Lecture Notes in Computer Science 9129

More information about this series at http://www.springer.com/series/7408

Ralf Hinze · Janis Voigtländer (Eds.)

Mathematics of Program Construction

12th International Conference, MPC 2015
Königswinter, Germany, June 29 – July 1, 2015
Proceedings

Springer

Editors
Ralf Hinze
University of Oxford
Oxford
UK

Janis Voigtländer
University of Bonn
Bonn
Germany

ISSN 0302-9743 ISSN 1611-3349 (electronic)
Lecture Notes in Computer Science
ISBN 978-3-319-19796-8 ISBN 978-3-319-19797-5 (eBook)
DOI 10.1007/978-3-319-19797-5

Library of Congress Control Number: 2015939987

LNCS Sublibrary: SL2 – Programming and Software Engineering

Springer Cham Heidelberg New York Dordrecht London

Printed on acid-free paper

Springer International Publishing AG Switzerland is part of Springer Science+Business Media
(www.springer.com)

Preface

This volume contains the proceedings of MPC 2015, the 12th International Conference on the Mathematics of Program Construction. This conference series aims to promote the development of mathematical principles and techniques that are demonstrably practical and effective in the process of constructing computer programs, broadly interpreted. The focus is on techniques that combine precision with conciseness, enabling programs to be constructed by formal calculation.

The conference was held in Königswinter, Germany, during June 29–July 1, 2015. The previous eleven conferences were held in 1989 in Twente, The Netherlands (with proceedings published as LNCS 375); in 1992 in Oxford, UK (LNCS 669); in 1995 in Kloster Irsee, Germany (LNCS 947); in 1998 in Marstrand, Sweden (LNCS 1422); in 2000 in Ponte de Lima, Portugal (LNCS 1837); in 2002 in Dagstuhl, Germany (LNCS 2386); in 2004, in Stirling, UK (LNCS 3125); in 2006 in Kuressaare, Estonia (LNCS 4014); in 2008 in Marseille-Luminy, France (LNCS 5133); in 2010 in Lac-Beauport, Canada (LNCS 6120); and in 2012 in Madrid, Spain (LNCS 7342).

The volume contains the abstracts of two invited talks, and 15 papers selected for presentation by the Program Committee from 20 submissions. The quality of the papers submitted to the conference was in general very high. However, the number of submissions has decreased compared to the previous conferences in the series. Each paper was refereed by at least three reviewers, and on average by four. We are grateful to the members of the Program Committee and the external reviewers for their care and diligence in reviewing the submitted papers. The review process and compilation of the proceedings were greatly helped by Andrei Voronkov's EasyChair system, which we can highly recommend.

June 2015

Ralf Hinze
Janis Voigtländer

Organization

Program Committee

Eerke Boiten	University of Kent, UK
Jules Desharnais	Université Laval, Canada
Lindsay Groves	Victoria University of Wellington, New Zealand
Ralf Hinze	University of Oxford, UK
Zhenjiang Hu	National Institute of Informatics, Japan
Graham Hutton	University of Nottingham, UK
Johan Jeuring	Utrecht University and Open University, The Netherlands
Jay McCarthy	Vassar College, USA
Shin-Cheng Mu	Academia Sinica, Taiwan
Bernhard Möller	Universität Augsburg, Germany
Dave Naumann	Stevens Institute of Technology, USA
Pablo Nogueira	Universidad Politécnica de Madrid, Spain
Ulf Norell	University of Gothenburg, Sweden
Bruno Oliveira	The University of Hong Kong, Hong Kong, SAR China
José Nuno Oliveira	Universidade do Minho, Portugal
Alberto Pardo	Universidad de la República, Uruguay
Christine Paulin-Mohring	INRIA-Université Paris-Sud, France
Tom Schrijvers	KU Leuven, Belgium
Emil Sekerinski	McMaster University, Canada
Tim Sheard	Portland State University, USA
Anya Tafliovich	University of Toronto Scarborough, Canada
Tarmo Uustalu	Institute of Cybernetics, Estonia
Janis Voigtländer	Universität Bonn, Germany

Additional Reviewers

Bove, Ana	Keuchel, Steven	Roocks, Patrick
Gómez-Martínez, Elena	Kozen, Dexter	You, Shu-Hung
Karachalias, George	Panangaden, Prakash	Zelend, Andreas

Local Organizing Committee

Ralf Hinze	University of Oxford, UK (Co-chair)
Janis Voigtländer	Universität Bonn, Germany (Co-chair)
José Pedro Magalhães	Standard Chartered Bank, UK
Nicolas Wu	University of Bristol, UK
Maciej Piróg	University of Oxford, UK

Invited Talks

A Nondeterministic Lattice of Information

Carroll Morgan

University of New South Wales, NSW 2052 Australia
carrollm@cse.unsw.edu.au[1]

Abstract. In 1993 Landauer and Redmond [2] defined a "lattice of information," where a partition over the type of secret's possible values could express the security resilience of a sequential, deterministic program: values within the same cell of the partition are those that the programs does not allow an attacker to distinguish.

That simple, compelling model provided not only a refinement order for deterministic security (inverse refinement of set-partitions) but, since it is a lattice, allowed the construction of the "least-secure deterministic program more secure than these other deterministic programs", and its obvious dual. But Landauer treated neither demonic nor probabilistic choice.

Later work of our own, and independently of others, suggested a probabilistic generalisation of Landauer's lattice [1, 3]—although it turned out that the generalisation is only a partial order, not a lattice [5]. This talk looks between the two structures above: I will combine earlier qualitatitve ideas [6] with very recent quantitative results [4] in order to explore

- What an appropriate purely demonic lattice of information might be, the "meat in the sandwich" that lies between Landauer's deterministic, qualitative lattice and our probabilistic partial order.
- The importance of compositionality in determining its structure.
- That it is indeed a lattice, that it generalises [2] and that it is generalised by [1, 3].
- Its operational significance and, of course,
- Thoughts on how it might help with constructing (secure) programs.

References

1. Alvim, M.S., Chatzikokolakis, K., Palamidessi, C., Smith, G.: Measuring information leakage using generalized gain functions. In: Proceedings of the 25th IEEE Computer Security Foundations Symposium (CSF 2012), pp. 265–279, June 2012
2. Landauer, J., Redmond, T.: A lattice of information. In: Proceedings of the 6th IEEE Computer Security Foundations Workshop (CSFW 1993), pp. 65–70, June 1993

[1] I am grateful for the support of the Australian ARC via its grant DP120101413, and of NICTA, which is funded by the Australian Government through the Department of Communications and the Australian Research Council through the ICT Centre-of-Excellence Program.

3. McIver, A., Meinicke, L., Morgan, C.: Compositional closure for bayes risk in probabilistic noninterference. In: Abramsky, S., Gavoille, C., Kirchner, C., auf der Heide, F.M., Spirakis, P.G. (eds.) ICALP 2010, Part II. LNCS, vol. 6199, pp. 223–235. Springer, Heidelberg (2010)
4. McIver, A., Meinicke, L., Morgan, C.: Abstract Hidden Markov Models: a monadic account of quantitative information flow. In: Proceedings of the LiCS 2015 (2015, to appear)
5. McIver, A., Morgan, C., Meinicke, L., Smith, G., Espinoza, B.: Abstract channels, gain functions and the information order. In: FCS 2013 Workshop on Foundations of Computer Security (2013). http://prosecco.gforge.inria.fr/personal/bblanche/fcs13/fcs13proceedings.pdf.
6. Morgan, C.: *The Shadow Knows*: refinement of ignorance in sequential programs. In: Uustalu, T. (ed.) MPC 2006. LNCS, vol. 4014, pp. 359–378. Springer, Heidelberg (2006)

A Compilation of Compliments for a Compelling Companion: The Comprehension

Torsten Grust

Department of Computer Science, Universität Tübingen, Tübingen, Germany
torsten.grust@uni-tuebingen.de

Abstract. If I were to name the most versatile and effective tool in my box of abstractions, I could shout my answer: the *comprehension*. Disguised in a variety of syntactic forms, comprehensions are found at the core of most established and contemporary database languages. Comprehensions uniformly embrace iteration, filtering, aggregation, or quantification. Comprehensions concisely express query transformations that otherwise fill stacks of paper(s). Yet their semantics is sufficiently simple that it can be shaped to fit a variety of demands. Comprehensions fare as user-facing language constructs just as well as internal query representations that facilitate code generation. Most importantly, perhaps, comprehensions are found in the vocabulary of the database and programming language communities, serving as a much needed interpreter that connects both worlds.

The humble comprehension deserves a pat on the back—that is just what this talk will attempt to provide.

Contents

Exploring an Interface Model for CKA

Bernhard Möller[1(✉)] and Tony Hoare[2]

[1] Institut für Informatik, Universität Augsburg, Augsburg, Germany
bernhard.moeller@informatik.uni-augsburg.de
[2] Microsoft Research, Cambridge, UK

Abstract. Concurrent Kleene Algebras (CKAs) serve to describe general concurrent systems in a unified way at an abstract algebraic level. Recently, a graph-based model for CKA has been defined in which the incoming and outgoing edges of a graph define its input/output interface. The present paper provides a simplification and a significant extension of the original model to cover notions of states, predicates and assertions in the vein of algebraic treatments using modal semirings. Moreover, it uses the extension to set up a variant of the temporal logic CTL^* for the interface model.

Keywords: Concurrency · Temporal logic · Algebra · Formal methods

1 Introduction

Concurrent Kleene Algebra (CKAs) is intended to describe general concurrent systems in a unified way at an abstract algebraic level. It may be used to define and explore the behaviour of a computer system embedded in its wider environment, including even human users. Moreover, it provides an algebraic presentation of common-sense (non-metric) spatial and temporal reasoning about the real world.

Its basic ingredients are *events*, i.e., occurrences of certain primitive actions, such as assignments or sending/receiving on channels. The different occurrences of an action are thought as being distinguished, for instance, by time or space coordinates. A *trace* is a set of events, and a *program* or *specification* is a set of traces. Programs and specifications are distinguished by the operators that are applicable to them. For instance, while set-theoretic intersection or complement make sense for specifications, they do not for programs. Another essential concept is that of a *dependence* relation between events which expresses causal or temporal succession. The events of a trace and their dependence therefore define a graph. The fundamental connection operators on traces and programs are sequential composition ; and concurrent composition |, governed by characteristic laws.

Various models of these laws have been proposed. In [18] an interface model for CKA was sketched. It considers *graphlets*, i.e., subgraphs of an overall trace, as "events" of their own. The *interface* of a graphlet consists of the arrows that

R. Hinze and J. Voigtländer (Eds.): MPC 2015, LNCS 9129, pp. 1–29, 2015.
DOI:10.1007/978-3-319-19797-5_1

connect it to neighbouring ones. In the present paper we simplify and significantly extend that model. Moreover, we show how the interface operators can be used to cover notions of states, predicates and assertions in the vein of algebraic treatments using modal semirings. Finally we set up a variant of the temporal logic CTL^* for the interface model.

Temporal logics, even for so-called branching time, were inspired by interleaving semantics and hence linear traces. It has been shown e.g. in [23] that this can be abstracted to more general algebraic structures such as quantales. We use this idea to set up a temporal logic for the interface model. The resulting semantics mainly covers sequential aspects. We think, though, that it can be useful for analysing quasilinear threads of CKA programs, i.e., subgraphs which do not consider incoming or outgoing arrows from the environment but only their own "inner flow" of causal or temporal succession.

Besides these main topics, the paper also sets up links to other semantic notions, in particular, to the traditional assertional interpretation of Hoare Logic and to the modal operators diamond and box. It is structured as follows.

Section 2 repeats the most important notions of CKA and its overall approach to semantics.

In Sect. 3 the interface model is presented. Using restriction operators, simple while programs and a counterpart of standard Hoare triples are defined. A variant of the sequential composition operator allows a connection to the more recent type of Hoare triples from [15,30]. Finally, restriction operators are used to define modal box and diamond operators for the interface model.

After that, Sect. 4 deals with the mentioned variant of CTL^*. After some basic material on temporal logics and their algebraic semantics in general, based on earlier work in [23], an adaptation for the interface model is achieved and shown at work in some small examples.

The paper finishes with a short conclusion and outlook; an appendix provides deferred detailed proofs.

2 Basic Concepts of CKA

To make the paper self-contained, we repeat some basic facts about CKA.

2.1 Traces

Assume a set of *events*. As mentioned in the Introduction, these might be occurrences of basic actions, distinguished by time or space stamps or the like. A *trace* is a set of such events. Traces will be denoted by p, q, r and primed versions of these. CKA assumes a *refinement relation* $\Rightarrow$ between traces which is at least a preorder. $p \Rightarrow q$ means that every event of p is also possible for q where q may be a specification of all desirable events of p.

The trace *skip* (also denoted by 1) describes doing nothing, for example because the task at hand has been accomplished. *Bottom* ($\bot$) is an error element standing for a trace which is physically impossible, for example because it requires an event to occur before an event which caused it. It is the least element w.r.t. $\Rightarrow$. *Top* ($\top$) stands for a trace which contains a generic programming error, for example a null dereference, a race condition or a deadlock. It is the greatest element w.r.t. $\Rightarrow$.

The two essential binary composition operators ; and | require disjointness of their argument traces to achieve modularisation in the sense of Separation Logic [24]. The *sequential composition* $p\,;q$ describes execution of both p and q, where p can finish before q starts. *Concurrent composition* $p\,|\,q$ describes execution of both p and q, where p and q can start together and can finish together. In between, they can interact with each other and with their common environment.

Both ; and | are assumed to be associative with unit 1 and zero $\bot$; in addition | has to be commutative. Moreover, ; and | have to be *covariant*[1] in both arguments. For example, $p \Rightarrow q$ implies $p;r \Rightarrow q;r$. This allows refinement in larger contexts composed using | and ;.

Sequential and concurrent composition need to satisfy the following inequational analogue of the *exchange (or interchange) law* of Category Theory, also known as subsumption or subdistribution:

$$(p\,|\,q)\,;\,(p'\,|\,q') \Rightarrow (p\,;\,p')\,|\,(q\,;\,q'). \qquad \text{(exchange)}$$

Since 1 is a shared unit of ; and |, specialising q or/and p' to 1 and use of commutativity of | yields the *frame laws*

$$p\,;\,(p'\,|\,q') \Rightarrow (p\,;\,p')\,|\,q', \qquad \text{(frame I)}$$

$$(p\,|\,q)\,;\,q' \Rightarrow p\,|\,(q\,;\,q'), \qquad \text{(frame II)}$$

$$p\,;\,q' \Rightarrow p\,|\,q'. \qquad \text{(frame III)}$$

2.2 Programs and Specifications

A *program* or a *specification* is a non-empty set P of traces that is downward closed w.r.t. $\Rightarrow$, i.e., satisfies $p \in P \wedge p' \Rightarrow p \implies p' \in P$. As mentioned above, specifications and programs are distinguished by the operators that are admissible for them. Programs will be denoted by P, Q, R and primed versions of these. By $\bot$ we also denote the program/specification consisting just of the trace $\bot$; it corresponds to a contradictory specifying predicate, i.e., to false. The set of all traces will be denoted by U.

The function dc forms the downward closure of a set S of traces, i.e.,

$$dc(S) =_{df} \{p' \mid \exists\, p \in S : p' \Rightarrow p\}. \qquad (1)$$

For a single element $t \in U$ we abbreviate $dc(\{t\})$ by $dc(t)$. Hence a program is a set P of traces with $P = dc(P)$. For instance, $\mathsf{skip} =_{df} dc(1) = \{\bot, 1\}$ lifts 1 to the level of programs.

[1] Also called *monotone* or *isotone*.

The *choice* $P \mathbin{[\!]} Q$ between programs is the union $P \cup Q$ and describes the execution of a P-trace or a Q-trace. The choice may be determined or influenced by the environment, or it may be left wholly indeterminate. The operator is associative, commutative and idempotent, with $\perp$ as unit and U as zero. Finally, it can be used to define the refinement relation between programs/specifications:

$$P \Rightarrow Q \quad \text{iff} \quad P \mathbin{[\!]} Q = Q.$$

By downward closure, $\Rightarrow$ coincides with inclusion on programs. Pointwise one has

$$P \Rightarrow P' \iff_{df} \forall p \in P : \exists p' \in P' : p \Rightarrow p'. \tag{2}$$

By this, a program P refines a specification P' if each of its traces refines a trace admitted by the specification. $\perp$ and U are the least and greatest elements w.r.t. $\Rightarrow$.

Occasionally we also use the intersection operator $\cap$ on specifications; it is associative, commutative and idempotent with unit U and zero $\perp$.

2.3 Lifting and Laws

We now present the important principle of lifting operators and their laws from the level of traces to that of programs.

Definition 2.1. When $\circ : U \times U \to U$ is a, possibly partial, binary operator on U, its *pointwise lifting* to programs P, P' is defined as

$$P \circ P' =_{df} dc(\{t \circ t' \mid t \in P, t' \in P' \text{ and } t \circ t' \text{ is defined}\}).$$

A sufficient condition for an inequational law $p \Rightarrow p'$ to lift from traces to programs is *linearity*, viz. that every variable occurs at most once on both sides of the law and that all variables in the left hand side P also occur in the right hand side P'. Examples are the exchange and frame laws. For equations a sufficient condition is *bilinearity*, meaning that both constituting inequations are linear. Examples are associativity, commutativity and neutrality. The main result is as follows.

Theorem 2.2. *If a linear law $p \Rightarrow p'$ holds for traces then it also holds when all variables in p, p' are replaced by variables for programs and the operators are interpreted as the liftings of the corresponding trace operators.*

We illustrate the gist of the proof for the case of the frame law $P\,;\,P' \Rightarrow P \mid P'$.

$$\begin{array}{ll}
 & r \in P\,;\,P' \\
\Leftrightarrow & \quad \{\!\!\{ \text{ by Definition 2.1 and (1) } \}\!\!\} \\
 & \exists\, t \in P, t' \in P' : r \Rightarrow t\,;\,t' \\
\Rightarrow & \quad \{\!\!\{ \text{ by } t\,;\,t' \Rightarrow t \mid t' \text{ (frame III) and transitivity of } \Rightarrow \}\!\!\} \\
 & \exists\, t \in P, t' \in P' : r \Rightarrow t \mid t' \\
\Rightarrow & \quad \{\!\!\{ \text{ by Definition 2.1 and (1) } \}\!\!\} \\
 & r \in P \mid P'.
\end{array}$$

The full proof for general preorders can be found in [17].

Corollary 2.3 (Laws of Trace Algebra for Programs). *The liftings of the operators ; and | to programs are associative and have* 1 *as a shared unit. Moreover,* | *is commutative and the exchange law and therefore also the frame laws hold.*

There are further useful consequences of our definition of programs. The set $\mathcal{P}$ of all programs forms a complete lattice w.r.t. the inclusion ordering; it has been called the *Hoare power domain* in the theory of denotational semantics (e.g. [4,22,31]). The least element is the program $\{\bot\}$, again denoted by $\bot$, while the greatest element is the program U consisting of all traces. Infimum and supremum coincide with intersection and union, since downward closed sets are also closed under these operations. Also, refinement coincides with inclusion, i.e., $P \Rightarrow Q \iff P \subseteq Q$. We will use this latter notation throughout the rest of the paper.

By completeness of the lattice we can define (unbounded) choice between a set $\mathcal{Q} \subseteq \mathcal{P}$ of programs as

$$[\!]\,\mathcal{Q} =_{df} \bigcup \mathcal{Q}.$$

The lifted versions of covariant trace operators are covariant again, but even distribute through arbitrary choices between programs. This means that the set of all programs forms a *quantale* (e.g. [25]) w.r.t. to the lifted versions of both ; and |. This will be used in Sect. 4 to set up a connection with temporal logics.

Covariance of the lifted operators, together with completeness of the lattice of programs and the Tarski-Knaster fixed point theorem guarantees that recursion equations have least and greatest solutions. More precisely, let $f : \mathcal{P} \to \mathcal{P}$ be a covariant function. Then f has a least fixed point μf and a greatest fixed point νf, given by the formulas

$$\mu f = \bigcap \{P \mid f(P) \subseteq P\}, \qquad \nu f = \bigcup \{P \mid P \subseteq f(P)\}. \tag{3}$$

With the operator ;, this can be used to define the Kleene star (see e.g. [5]), i.e., unbounded finite sequential iteration, of a program P as $P^* =_{df} \mu f_P$, where

$$f_P(X) =_{df} \mathsf{skip} \;[\!]\; P \;[\!]\; X \,;\, X.$$

Equivalently, $P^* = \mu g = \mu h$, where

$$g_P(X) =_{df} \mathsf{skip} \;[\!]\; P \,;\, X, \qquad h_P(X) =_{df} \mathsf{skip} \;[\!]\; X \,;\, P. \tag{4}$$

Since f_P, by the above remark, distributes through arbitrary choices between programs, it is even continuous and Kleene's fixed point theorem tells us that $P^* = \mu f_P$ has the iterative representation

$$P^* = \bigcup \{f_P^i(\emptyset) \mid i \in \mathbb{N}\}, \tag{5}$$

which transforms into the well known representation of star, viz.

$$P^* = \bigcup \{P^i \mid i \in \mathbb{N}\}$$

with $P^0 =_{df} \mathsf{skip}$ and $P^{i+1} =_{df} P \,;\, P^i$.

We show an example for the interplay of Kleene star and the exchange law.

Lemma 2.4. $(P \mid Q)^* \subseteq P^* \mid Q^*$.

Proof. Given the representation of least fixed points in Eq. (3) it suffices to show that $P^* \mid Q^*$ is contracted by the generating function $g_{P \mid Q}$ of $(P \mid Q)^*$. By the fixed point property of star, distributivity of lifted operators, neutrality of skip and omitting choices, exchange law and covariance, and definition of $g_{P \mid Q}$:

$$
\begin{array}{ll}
 & P^* \mid Q^* \\
= & (\mathsf{skip} \,[\!]\, P\,;\, P^*) \mid (\mathsf{skip} \,[\!]\, Q\,;\, Q^*) \\
= & (\mathsf{skip} \mid \mathsf{skip}) \,[\!]\, (\mathsf{skip} \mid (Q\,;\, Q^*)) \,[\!]\, ((P\,;\, P^*) \mid \mathsf{skip}) \,[\!]\, ((P\,;\, P^*) \mid (Q\,;\, Q^*)) \\
\supseteq & \mathsf{skip} \,[\!]\, ((P\,;\, P^*) \mid (Q\,;\, Q^*)) \\
\supseteq & \mathsf{skip} \,[\!]\, ((P \mid Q)\,;\, (P^* \mid Q^*)) \\
= & g_{P \mid Q}(P^* \mid Q^*).
\end{array}
$$

□

Infinite iteration P^ω can be defined as the greatest fixed point νk_P where

$$k_P(X) =_{df} P\,;\, X.$$

However, there is no representation of P^ω similar to (5) above, because semicolon does not distribute through intersection and hence is not co-continuous; we only have the inequation

$$P^\omega \subseteq \bigcap \{P^i\,;\, U \mid i \in \mathbb{N}\}.$$

To achieve equality, in general the iteration and intersection would need to be transfinite.

Sometimes it is convenient to work with an iteration operator that leaves it open whether the iteration is finite or infinite; this can be achieved by Back and von Wright's operator [1]

$$P^{\widehat{\omega}} =_{df} P^\omega \,[\!]\, P^*,$$

which is the greatest fixed point of the function g_P above.

Along the same lines, unbounded finite and infinite concurrent iteration of a program can be defined.

3 An Interface Model for CKA

3.1 Motivation and Basic Notions

This section presents a simplification and substantial extension of a CKA model given in [15]. It supports the process of developing a system architecture and design at any desired level of granularity. It abstracts from the plethora of internal events of a trace, and models only the external interface of each component. The interface is described as a set of arrows, each of which connects a pair of events, one of them inside the component, and one of them outside.

The set of all arrows defines a desired minimum level of granularity of the global dependence relation which is induced as its reflexive-transitive closure. Events and arrows together define a graph. A part of that graph will be called a graphlet below.

3.2 Graphlets

Assume, as in Sect. 2.1, a set EV of *events* and additionally a set AR of *arrows*. Moreover, assume total functions $s, t : \mathrm{AR} \rightarrow \mathrm{EV}$ that yield the *source* and *target* of each arrow. For event sets $E, E' \subseteq \mathrm{EV}$ we define, by a slight abuse of notation, the set $E \times E'$ of arrows by

$$a \in E \times E' \Longleftrightarrow_{df} s(a) \in E \wedge t(a) \in E'.$$

We choose this notation, since $\times$ will play the role of the conventional Cartesian product of sets of events as used in relationally based models such as the trace model of [17]. The operator $\times$ distributes through *disjoint union* $+$ and hence is covariant in both arguments. We let $\times$ bind tighter than $+$ and $\cap$. This gives the following properties (with $\overline{E} =_{df} \mathrm{EV} - E$) which are immediate by Boolean algebra:

$$\begin{aligned} \overline{E \times E'} &= \mathrm{EV} \times \overline{E'} + \overline{E} \times E' = E \times \overline{E'} + \overline{E} \times \mathrm{EV} \\ &= E \times \overline{E'} + \overline{E} \times E' + \overline{E} \times \overline{E'}, \\ (E \cap E') \times (E'' \cap E''') &= (E \times E'') \cap (E' \times E''') = (E \times E''') \cap (E' \times E''), \\ \emptyset \times E &= \emptyset = E \times \emptyset. \end{aligned} \tag{6}$$

Definition 3.1. A *graphlet* is a pair $G = (E, A)$ with $E \subseteq \mathrm{EV}, A \subseteq \mathrm{AR}$ satisfying the following healthiness condition: there are no "loose" arrows, i.e.,

$$A \cap \overline{E} \times \overline{E} = \emptyset, \tag{7}$$

In words, every arrow in A has at least one event in E.

The healthiness condition is mandatory for validating the associativity and exchange laws for the trace algebra operators on graphlets, as will become manifest in their proof.

Since arrows have identity (and do not just record whether or not a connection exists between two events, as is done in relational models), we can associate values, labels, colours etc. with them without having to include extra functions for that into the model.

Example 3.2. We specify a graph H that models a thread corresponding to a single natural-number variable x, already initialised to 0, with increment as the only operation. Let $\mathbb{N}$ be the set of natural numbers and $\mathbb{N}_+ =_{df} \mathbb{N} - \{0\}$. We use $\mathrm{EV} =_{df} \{inc_i \mid i \in \mathbb{N}\}$ and $A =_{df} \mathbb{N}_+$ with the source and target maps

$$s(i) =_{df} inc_{i-1}, \qquad t(i) =_{df} inc_i.$$

Pictorially,

$$H: \quad inc_0 \xrightarrow{\;1\;} inc_1 \xrightarrow{\;2\;} \cdots$$

Some graphlets of H then are the following:

$$inc_0 \xrightarrow{1}$$

$$\xrightarrow{2} inc_2 \xrightarrow{3}$$

$$inc_0 \xrightarrow{1} inc_1 \xrightarrow{2} inc_2 \xrightarrow{3}$$

□

Definition 3.3. The sets of *input, output* and *internal* arrows of graphlet G are

$$\begin{aligned} in(G) &=_{df} A \cap \overline{E} \times E, \\ out(G) &=_{df} A \cap E \times \overline{E}, \\ int(G) &=_{df} A \cap E \times E. \end{aligned}$$

These sets are pairwise disjoint, and by the healthiness condition Eq. (7) in Definition 3.1 we have $A = in(G) \cup out(G) \cup int(G)$. The sets $in(G)$ and $out(G)$ together constitute the *interface* of G to its environment. The set of all graphlets is denoted by G.

We want to compose graphlets G and G' by connecting output arrows of G to input arrows of G'. If these arrows carry values of some kind, we view the composition as transferring these values from the source events of these arrows in G to their target events in G'. To achieve separation, we require that the event sets of G and G' are disjoint. While concurrent composition $G \,|\, G'$ does not place any further restriction, in sequential composition $G \,;\, G'$ we forbid "backward" arrows from G' to G.

Definition 3.4. Let G, G' be graphlets with disjoint event sets. We set

$$\begin{aligned} G \,|\, G' &=_{df} (E + E', A \cup A'), \\ G \,;\, G' &=_{df} \begin{cases} G \,|\, G' & \text{if } \mathrm{CS}(G, G'), \\ \text{undefined otherwise,} \end{cases} \end{aligned}$$

where

$$\begin{aligned} \mathrm{CS}(G, G') &\Longleftrightarrow_{df} A \cap A' \cap E' \times E = \emptyset \\ &\Longleftrightarrow \quad A \cap A' \subseteq E \times E' \end{aligned}$$

formalises the above requirement of no backward arrows from G' to G.

The "connection" of output and input arrows takes place as follows: if a is an arrow in $out(G) \cap in(G')$ then it will also be in $A \cup A'$ and both its end points will be in $E + E'$, so that it is an internal arrow of $G \,|\, G'$. Clearly, $G \,|\, G'$ is a graphlet again. Since disjoint union and union are associative, also $|$ is associative and has the *empty graphlet* $\square =_{df} (\emptyset, \emptyset)$ as its unit.

Assume that in a term $(G \,|\, G') \,;\, (H \,|\, H')$ the condition CS is satisfied for the operands of ;. Then it also holds for the sequential compositions $G \,;\, H$ and $G' \,;\, G''$ that occur on the right hand side of the corresponding exchange law, so that then all three sequential compositions are defined.

Lemma 3.5

1. $in(G \,|\, G') = (in(G) \cap \overline{E'} \times E) \cup (in(G') \cap \overline{E} \times E')$.

2. $out(G \mid G') = (out(G) \cap E \times \overline{E'}) \cup (out(G') \cap E' \times \overline{E})$.

The somewhat tedious proof is deferred to the Appendix.

As the refinement relation between graphlet terms p, p' we use what is known as the *flat order* in denotational semantics:

$$p \Rightarrow p' \iff_{df} p \text{ is undefined or } p, p' \text{ are both defined and have equal values.}$$

Theorem 3.6. *The operators ; and* | *are associative and obey the exchange and frame laws. Moreover,* | *is commutative and* $\square$ *is a shared unit of ; and* |.

See again the Appendix for details of the proof. We note, however, that absence of "loose" arrows from graphlets is essential for it.

Definition 3.7. We now use graphlets as traces. Hence in the interface model a program is a set of graphlets. The program that does nothing is $\mathsf{skip} =_{df} \{\square\}$, while U is the program consisting of all graphlets.

Since we use the flat refinement order between graphlets, the set of programs is isomorphic to the power set of the program U that consists of all graphlets. By the general lifting result quoted in Theorem 2.2, the above Theorem 3.6 extends to graphlet programs as well.

Example 3.8. Consider again the graph H from Example 3.2 and the program P consisting of all singleton graphlets of H:

$$P =_{df} \{\, inc_0 \xrightarrow{1} \,\} \cup \{\, \xrightarrow{i} inc_i \xrightarrow{i+1} \mid i \in \mathbb{N}_+\}.$$

It represents the single-step transition relation associated with the statement `x := x + 1`. Then $P^2 = P\,;\,P = P \mid P$ contains

$$\xrightarrow{i} inc_i \xrightarrow{i+1} inc_{i+1} \xrightarrow{i+2}$$

for all $i \in \mathbb{N}$, but also non-contiguous graphlets such as

$$\xrightarrow{i} inc_i \xrightarrow{i+1} \qquad \xrightarrow{j} inc_{i+1} \xrightarrow{j+1}$$

for $j + 1 \neq i \neq j \neq i + 1$. We will later define a variant of; that excludes non-contiguous graphlets. $\square$

3.3 Restriction and while Programs

As mentioned in the motivation in Sect. 3.1, a set of arrows can be viewed as a description of a set of variable/value associations. We can use such a set to characterise "admissible" or "desired" associations. Usually one will require such a set to be *coherent* in some sense, like not having contradictory associations. We give no precise definition of coherence, but leave it a parameter to our treatment.

A minimum healthiness requirement is that the empty arrow set $\emptyset$ and each of the interface sets $in(G)$ and $out(G)$ of every graphlet G should be coherent. The set $int(G)$ of internal arrows of G will usually not be coherent, since the variable/value associations may change during the internal flow of control. For brevity we refer to coherent arrow sets as *states* in the sequel.

First we introduce restriction operators and, based on these, while programs.

Definition 3.9. The *input restriction* of graphlet G to set $C \subseteq \mathrm{AR}$ is

$$C \rfloor G =_{df} \begin{cases} G & \text{if } in(G) = C, \\ \text{undefined} & \text{otherwise.} \end{cases}$$

Output restriction $G \lfloor C$ is defined symmetrically.

Definition 3.10. A *predicate* is a set of arrow sets that satisfy the coherence requirement. Restriction is lifted pointwise to predicates and programs.

Hence the input restriction $S \rfloor Q$ of program Q to a predicate S retains only those graphlets $G \in Q$ for which $in(G) \in S$. An analogous remark holds for output restriction.

Predicates will mimic the role of tests, i.e., of elements below the multiplicative unit, in semirings. In the standard model of CKA and also the interface model there are, however, only the tests $\perp$ and skip. So internalising restriction as pre- or post-multiplication by tests is not possible there in a reasonable way, which is why we resort to the above separate restriction operators. In the recent model of [19] non-trivial test elements exist, but the model is restricted in a number of ways.

Restriction obeys useful laws, quite analogous to the case of semirings with tests; for the proofs see the corresponding ones for the *mono modules* introduced in [7] (Lemmas 7.1 and 7.2 there).

Lemma 3.11. *For predicate S and programs P, Q,*

$$\begin{aligned} S \rfloor (P \cap Q) &= (S \rfloor P) \cap Q = P \cap (S \rfloor Q) = (S \rfloor P) \cap (S \rfloor Q), \\ (P \cap Q) \lfloor S &= (P \lfloor S) \cap Q = P \cap (Q \lfloor S) = (P \lfloor S) \cap (Q \lfloor S). \end{aligned}$$

In particular, for $Q = U$ with U being the program consisting of all graphlets, $S \rfloor P = P \cap S \rfloor U$ and $P \lfloor S = P \cap U \lfloor S$.

Now we can define an if then else and a while do construct.

Definition 3.12. Given a predicate S and programs P, Q we set

$$\begin{aligned} \text{if } S \text{ then } P \text{ else } Q &=_{df} (S \rfloor P) \mathbin{[\![]\!]} (\overline{S} \rfloor Q), \\ \text{while } S \text{ do } P &=_{df} (S \rfloor P)^* \lfloor \overline{S}, \end{aligned}$$

where $\overline{S}$ is the complement of S in the set of predicates.

Since this definition is analogous to the one for semirings with tests, all the usual algebraic rules for while programs carry over to the present setting.

If one wants to include the possibility of loops with infinite iteration one can use the program (while S do P) $[\![]\!]$ $(S \rfloor P)^\omega$.

3.4 Hoare Triples

We show now how predicates can be used to define Hoare triples for graphlet programs.

Definition 3.13. The *standard Hoare triple* $\{S\}\,P\,\{S'\}$ with predicates S, S' and program P is defined by

$$\{S\}\,P\,\{S'\} \iff_{df} S\rfloor P \subseteq P\lfloor S'.$$

This means that, starting with a variable/value association in S, execution of P is guaranteed to "produce" a variable/value association in S'. See [2] for a closely related early relational definition. We give two equivalent formulations of the standard Hoare triple.

Lemma 3.14. *Let U, as in Definition 3.7, be the program consisting of all graphlets.*

$$\{S\}\,P\,\{S'\} \iff S\rfloor P \subseteq U\lfloor S' \iff S\rfloor P = (S\rfloor P)\lfloor S'.$$

The formula $S\rfloor P \subseteq U\lfloor S'$ expresses more directly that the "outputs" of $S\rfloor P$ satisfy the predicate S' and is calculationally advantageous, since the variable P is not repeated. The formula $S\rfloor P = (S\rfloor P)\lfloor S'$ says that post-restriction of $S\rfloor P$ to S' is no proper restriction, since its "outputs" already satisfy S'.

Proof. For the first equivalence let $Q = S\rfloor P$. By Boolean algebra, Lemma 3.11, definition of intersection, and since $Q \subseteq P$ is true by the definition of restriction:

$$\begin{aligned} Q \subseteq P\lfloor S' &\iff Q \subseteq U.\cap(P\lfloor S') \iff Q \subseteq (U\lfloor S')\cap P \\ &\iff Q \subseteq U\lfloor S' \wedge Q \subseteq P \iff Q \subseteq U\lfloor S'. \end{aligned}$$

For the second one we have by Lemma 3.11, and by the just shown first equivalence:

$$(S\rfloor P)\lfloor S' = (S\rfloor P) \cap (U\lfloor S') = S\rfloor P.$$

□

Definition 3.13 entails the well known Hoare calculus with all its inference rules, in particular the if then else and while rules.

Let us connect this to another, more recent view of Hoare triples [15].

Definition 3.15. For programs P, P' and Q one sets

$$P\ \{\!\!\{Q\}\!\!\}\ P' \iff_{df} P\,;\,Q \subseteq P'.$$

This expresses that, after any graphlet in "pre-history" P, execution of Q is guaranteed to yield an overall graphlet in P'. For the case where programs are relations between states this definition appears already in [29]. These new triples enjoy many pleasant properties; see again [15] for details. We show two samples which will be taken up again in the next section.

Lemma 3.16. *Let program P be an invariant of program Q, i.e., assume P* $\{Q\}$ P.

1. *P is also an invariant of program Q^*, i.e., P* $\{Q^*\}$ P.
2. $(P\,;\,Q)^\omega \subseteq P^\omega$ *and* P $\{(Q\,;\,P)^\omega\}$ P^ω.

Proof.

1. By standard Kleene algebra, in generalisation of Eq. (4) one has $P\,;\,Q^* = \mu k_{PQ}$, where $k_{PQ}(X) =_{df} P \;[\!]\; X\,;Q$. Thus the claim is shown if we can prove that also P is a fixed point of k_{PQ}. Indeed, by the assumption $P\,;\,Q \subseteq P$,

$$k_{PQ}(P) \;=\; P \;[\!]\; P\,;\,Q = P.$$

2. The first claim is immediate from the assumption and covariance of the $^\omega$ operator. The second claim follows from the first one by the so-called rolling rule for the omega operator: $(P\,;\,Q)^\omega = P\,;\,(Q\,;\,P)^\omega$. □

A relationship between standard Hoare triples and a variant of the new ones will be set up in Lemmas 3.19 of the next section.

3.5 Quasilinear Sequential Composition

We define a strengthened form of sequential composition.

Definition 3.17. For graphlets G, G' the *quasilinear sequential composition* is defined by

$$G \mathbin{\boldsymbol{;}} G' =_{df} \begin{cases} G' & \text{if } G = \Box, \\ G & \text{if } G' = \Box, \\ G\,;\,G' & \text{if } G, G' \neq \Box \text{ and } out(G) = in(G'), \\ \text{undefined} & \text{otherwise.} \end{cases}$$

We call it quasilinear, since it does not allow "in-branching" or "out-branching" at the border between the composed graphlets and therefore leads to quasi linear thread-like graphlets when iterated finitely or infinitely. Note, however, that in the overall graph there may still be arrows to or from the environment which have been disregarded in selecting the arrows belonging to the graphlets in question.

The definition could be simplified if graphlets were allowed to have "loose" arrows; however, as remarked after Theorem 3.6, this would destroy associativity of | and **;**.

Lemma 3.18. *Quasilinear sequential composition is associative and satisfies the following laws (that do not hold for ;), assuming definedness of the terms involved.*

$$\begin{array}{rcl} in(G \mathbin{\boldsymbol{;}} G') = in(G) = in(G \,\llcorner\, in(G')) & \Leftarrow & G \neq \Box, \\ out(G \mathbin{\boldsymbol{;}} G') = out(G') = out(out(G) \,\lrcorner\, G') & \Leftarrow & G' \neq \Box, \\ (G \,\llcorner\, C) \mathbin{\boldsymbol{;}} G' = G \mathbin{\boldsymbol{;}} (C \,\lrcorner\, G|) & \Leftarrow & G, G' \neq \Box, \\ C \,\lrcorner\, (G \mathbin{\boldsymbol{;}} G') = (C \,\lrcorner\, G) \mathbin{\boldsymbol{;}} G' & \Leftarrow & G \neq \Box, \\ (G \mathbin{\boldsymbol{;}} G') \,\llcorner\, C = G \mathbin{\boldsymbol{;}} (G' \,\llcorner\, C) & \Leftarrow & G' \neq \Box. \end{array}$$

The proof is again somewhat tedious and hence deferred to the Appendix.

We lift ; to programs as usual. The lifted operator has skip as its unit.

Let again U denote the universal program consisting of all graphlets. Then we have the following connection between standard Hoare triples and a variant of the new ones.

Lemma 3.19. *Assume that* $\square \in P \iff in(\square) = \emptyset \in S \iff \emptyset \in S'$. *Then*

$$\{S\}\,P\,\{S'\} \iff U \downharpoonright S\ \boldsymbol{\{}P\boldsymbol{\}}\ U \downharpoonright S'.$$

The right hand side means that an arbitrary "pre-history" with a result that satisfies S, followed by P, makes an overall history with a result satisfying S'.

Proof. According to the definitions, $U \downharpoonright S\ \boldsymbol{\{}P\boldsymbol{\}}\ U \downharpoonright S'$ spells out to $(U \downharpoonright S) \,;\, P \subseteq U \downharpoonright S'$. By the assumption, the laws of Lemma 3.18 lift to the programs and predicates involved.

($\Rightarrow$) By Lemma 3.14, isotony of lifted operators, Lemma 3.18, and $U \,;\, U \subseteq U$ with isotony of restriction:

$$\begin{aligned}\{S\}\,P\,\{S'\} &\iff S \downharpoonleft P \subseteq U \downharpoonright S' \implies U \,;\, (S \downharpoonleft P) \subseteq U \,;\, (U \downharpoonright S') \\ &\iff (U \downharpoonright S) \,;\, P \subseteq (U \,;\, U) \downharpoonright S' \implies (U \downharpoonright S) \,;\, P \subseteq U \downharpoonright S'.\end{aligned}$$

($\Leftarrow$) By neutrality of skip, isotony of lifted operators, Lemma 3.18, and the assumption:

$$S \downharpoonleft P = \mathsf{skip} \,;\, (S \downharpoonleft P) \subseteq U \,;\, (S \downharpoonleft P) = (U \downharpoonright S) \,;\, P \subseteq U \downharpoonright S'.$$

$\square$

A dual construction using input restriction gives a connection between standard Hoare triples and the analogous variant of *Milner triples*, see [16]. These take the form $P \rightarrow_Q R$ and are defined by

$$P \rightarrow_Q R \iff_{df} P \supseteq Q; R.$$

Although the Milner triple modulo renaming has the same definition as the new Hoare triple, its informal interpretation is more that of operational semantics: it says that one way of executing P is to first execute Q (or to do a Q-transition), leading to the residual program R.

Example 3.20. Using ; in place of ; in forming powers and star of a program eliminates non-contiguous graphlets like the one shown in Example 3.8. Considering again the program P from Example 3.2 we can write a loop that increments the variable x until it becomes 10:

$$\mathsf{while}\ [0, 9]\ \mathsf{do}\ P,$$

where $[0, 9]$ is the interval from 0 to 9.

It consists of the graphlets

$$\xrightarrow{\ i\ } inc_i \xrightarrow{\ i+1\ } \cdots \xrightarrow{\ 9\ } inc_9 \xrightarrow{\ 10\ }$$

for $i \in [1, 9]$ and

$$inc_0 \xrightarrow{\ 1\ } inc_1 \xrightarrow{\ 2\ } \cdots \xrightarrow{\ 9\ } inc_9 \xrightarrow{\ 10\ }$$

$\square$

We conclude this section with an invariance property, related to that of Lemma 3.16.2.

Lemma 3.21. *If* $\Box \notin P$ *and* S *is an invariant of* P*, i.e., if* $\{S\}\,P\,\{S\}$*, then it is also preserved throughout the infinite iteration of* P*, i.e.,*

$$S \rfloor P^\omega \;=\; (S \rfloor P)^\omega,$$

where $^\omega$ *is taken w.r.t.* ;.

Proof. ($\supseteq$) We do not even need the assumption: by the fixed point definition of $(S \rfloor P)^\omega$, the definition of restriction, Lemma 3.18, the fixed point definition again, $S \rfloor P \subseteq S$ and isotony of $^\omega$,

$$\begin{aligned}(S \rfloor P)^\omega = (S \rfloor P)\,;(S \rfloor P)^\omega = (S \rfloor (S \rfloor P))\,;(S \rfloor P)^\omega = S \rfloor ((S \rfloor P)\,;(S \rfloor P)^\omega)\\ = S \rfloor (S \rfloor P)^\omega \supseteq S \rfloor P^\omega.\end{aligned}$$

($\subseteq$) By the fixed point definition, Lemma 3.18, the assumption with Lemmas 3.14, and 3.18:

$$S \rfloor P^\omega = S \rfloor (P\,;P^\omega) = (S \rfloor P)\,;P^\omega = ((S \rfloor P) \lfloor S)\,;P^\omega = (S \rfloor P)\,;(S \rfloor P^\omega),$$

which means that $S \rfloor P^\omega$ is a fixed point of $\lambda x\,.\,(S \rfloor P)\,;x$ and hence below the greatest fixed point $(S \rfloor P)^\omega$ of that function. □

Example 3.22. Consider again the program P from Example 3.8. If the powers of P as well as P^* and P^ω are taken w.r.t ; rather than ; then the non-contiguous subsequences of the overall graph are ruled out from them. Moreover, letting again $\mathbb{N}_+$ be the predicate consisting of all positive natural numbers, we have $\{\mathbb{N}_+\}\;P\;\{\mathbb{N}_+\}$ and hence

$$\mathbb{N}_+ \rfloor P^\omega \;=\; (\mathbb{N}_+ \rfloor P)^\omega.$$

This example will be taken up later in connection with the temporal logic CTL*. □

3.6 Disjoint Concurrent Composition

We now briefly deal with *interaction-free* or *disjoint concurrent composition* $|||$, a special variant of concurrent composition $|$ that does not admit any arrows between its operands and hence does not prescribe any particular interleavings of their events.

Definition 3.23

$$G \,|||\, G' =_{df} \begin{cases} G \,|\, G' & \text{if } \mathrm{CD}(G,G'),\\ \text{undefined} & \text{otherwise,}\end{cases}$$

where

$$\mathrm{CD}(G,G') \iff_{df} A \cap E' \times E = \emptyset = A' \cap E \times E'$$

excludes any arrow between G and G'.

Lemma 3.24. *Recall that + means disjoint union.*

$$in(G \,|||\, G') = in(G) + in(G'),$$
$$out(G \,|||\, G') = out(G) + out(G').$$

The proof can be found in the Appendix.

Lemma 3.25. *The operators* ; *and* $|||$ *satisfy the reverse exchange law*

$$(G \,;\, G') \,|||\, (G'' \,;\, G''') \Rightarrow (G \,|||\, G'') \,;\, (G' \,|||\, G'''),$$

in which the order of refinement is the reverse of that in (exchange).

Proof. Straightforward from Lemma 3.24 and the definition of ;. □

The operator $|||$ is closely related to the disjoint concurrent composition operator $*$ in [6], for which also a reverse exchange law holds.

We conclude this section by an inference rule for Hoare triples involving $|||$: for graphlets G, G' and states C, C', D, D' we have

$$\frac{\{C\}\,G\,\{D\} \qquad \{C'\}\,G'\,\{D'\}}{\{C+D\}\,G \,|||\, G'\,\{D+D'\}},$$

where we have identified graphlets and singleton programs. The proof is again straightforward from Lemma 3.24 and the definitions.

3.7 Modal Operators

We continue with another interesting connection, namely to the theory of modal semirings (e.g. [9]), which will be useful for the variant of CTL presented in Sect. 4.4.

Input restriction obeys the following laws, where false is the empty predicate and true is the predicate containing all states:

$$\begin{aligned}
&\mathsf{false}\rfloor P = \bot,\\
&\mathsf{true}\rfloor P = P\\
&(S \cup S')\rfloor P = (S\rfloor P) \;[\!]\; (S'\rfloor P),\\
&(S \cap S')\rfloor P = S\rfloor (S'\rfloor P),\\
&P = S\rfloor P \iff in(P) \subseteq S.
\end{aligned}$$

The first four of these say that restriction acts as a kind of module operator between states and graphlets. Symmetric laws hold for output restriction.

The last law means that in satisfies the characteristic property of an abstract domain operator as known from modal semirings [9], namely that $in(G)$ is the least preserver of G under input restriction. The law is equivalent to the following pair of laws:

$$in(G)\rfloor G = G, \qquad in(S\rfloor G) \subseteq S.$$

The domain operator can also be viewed as an "enabledness" predicate (cf. [28]).

Lifting the in function to programs P and predicates S we can now define modal operators.

Definition 3.26. The forward diamond operator $|P\rangle$ and box operator $|P]$ are given by

$$|P\rangle S =_{df} in(P \lfloor S),$$
$$|P]S =_{df} \neg|P\rangle\neg S.$$

The backward operators $\langle P|$ and $[P|$ are defined symmetrically using the lifted *out* function.

The forward diamond $|P\rangle S$ calculates all immediate predecessor states of S-states under program P, i.e., all states from which an S state can be reached by one P-step. The forward box operator $|P]S$ calculates all states from which every P-transition is guaranteed to satisfy S; it corresponds to Dijkstra's wlp operator, all of whose laws can be derived algebraically from these definitions. For the relational case analogous definitions appear already in [3,26]. Diamond distributes through union and is strict and covariant in both arguments. Box antidistributes through union and hence is contravariant in its first argument, while distributes through intersection and hence is covariant in its second argument.

4 CKA and Temporal Logics

The temporal logic CTL^* and its sublogics CTL and LTL are prominent tools in the analysis of concurrent and reactive systems. First algebraic treatments of these logics were obtained by von Karger and Berghammer [20,21]. A partial treatment in the framework of fork algebras was also given by Frías and Lopez Pombo [13]. For LTL compact closed expressions could be obtained by Desharnais, Möller and Struth in [8]. This was extended in [23] to a semantics for full CTL^* and its sublogics in the framework of quantales. Since, as mentioned in Sect. 2.2, CKA has a quantale semantics, that approach can be applied to the interface model as well.

In this, sets of states and hence the semantics of state formulas can be represented as predicates, while general programs represent the semantics of path formulas.

4.1 An Algebraic Semantics of CTL^*

To make the paper self-contained, we repeat some basic facts about CTL^* and its algebraic semantics.

The language Ψ of CTL^* *formulas* (see e.g. [12]) over a set Φ of atomic propositions is defined by the grammar

$$\Psi \ ::= \ \bot \mid \Phi \mid \Psi \rightarrow \Psi \mid \mathsf{X}\Psi \mid \Psi\,\mathsf{U}\,\Psi \mid \mathsf{E}\Psi,$$

where $\bot$ denotes falsity, $\rightarrow$ is logical implication, X and U are the next-time and until operators and E is the existential quantifier on paths.

The use of the next-time operator X means that implicitly a small-step semantics for the specified or analysed transition system is used. Informally, the formula

$\mathsf{X}\varphi$ is true for a trace p if φ is true for the remainder of p after one such step. We will briefly discuss below which algebraic properties the semantic element, a program, corresponding to X should have.

The formula $\varphi \,\mathsf{U}\, \psi$ is true for a trace p if

- after zero or more X steps within p the formula ψ holds for the remaining trace and
- for all intermediate trace pieces for which ψ does not yet hold the formula φ is true.

The logical connectives $\neg, \wedge, \vee, \mathsf{A}$ are defined, as usual, by $\neg\varphi =_{df} \varphi \rightarrow \bot$, $\top =_{df} \neg\bot$, $\varphi \wedge \psi =_{df} \neg(\varphi \rightarrow \neg\psi)$, $\varphi \vee \psi =_{df} \neg\varphi \rightarrow \psi$ and $\mathsf{A}\varphi =_{df} \neg\,\mathsf{E}\neg\varphi$. Moreover, the "finally" operator F and the "globally" operator G are defined by

$$\mathsf{F}\psi =_{df} \top \mathsf{U} \psi \qquad \text{and} \qquad \mathsf{G}\psi =_{df} \neg\mathsf{F}\neg\psi .$$

Informally, $\mathsf{F}\psi$ holds if after a finite number of steps the remainder of the trace satisfies ψ, while $\mathsf{G}\psi$ holds if after every finite number of steps ψ still holds.

The sublanguages Σ of *state formulas* that denote sets of states and Π of *path formulas* that denote sets of traces are given by

$$\Sigma ::= \bot \mid \Phi \mid \Sigma \rightarrow \Sigma \mid \mathsf{E}\Pi,$$
$$\Pi ::= \Sigma \mid \Pi \rightarrow \Pi \mid \mathsf{X}\Pi \mid \Pi \,\mathsf{U}\, \Pi .$$

To motivate our algebraic semantics, we briefly recapitulate the standard CTL^* semantics formulas. Its basic objects are traces σ from T^+ or T^ω, the sets of finite non-empty or infinite words over some set T of states. The i-th element of σ (indices starting with 0) is denoted σ_i, and σ^i is the trace that results from σ by removing its first i elements.

Each atomic proposition $\pi \in \Phi$ is associated with the set $T_\pi \subseteq T$ of states for which π is true. The relation $\sigma \models \varphi$ of *satisfaction* of a formula φ by a trace is defined inductively (see e.g. [12]) by

$$\begin{array}{lll}
\sigma & \not\models \bot, & \\
\sigma & \models \pi & \text{iff } \sigma_0 \in T_\pi, \\
\sigma & \models \varphi \rightarrow \psi & \text{iff } \sigma \models \varphi \text{ implies } \sigma \models \psi, \\
\sigma & \models \mathsf{X}\varphi & \text{iff } \sigma^1 \models \varphi, \\
\sigma & \models \varphi \,\mathsf{U}\, \psi & \text{iff } \exists j \geq 0 .\ \sigma^j \models \psi \text{ and } \forall k < j .\ \sigma^k \models \varphi, \\
\sigma & \models \mathsf{E}\varphi & \text{iff } \exists \tau .\ \tau_0 = \sigma_0 \text{ and } \tau \models \varphi .
\end{array}$$

In particular, $\sigma \models \neg\varphi$ iff $\sigma \not\models \varphi$.

From this semantics one can extract a set-based one by assigning to each formula φ the set $[\![\varphi]\!] =_{df} \{\sigma \mid \sigma \models \varphi\}$ of paths that satisfy it. This is the basis of the algebraic semantics in terms of quantales.

We now repeat the algebraic interpretation of CTL^* over a Boolean left quantale B from [23]. To save some notation we set Φ equal to the set of *tests*, i.e., elements below the multiplicative unit 1, of that quantale. These abstractly represent sets of states of the modelled transition system. Moreover, we fix an element X that represents the transition system underlying the logic. The precise requirements for X are discussed in Sect. 4.3. Then the concrete semantics above generalises to a function $[\![\ ]\!] : \Psi \to B$, where $[\![\varphi]\!]$ abstractly represents the set of paths satisfying formula φ and $+$ and $\cdot$ are now the general quantale operators of choice and sequential composition. We note that in quantales with non-trivial test sets, left multiplication $t \cdot a$ for test t and general element a corresponds to input restriction: it leaves only those transitions of a that start with a state in t. With this, we can transform the above concrete semantics into the abstract algebraic one.

Definition 4.1. The *general quantale semantics* of CTL^* formula φ is defined inductively over the structure of φ, where $\bot$ and $\top$ are the least and greatest elements, resp., and $\overline{}$ is the complement operator:

$$\begin{aligned}
[\![\bot]\!] &= \bot, \\
[\![t]\!] &= t \cdot \top, \\
[\![\varphi \to \psi]\!] &= \overline{[\![\varphi]\!]} + [\![\psi]\!], \\
[\![\mathsf{X}\,\varphi]\!] &= \mathsf{X} \cdot [\![\varphi]\!], \\
[\![\varphi\,\mathsf{U}\,\psi]\!] &= \bigsqcup_{j \geq 0} (\mathsf{X}^j \cdot [\![\psi]\!] \sqcap \bigsqcap_{k<j} \mathsf{X}^k \cdot [\![\varphi]\!]), \\
[\![\mathsf{E}\varphi]\!] &= {}^{\ulcorner}[\![\varphi]\!] \cdot \top.
\end{aligned}$$

Here the domain operator $^{\ulcorner}$ is the algebraic abstraction of the operator that yields the set of starting states τ_0 of a set of CTL^* traces τ.

4.2 CTL* for the Interface Model

We now concretise this semantics for the CKA interface model. In fact, we can do so in two ways, namely by interpreting quantale multiplication $\cdot$ by sequential composition ; or quasilinear composition **;**. Choice $+$ is in both cases interpreted as $[\!]$. Since we have used the flat refinement order on graphlets, the set of programs, i.e., of downward closed sets of graphlets, is isomorphic to the power set of the set U of all graphlets, so that the complement operator is well defined.

To model the next-step operator X of CTL^* we assume a fixed graphlet program, denoted again by X, that reflects the small-step semantics of the system to be analysed. An example for such a program would be P from Example 3.8. The other operators of CTL^* deal with the iteration (both finite and infinite) of the single-step semantics, i.e., with the large-step semantics. To make CTL^* work in the sense that the standard properties are obtained, the small-step program X needs to satisfy a number of assumptions that are reflected by certain CTL^* axioms. This is discussed in detail in Sect. 4.3.

To give a CTL^* semantics based on the interface model, we want to assign to every CTL^* formula φ a program, i.e., a set of graphlets, $[\![\varphi]\!]$ that describes admissible iterations of the single-step program X. In adapting the general quantale semantics of Definition 4.1, we replace tests by predicates, left multiplication by tests by restriction, $\top$ by the universal program U and the operator $^{\ulcorner}$ by *in*.

Definition 4.2. The semantics of CTL^* in the interface model is given as follows, where $\cdot \in \{;\,,\,;\}$ and the powers of X are taken w.r.t. $\cdot$.

$$
\begin{aligned}
[\![\bot]\!] &= \bot, \\
[\![S]\!] &= S \downharpoonleft U, \\
[\![\varphi \to \psi]\!] &= \overline{[\![\varphi]\!]} \mathbin{[\,]} [\![\psi]\!], \\
[\![\mathsf{X}\,\varphi]\!] &= \mathsf{X} \cdot [\![\varphi]\!], \\
[\![\varphi \mathbin{\mathsf{U}} \psi]\!] &= \bigcup_{j \geq 0} (\mathsf{X}^j \cdot [\![\psi]\!] \cap \bigcap_{k<j} \mathsf{X}^k \cdot [\![\varphi]\!]), \\
[\![\mathsf{E}\varphi]\!] &= in([\![\varphi]\!]) \downharpoonleft U.
\end{aligned}
$$

Using these definitions, it is straightforward to check that

$$[\![\varphi \vee \psi]\!] = [\![\varphi]\!] \mathbin{[\,]} [\![\psi]\!], \qquad [\![\varphi \wedge \psi]\!] = [\![\varphi]\!] \cap [\![\psi]\!], \qquad [\![\neg\varphi]\!] = \overline{[\![\varphi]\!]}. \tag{8}$$

We have the following important property.

Theorem 4.3. *Assume that multiplication with* X *from the left distributes through arbitrary joins and binary meets, i.e.,* $\mathsf{X} \cdot (P \cap Q) = \mathsf{X} \cdot P \cap \mathsf{X} \cdot Q$ *for all programs* P, Q.

1. $[\![\varphi \mathbin{\mathsf{U}} \psi]\!]$ *is the least fixed point* μf *of the function* $f(Y) =_{df} [\![\psi]\!] \mathbin{[\,]} ([\![\varphi]\!] \cap \mathsf{X} \cdot Y)$.
2. $[\![\mathsf{F}\psi]\!] = \mathsf{X}^* \cdot [\![\psi]\!]$. *In particular,* $[\![\mathsf{F}\top]\!] = U$.

The proof is a direct translation of the one for general quantales given in [23].

Example 4.4. Our program P from Example 3.8 satisfies the assumption of that theorem. □

4.3 Requirements for Small-Step Programs

We now want to find suitable requirements on X. A fundamental requirement in standard CTL^* is validity of the axiom

$$\neg\mathsf{X}\varphi \leftrightarrow \mathsf{X}\neg\varphi. \tag{9}$$

To satisfy it in the algebraic setting, we need to have for all formulas φ and their semantic values $Q =_{df} [\![\varphi]\!]$,

$$\overline{\mathsf{X} \cdot Q} = [\![\neg\mathsf{X}\varphi]\!] = [\![\mathsf{X}\neg\varphi]\!] = \mathsf{X} \cdot \overline{Q}. \tag{10}$$

This semantic property can equivalently be characterised as follows (see again [23]).

Lemma 4.5. *Assume the same properties of* X *as in Theorem 4.3.*

1. $\forall Q : \mathsf{X} \cdot \overline{Q} \subseteq \overline{\mathsf{X} \cdot Q} \iff \forall Q, R : \mathsf{X} \cdot (Q \cap R) = \mathsf{X} \cdot Q \cap \mathsf{X} \cdot R.$
2. $\forall Q : \overline{\mathsf{X} \cdot Q} \subseteq \mathsf{X} \cdot \overline{Q} \iff \mathsf{X} \cdot U = U \iff \mathsf{X}^\omega = U.$

Let us explain the meaning of these formulas. In relation algebra, the special case $\mathsf{X} \cdot \overline{\mathsf{skip}} \subseteq \overline{\mathsf{X}}$ of the property in Lemma 4.5.1 characterises X as a partial function and is equivalent to the full property (10) (see [27]). But in general quantales the special and the full case are not equivalent [10]. Moreover, again from [10], we know that in quantales such as that of formal languages under concatenation, left composition with an element X distributes over meet iff X is prefix-free, i.e. if no member of X is a prefix of another member. This holds in particular if all words in X have equal length, which is the case if X models a transition relation and hence consists only of words of length 2. The program P from Example 3.8 has analogous character.

The equivalent condition $\forall b : \mathsf{X} \cdot Q \cap \mathsf{X} \cdot \overline{Q} = 0$ was used in the computation calculus of R.M. Dijkstra [11].

But what about the property in Lemma 4.5.2? Only rarely will the set of all graphlets be "generated" by an element X in the sense that $\mathsf{X}^\omega = U$. The solution is to choose a left-distributive and right-strict element X and restrict the set of semantic values to the subset $\mathrm{SEM}(\mathsf{X}) =_{df} \{Q : Q \subseteq \mathsf{X}^\omega\}$, taking complements relative to X^ω. This set is clearly closed under $[\!]$ and $\cap$ and under prefixing by X, since by isotony

$$\mathsf{X} \cdot Q \subseteq \mathsf{X} \cdot \mathsf{X}^\omega = \mathsf{X}^\omega .$$

Finally, it also contains all elements $S \rfloor \mathsf{X}^\omega$ for predicates S, since $S \rfloor Q \subseteq Q$ for all Q. Hence the above semantics is well-defined in $\mathrm{SEM}(\mathsf{X})$ if we replace U by X^ω. This entails

$$[\![\mathsf{G}\psi]\!] = \bigcap_{i \in \mathbb{N}} \mathsf{X}^i \cdot [\![\psi]\!],$$

in pleasant symmetry to the property $[\![\mathsf{F}\psi]\!] = \mathsf{X}^* \cdot [\![\psi]\!] = \bigcup_{i \in \mathbb{N}} \mathsf{X}^i \cdot [\![\psi]\!]$.

We conclude this section by showing that one can transform the semantics of $\mathsf{G}\psi$ into closed form.

Theorem 4.6. *Assume Eq.* (10)*, i.e.,* $\forall Q : \overline{\mathsf{X} \cdot Q} = \mathsf{X} \cdot \overline{Q}$.

1. *Left multiplication by* X *distributes through arbitrary intersections.*
2. *If* $\cdot$ *is interpreted as* ; *, for a state formula* ψ *with* $[\![\psi]\!] = S \rfloor U$ *for some predicate* S *we have*

$$[\![\mathsf{G}\psi]\!] = (S \rfloor \mathsf{X})^\omega .$$

The lengthy proof is presented in the Appendix.

Example 4.7. For the program P from Example 3.8 we obtain, using Example 3.22,

$$1 \rfloor P^\omega \subseteq \mathbb{N}_+ \rfloor P^\omega \subseteq (\mathbb{N}_+ \rfloor P)^\omega = [\![\mathsf{G}\mathbb{N}_+]\!].$$

This means that if the variable x starts with value 1 then its contents will always remain positive under infinite iteration of the increment program P. □

A more elaborate example is presented in Sect. 4.6.

4.4 From CTL* to CTL

For a number of applications the sublogic CTL of CTL* suffices. Syntactically, it consists of those CTL* state formulas that only use path formulas of the restricted form $\Pi ::= \mathsf{X}\Sigma \mid \Sigma \mathsf{U} \Sigma$.

It can be shown (in analogy to [23]) that the semantics of every CTL formula has the form $S \rfloor U$ for some predicate S; moreover, S can be retrieved as $S = in(S \rfloor U)$. This is reflected by the simplified semantics

$$[\![\varphi]\!]_s =_{df} in([\![\varphi]\!]),$$

which enables us to calculate solely with predicates in the case of CTL.

First, for the Boolean connectives we obtain

$$[\![\varphi \vee \psi]\!]_s = [\![\varphi]\!]_s \cup [\![\psi]\!]_s, \qquad [\![\varphi \wedge \psi]\!]_s = [\![\varphi]\!]_s \cap [\![\psi]\!]_s, \qquad [\![\neg\varphi]\!]_s = \overline{[\![\varphi]\!]_s}.$$

To make this simplified semantics work well, we need strong properties of restriction and the *in* operator. Therefore, to use Lemma 3.18, we now choose the interpretation $\cdot = ;$. Then the inductive behaviour of $[\![\_]\!]_s$ for all CTL formulas is as follows; it involves now the modal operators from Sect. 3.7.

Theorem 4.8

1. $[\![\bot]\!]_s = \emptyset,$
2. $[\![S]\!]_s = S,$
3. $[\![\varphi \rightarrow \psi]\!]_s = \overline{[\![\varphi]\!]_s} \cup [\![\psi]\!]_s,$
4. $[\![\mathsf{EX}\varphi]\!]_s = |\mathsf{X}\rangle [\![\varphi]\!]_s,$
5. $[\![\mathsf{AX}\varphi]\!]_s = |\mathsf{X}] [\![\varphi]\!]_s,$
6. $[\![\mathsf{E}(\varphi \mathsf{U} \psi)]\!]_s = |([\![\varphi]\!]_s ; \mathsf{X})^*\rangle [\![\psi]\!]_s,$
7. $[\![\mathsf{A}(\varphi \mathsf{U} \psi)]\!]_s = [\![\mathsf{F}\psi]\!]_s \cap |(\overline{[\![\psi]\!]_s} ; \mathsf{X})^*]([\![\varphi]\!]_s + [\![\psi]\!]_s).$

Parts 4 and 5 mean that the existential and universal quantifiers of CTL are semantically reflected as the existential and universal modal operators diamond and box. Part 6 means that the starting states of the traces in $[\![\mathsf{E}(\varphi \mathsf{U} \psi)]\!]_s$ are precisely those from which after finitely many X steps through φ states a ψ state can be reached. Part 7 characterises $[\![\mathsf{A}(\varphi \mathsf{U} \psi)]\!]_s$ as the set of those states from which eventually a ψ state must be reached and for which iteration through non-ψ states must lead to a φ or a ψ state. The proofs are again direct translations of the corresponding ones in [23].

This theorem entails the following pleasant characterisations.

Corollary 4.9

$$[\![\mathsf{EF}\psi]\!]_s = |\mathsf{X}^*\rangle [\![\psi]\!]_s, \qquad [\![\mathsf{EG}\psi]\!]_s = in(([\![\psi]\!]_s ; \mathsf{X})^\omega),$$
$$[\![\mathsf{AG}\psi]\!]_s = |\mathsf{X}^*] [\![\psi]\!]_s, \qquad [\![\mathsf{AF}\psi]\!]_s = \overline{in((\overline{[\![\psi]\!]_s} ; \mathsf{X})^\omega)}.$$

4.5 CTL* and LTL

The logic LTL is the fragment of CTL* in which only A may occur, once and outermost only, as path quantifier. More precisely, the LTL path formulas are given by

$$\Pi ::= \Phi \mid \bot \mid \Pi \rightarrow \Pi \mid \mathsf{X}\Pi \mid \Pi \mathsf{U} \Pi.$$

The LTL semantics is embedded into the CTL^* one by assigning to $\varphi \in \Pi$ the semantic value $[\![\mathsf{A}\varphi]\!]$.

In so far, LTL for the interface model is covered by the semantics in Sect. 4.2. If instead of the graphlet quantale the quantale of finite and infinite sequences is used, by a slight twist of the general semantics one can obtain a simplified version using only *finite* iteration and the modal operators [23]. However, that twist is possible only, since the generating element X in that algebra is left cancellative w.r.t. sequential composition. Since this in general will not hold in the interface model, we just stay with the LTL semantics given above.

4.6 A Somewhat Larger Example

We consider a program that works on two variables `x` and `y` that are allocated and initialised to 0 by the actions *allocx* and *allocy*. Then the program uses two actions *stepy* and *addxy* given by `y := y+2` and `x := x+y`, respectively. These actions are performed repeatedly. For `x` only 8 such actions take place, after which `x` is deallocated; the action on `y` is repeated forever.

We use arrows of the form $\xrightarrow[v]{z}$, where z is a variable and v a value. To avoid excessive indexing we assume that arrows are, in addition to their labelling, distinguished by their position within the diagram. A set of arrows is considered coherent, i.e., a state, if for every variable z it contains at most one arrow $\xrightarrow[v]{z}$. The overall graph H of the program looks as follows.

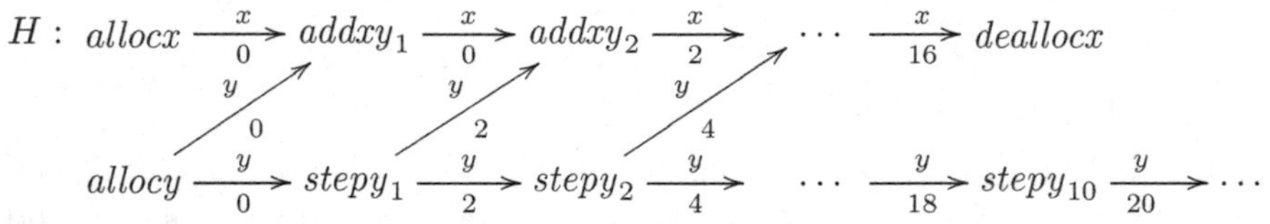

This example exhibits the following phenomena.

- The graph contains special subgraphs the arrows of which have the shape of the letter N rotated clockwise by 90°. Each of these is induced by events $addxy_i, addxy_{i+1}, stepy_i, stepy_{i+1}$ $(1 \leq i \leq 7)$ or by $allocx, addxy_1, allocy, stepy_1$. According to a result by Gischer [14], a graph containing such N-shaped subgraphs, hence in particular the graph H, cannot be expressed as a composition of singleton graphlets using only the ; and | operators.
- The graph can, however, be decomposed into a sequential composition of the slices $\{allocx, allocy\}$, $\{addxy_i, stepy_i\}$ $(1 \leq i \leq 9)$, $\{deallocx, stepy_{10}\}$ and $\{stepy_j\}$ $(j \geq 11)$. These slices can be taken as the elements of a small-step program X that corresponds to the CTL^* next-time operator. Then the overall graph is the "longest" element of X^ω, selected from X^ω by the restriction $\{\emptyset\} \rfloor \mathsf{X}^\omega$, i.e., by taking the only trace in X^ω that has an empty input interface.
- We can express in CTL^* and show that y remains available forever, whereas x does not. A possible cause for this may be that deallocation of y has just been

forgotten, since it is not discovered that y never is actually used any more. Such an analysis can be used to prevent possible memory leaks. To formalise that we say that a state A satisfies the predicate $has(z)$ for a variable z iff for some value v there is an arrow $\xrightarrow[v]{z}$ in A. Then we can express the above properties as

$$\mathsf{F}(\overline{has(z)}), \qquad \mathsf{XG}(has(y)),$$

which indeed hold for our program.

- We can also describe and show the joint behaviour of x and y, notably their coupling by an invariant: our program satisfies $\mathsf{G}(y > 0 \implies y = x + 2)$, with the obvious semantics for the predicate following the G operator.
- If we choose a different small-step program consisting only of the elementary *stepy* events we can specify and analyse the y subthread by itself in an analogous fashion: for instance, we have $\mathsf{G}(even(y))$.

5 Conclusion and Outlook

We have presented an interface model for CKA and have shown how its operators can be used to formalise variants of the temporal logics CTL^* and CTL that are suitable for specifying and reasoning about temporal subthreads of CKA programs. This was done using the fact that CKA induces a quantale w.r.t. to each of its composition operators. We have, however, only exploited the quantales dealing with sequential compositions ; and **;**. The quantale induced by the concurrent composition operator $|$ could be used in a similar manner to set up analogous "spatial" logics, again based on finite or infinite iteration. Also, links between $\mathsf{CTL}, \mathsf{CTL}^*, \mathsf{LTL}$, the algebraic temporal logics of von Karger [20] and the Duration Calculus of Zhou [32] need to be established. The details of that as well as a combination of temporal and spatial logic constructs to deal properly with truly concurrent aspects will be the subject of further research.

Acknowledgments. We are grateful to Jules Desharnais, Peter Höfner, Martin E. Müller, Patrick Roocks, Stephan van Staden and the anonymous referees for thorough proofreading and valuable comments.

Appendix: Deferred Proofs

Proof of Lemma 3.5.
We only show Part 1, since Part 2 is analogous. First,

$$\begin{array}{ll} & in(G\,|\,G') \\ = & \quad \{\!\!\{ \text{ by Definition 3.3 } \}\!\!\} \\ & (A \cup A') \cap \overline{E + E'} \times (E + E') \\ = & \quad \{\!\!\{ \text{ distributivity } \}\!\!\} \\ & (A \cup A') \cap (\overline{E + E'} \times E + \overline{E + E'} \times E') \end{array}$$

$$\begin{array}{ll}
= & \{\!\![\text{ distributivity }]\!\!\} \\
& ((A \cap \overline{E+E'} \times E) + (A \cap \overline{E+E'} \times E')) \cup \\
& ((A' \cap \overline{E+E'} \times E) + (A' \cap \overline{E+E'} \times E')).
\end{array}$$

We simplify the first two summands; the other two are analogous. For summand number one we have

$$\begin{array}{ll}
& A \cap \overline{E+E'} \times E \\
= & \{\!\![\text{ De Morgan }]\!\!\} \\
& A \cap (\overline{E} \cap \overline{E'}) \times E \\
= & \{\!\![\cap/\times\text{-exchange using } E = E \cap E \ ]\!\!\} \\
& A \cap \overline{E} \times E \cap \overline{E'} \times E \\
= & \{\!\![\text{ by Definitions 3.3 }]\!\!\} \\
& in(G) \cap \overline{E'} \times E.
\end{array}$$

For summand number two we have

$$\begin{array}{ll}
& A \cap \overline{E+E'} \times E' \\
= & \{\!\![\text{ De Morgan, distributivity }]\!\!\} \\
& A \cap \overline{E} \times E' \cap \overline{E'} \times E' \\
\subseteq & \{\!\![\text{ by } E \cap E' = \emptyset \text{, hence } E' \subseteq \overline{E} \text{ and covariance of } \times \ ]\!\!\} \\
& A \cap \overline{E} \times \overline{E} \cap \overline{E'} \times E' \\
= & \{\!\![\text{ by Definition 3.1.7 }]\!\!\} \\
& \emptyset \cap \overline{E'} \times E' \\
= & \{\!\![\text{ set algebra }]\!\!\} \\
& \emptyset.
\end{array}$$

Altogether, the claim is shown. □

Proof of Theorem 3.6.
First we observe that $(\mathsf{G}, |)$ is an aggregation algebra in the sense of [15] and CS and $\mathrm{CC}(G, G') =_{df} \mathsf{TRUE}$ can be viewed as independence relations with CS $\subseteq$ CC. With this, the claim follows from Lemmas 3.4 and 3.5 and Proposition 3.6 of [15] if we can show that the restricting predicates CS and CC are bilinear, i.e., satisfy.

$$\begin{array}{lll}
\mathrm{CS}(G \,|\, G', G'') \iff \mathrm{CS}(G, G'') & \wedge \ \mathrm{CS}(G', G''), & \\
\mathrm{CS}(G, G' \,|\, G'') \iff \mathrm{CS}(G, G') & \wedge \ \mathrm{CS}(G, G'') & \quad (11)
\end{array}$$

and the analogous property for CC, and CC is symmetric. For CC both claims are trivial. To show (11), we exploit that the definition of CS is symmetric in both arguments, so that it suffices to consider the first argument.

$$\begin{array}{ll}
& \mathrm{CS}(G \,|\, G', G'') \\
\Leftrightarrow & \{\!\![\text{ by Definitions 3.4 }]\!\!\} \\
& (A \cup A') \cap A'' \cap E'' \times (E + E') = \emptyset
\end{array}$$

$$
\begin{array}{ll}
\Leftrightarrow & \{\!\!\{ \text{ distributivity, set algebra } \}\!\!\} \\
& A \cap A'' \cap E'' \times E = \emptyset \ \wedge \\
& A \cap A'' \cap E'' \times E' = \emptyset \ \wedge \\
& A' \cap A'' \cap E'' \times E = \emptyset \ \wedge \\
& A' \cap A'' \cap E'' \times E' = \emptyset \\
\Leftrightarrow & \{\!\!\{ \text{ by Definition 3.4, and since } E' \cap E = E \cap E'' = E' \cap E'' = \emptyset, \\
& \quad \text{hence } E'', E' \subseteq \overline{E} \text{ and } E'', E \subseteq \overline{E'}, \text{ therefore} \\
& \quad A \cap E'' \times E' = A' \cap E'' \times E = \emptyset \text{ by condition (7) of Definition 3.1} \}\!\!\} \\
& \mathsf{CS}(G, G'') \wedge \mathsf{TRUE} \wedge \mathsf{TRUE} \wedge \mathsf{CS}(G', G'').
\end{array}
$$

Proof of Lemma 3.18.
By the definitions of; and ; their result, and its interfaces, coincides with that of parallel composition whenever it is defined.

We start by showing the first equation. By the above remark and Lemma 3.5,

$$in(G\,;G') \;=\; (in(G) \cap \overline{E'} \times E) \cup (in(G') \cap \overline{E} \times E').$$

For the second summand we calculate by $in(G') = out(G)$ and the definition of out, Eq. (6), and disjointness of E, E' with (6) and Boolean algebra:

$$in(G') \cap \overline{E} \times E' = A \cap \overline{E} \times E \cap \overline{E} \times E' = A \cap \overline{E} \times (E \cap E') = \emptyset.$$

Concerning the first summand we have by by $in(G') = out(G)$ and the definition of in, out, definition of CS in Definition 3.4 and shunting, Eq. (6), distributivity, second summand $= \emptyset$ by Eq. (6), definition of $\cap$,

$$
\begin{array}{l}
in(G) \subseteq A \cap A' \cap \overline{E} \times E \subseteq \overline{E'} \times \overline{E} \cap \overline{E} \times E = (\overline{E'} \times E + \mathrm{EV} \times \overline{E}) \cap \overline{E} \times E = \\
\quad (\overline{E'} \times E \cap \overline{E} \times E) + (\mathrm{EV} \times \overline{E} \cap \overline{E} \times E) = \overline{E'} \times E \cap \overline{E} \times E \subseteq \overline{E'} \times E.
\end{array}
$$

Therefore the first summand reduces to $in(G)$, as required.

The equation $in(G) = in(G \downharpoonleft in(G'))$ is immediate, since the definition of $\downharpoonleft$ and $in(G') = out(G)$ entail $G \downharpoonleft in(G') = G$.

The equations for out are proved completely symmetrically.

Now we can show associativity of ;. Assume graphlets G, G, G''. If any of them is $\square$ then the associativity equation is immediate from the definition of ;. Otherwise we only need to check the case where $in(G') = out(G)$ and $in(G'') = out(G')$ so that both $G;G'$ and $G';G''$ are defined. By the just proved equations for in and out then also $in(G'') = out(G\,;G')$ and $in(G'\,;G'') = out(G)$, so that also $(G;G');G''$ and $G;(G'\,;G'')$ are defined; by definition of ; and associativity of ; they coincide.

Next, by the definition of the restriction operators, $G \downharpoonleft C$ and $C \downharpoonright G'$ are defined iff $out(G) \subseteq C$ and $in(G') \subseteq C$. In that case $G \downharpoonleft C = G$ and $C \downharpoonright G' = G'$ and therefore

$$(G \downharpoonleft C)\,;G' = G\,;G' = G\,;\ (C \downharpoonright G').$$

Finally, the last two claims are immediate from the definition of restriction and the first two claims. $\square$

Proof of Lemma 3.24.
We only show the first equation, since the second is analogous. By the definition of $|||$ and Lemma 3.5 we have

$$in(G\,|\,G') \;=\; (in(G) \cap \overline{E'} \times E) \,\cup\, (in(G') \cap \overline{E} \times E').$$

The first summand reduces to $in(G)$ if we can show $in(G) \subseteq \overline{E'} \times E$, equivalently, $in(G) \cap \overline{\overline{E'} \times E} = \emptyset$. We calculate, by the definition of in with Eq. (6), distributivity with Eq. (6), and the definition of $|||$:

$$in(G) \cap \overline{\overline{E'} \times E} = A \cap \overline{E} \times E \cap (\mathrm{EV} \times \overline{E} + E' \times E) = A \cap \overline{E} \times E \cap E' \times E = \emptyset\ .$$

Symmetrically, the second summand reduces to $in(G')$. □

Proof of Theorem 4.6.

1.
$$\begin{aligned}
& \mathsf{X} \cdot \bigcap_i Q_i \\
= \quad & \{\!|\ \text{De Morgan}\ |\!\} \\
& \mathsf{X} \cdot \overline{\bigcup_i \overline{Q_i}} \\
= \quad & \{\!|\ \text{Eq. (10)}\ |\!\} \\
& \overline{\mathsf{X} \cdot \bigcup_i \overline{Q_i}} \\
= \quad & \{\!|\ \text{quantale distributivity}\ |\!\} \\
& \overline{\bigcup_i \mathsf{X} \cdot \overline{Q_i}} \\
= \quad & \{\!|\ \text{Eq. (10)}\ |\!\} \\
& \overline{\bigcup_i \overline{\mathsf{X} \cdot Q_i}} \\
= \quad & \{\!|\ \text{De Morgan}\ |\!\} \\
& \bigcap_i \mathsf{X} \cdot Q_i .
\end{aligned}$$

2. We show by induction on i that $\bigcap_{j \le i} \mathsf{X}^j \cdot [\![\psi]\!] = (S \downharpoonright \mathsf{X})^i \cdot [\![\psi]\!]$. The base case $i = 0$ is trivial. The induction step proceeds as follows:

$$\begin{aligned}
& \bigcap_{j \le i+1} \mathsf{X}^j \cdot [\![\psi]\!] \\
= \quad & \{\!|\ \text{set theory}\ |\!\} \\
& \bigcap_{0 < j \le i+1} \mathsf{X}^j \cdot [\![\psi]\!] \cap [\![\psi]\!] \\
= \quad & \{\!|\ \text{left distributivity of } \mathsf{X}\ |\!\} \\
& \mathsf{X} \cdot (\bigcap_{0 < j \le i+1} \mathsf{X}^{j-1} \cdot [\![\psi]\!]) \cap [\![\psi]\!] \\
= \quad & \{\!|\ \text{index transformation } k = j - 1\ |\!\} \\
& \mathsf{X} \cdot (\bigcap_{k \le i} \mathsf{X}^k \cdot [\![\psi]\!]) \cap [\![\psi]\!] \\
= \quad & \{\!|\ \text{induction hypothesis}\ |\!\}
\end{aligned}$$

$$
\begin{array}{ll}
 & \mathsf{X} \cdot (S \downharpoonleft \mathsf{X})^i \cdot [\![\psi]\!] \cap [\![\psi]\!] \\
= & \quad \{\![\text{ Lemma 3.11 }]\!\} \\
 & S \downharpoonleft (\mathsf{X} \cdot (S \downharpoonleft \mathsf{X})^i \cdot [\![\psi]\!]) \\
= & \quad \{\![\text{ Lemma 3.18 }]\!\} \\
 & (S \downharpoonleft \mathsf{X}) \cdot (S \downharpoonleft \mathsf{X})^i \cdot [\![\psi]\!]) \\
= & \quad \{\![\text{ definition of powers }]\!\} \\
 & (S \downharpoonleft \mathsf{X})^{i+1} \cdot [\![\psi]\!].
\end{array}
$$

By this,

$$[\![\mathsf{G}\psi]\!] = \bigcap_{i \in \mathbb{N}} \mathsf{X}^i \cdot [\![\psi]\!] = \bigcap_{i \in \mathbb{N}} \bigcap_{j \leq i} \mathsf{X}^j \cdot [\![\psi]\!] = \bigcap_{i \in \mathbb{N}} \bigcap_{j \leq i} (S \downharpoonleft \mathsf{X})^i \cdot [\![\psi]\!].$$

An easy induction shows $(S \downharpoonleft \mathsf{X})^\omega \subseteq (S \downharpoonleft \mathsf{X})^i \cdot [\![\psi]\!]$ for all i and hence $(S \downharpoonleft \mathsf{X})^\omega \subseteq [\![\mathsf{G}\psi]\!]$. For the reverse inclusion it suffices to show that $[\![\mathsf{G}\psi]\!] = (S \downharpoonleft \mathsf{X}) \cdot [\![\mathsf{G}\psi]\!]$. Indeed,

$$
\begin{array}{ll}
 & (S \downharpoonleft \mathsf{X}) \cdot [\![\mathsf{G}\psi]\!] \\
= & \quad \{\![\text{ Lemma 3.18 and Lemma 3.11 }]\!\} \\
 & \mathsf{X} \cdot [\![\mathsf{G}\psi]\!] \cap (S \downharpoonleft U) \\
= & \quad \{\![\text{ above representation }]\!\} \\
 & \mathsf{X} \cdot (\bigcap_{i \in \mathbb{N}} \mathsf{X}^i \cdot [\![\psi]\!]) \cap [\![\psi]\!] \\
= & \quad \{\![\text{ Part 1 }]\!\} \\
 & (\bigcap_{i \in \mathbb{N}} \mathsf{X} \cdot \mathsf{X}^i \cdot [\![\psi]\!]) \cap [\![\psi]\!] \\
= & \quad \{\![\text{ definition of powers }]\!\} \\
 & (\bigcap_{i \in \mathbb{N}} \mathsf{X}^{i+1} \cdot [\![\psi]\!]) \cap [\![\psi]\!] \\
= & \quad \{\![\text{ combining cases }]\!\} \\
 & \bigcap_{j \in \mathbb{N}} \mathsf{X}^j \cdot [\![\psi]\!] \\
= & \quad \{\![\text{ above representation }]\!\} \\
 & [\![\mathsf{G}\psi]\!].
\end{array}
$$

□

References

1. Back, R., von Wright, J.: Refinement Calculus - A Systematic Introduction. Graduate Texts in Computer Science. Springer, New York (1998)
2. de Bakker, J., Meertens, L.: On the completeness of the inductive assertion method. J. Comput. Syst. Sci. **11**(3), 323–357 (1975)
3. Blikle, A.: A comparative review of some program verification methods. In: Gruska, J. (ed.) Mathematical Foundations of Computer Science 1977. LNCS, vol. 53, pp. 17–33. Springer, Heidelberg (1977)
4. Brink, C., Rewitzky, I.: A Paradigm for Program Semantics: Power Structures and Duality. CSLI Publications, Stanford (2001)

5. Conway, J.: Regular Algebra and Finite Machines. Chapman and Hall, London (1971)
6. Dang, H.H., Möller, B.: Concurrency and local reasoning under reverse exchange. Sci. Comput. Prog. **85**(Part B), 204–223 (2013)
7. Dang, H., Glück, R., Möller, B., Roocks, P., Zelend, A.: Exploring modal worlds. J. Log. Algebr. Meth. Program. **83**(2), 135–153 (2014)
8. Desharnais, J., Möller, B., Struth, G.: Modal Kleene algebra and applications - a survey. J. Relational Methods Comput. Sci. **1**, 93–131 (2004)
9. Desharnais, J., Möller, B., Struth, G.: Kleene algebra with domain. ACM Trans. Comput. Log. **7**(4), 798–833 (2006)
10. Desharnais, J., Möller, B.: Characterizing determinacy in Kleene algebras. Inf. Sci. **139**(3–4), 253–273 (2001)
11. Dijkstra, R.M.: Computation calculus bridging a formalization gap. Sci. Comput. Program. **37**, 3–36 (2000)
12. Emerson, E.A.: Temporal and modal logic. In: van Leeuwen, J. (ed.) Handbook of Theoretical Computer Science. Formal Models and Semantics (B), vol. B, pp. 995–1072. Elsevier, Amsterdam (1990)
13. Frias, M.F., Pombo, C.L.: Interpretability of first-order linear temporal logics in fork algebras. J. Log. Algebr. Program. **66**(2), 161–184 (2006)
14. Gischer, J.L.: The equational theory of pomsets. Theoret. Comput. Sci. **61**(2–3), 199–224 (1988)
15. Hoare, T., Möller, B., Struth, G., Wehrman, I.: Concurrent Kleene algebra and its foundations. J. Log. Algebr. Program. **80**(6), 266–296 (2011)
16. Hoare, T., van Staden, S.: The laws of programming unify process calculi. Sci. Comput. Program. **85**, 102–114 (2014)
17. Hoare, T., van Staden, S., Möller, B., Struth, G., Villard, J., Zhu, H., O'Hearn, P.: Developments in concurrent Kleene algebra. In: Höfner, P., Jipsen, P., Kahl, W., Müller, M.E. (eds.) RAMiCS 2014. LNCS, vol. 8428, pp. 1–18. Springer, Heidelberg (2014)
18. Hoare, T., van Staden, S., Möller, B., Struth, G., Villard, J., Zhu, H., O'Hearn, P.: Developments in concurrent Kleene algebra. In: Höfner, P., Jipsen, P., Kahl, W., Müller, M.E. (eds.) RAMiCS 2014. LNCS, vol. 8428, pp. 1–18. Springer, Heidelberg (2014)
19. Jipsen, P.: Concurrent Kleene algebra with tests. In: Höfner, P., Jipsen, P., Kahl, W., Müller, M.E. (eds.) RAMiCS 2014. LNCS, vol. 8428, pp. 37–48. Springer, Heidelberg (2014)
20. von Karger, B.: Temporal algebra. Math. Struct. Comput. Sci. **8**(3), 277–320 (1998)
21. von Karger, B., Berghammer, R.: A relational model for temporal logic. Logic J. IGPL **6**(2), 157–173 (1998)
22. Main, M.: A powerdomain primer – a tutorial for the Bulletin of the EATCS 33. Technical report, CU-CS-375-87 (1987). Paper 360, University Colorado at Boulder, Department of Computer Science (1987). http://scholar.colorado.edu/csci_techreports/360
23. Möller, B., Höfner, P., Struth, G.: Quantales and temporal logics. In: Johnson, M., Vene, V. (eds.) AMAST 2006. LNCS, vol. 4019, pp. 263–277. Springer, Heidelberg (2006)
24. O'Hearn, P.W., Reynolds, J.C., Yang, H.: Separation and information hiding. ACM Trans. Program. Lang. Syst. **31**(3), 1–50 (2009)
25. Rosenthal, K.: Quantales and Their Applications, Pitman Research Notes in Mathematics Series, vol. 234. Longman Scientific and Technical, Harlow (1990)

26. Schmidt, G.: Programme als partielle Graphen. TU Munich, FB Mathematatik. Habilitation Thesis (1977)
27. Schmidt, G., Ströhlein, T.: Relations and Graphs: Discrete Mathematics for Computer Scientists. Springer, Heidelberg (1993)
28. Solin, K., von Wright, J.: Enabledness and termination in refinement algebra. Sci. Comput. Program. **74**(8), 654–668 (2009)
29. Tarlecki, A.: A language of specified programs. Sci. Comput. Program. **5**(1), 59–81 (1985)
30. Wehrman, I., Hoare, C.A.R., O'Hearn, P.W.: Graphical models of separation logic. Inf. Process. Lett. **109**(17), 1001–1004 (2009)
31. Winskel, G.: On powerdomains and modality. Theor. Comput. Sci. **36**, 127–137 (1985)
32. Zhou, C., Hoare, C.A.R., Ravn, A.P.: A calculus of durations. Inf. Process. Lett. **40**(5), 269–276 (1991)

On Rely-Guarantee Reasoning

Stephan van Staden(✉)

University College London, London, UK
s.vanstaden@cs.ucl.ac.uk

Abstract. Many semantic models of rely-guarantee have been proposed in the literature. This paper proposes a new classification of the approaches into two groups based on their treatment of guarantee conditions. To allow a meaningful comparison, it constructs an abstract model for each group in a unified setting. The first model uses a weaker judgement and supports more general rules for atomic commands and disjunction. However, the stronger judgement of the second model permits the elegant separation of the rely from the guarantee due to Hayes et al. and allows refinement-style reasoning. The generalisation to models that use binary relations for postconditions is also investigated. An operational semantics is derived and both models are shown to be sound with respect to execution. All proofs have been checked with Isabelle/HOL and are available online.

Keywords: Rely-guarantee · Concurrency · Semantics · Soundness

1 Introduction

Rely-guarantee [8] is a well-established technique for reasoning about concurrent programs. It has been used to verify the correctness of tricky concurrent algorithms and inspired recent program logics such as RGSep, SAGL and LRG. It offers a compositional rule for concurrency by augmenting the usual pre- and postcondition specifications of Hoare logic with summaries of the interference of a program's concurrent environment and also of the program itself. The program can *rely* on its environment to behave according to the environment's interference specification, and must *guarantee* that it will adhere to its own interference constraints. Concretely, interference is summarised by a binary relation on states that over-approximates the effect of individual execution steps.

The judgements of rely-guarantee calculi are thus quintuples of the form:

$$Pre\ R\ \{Prog\}\ G\ Post$$

where Pre is the precondition (a set of states), R is the rely condition (a binary relation on states), $Prog$ is the program, G is the guarantee condition (a binary

This paper is dedicated to Ian Hayes.

S. van Staden—Currently affiliated with Google Switzerland.

R. Hinze and J. Voigtländer (Eds.): MPC 2015, LNCS 9129, pp. 30–49, 2015.
DOI:10.1007/978-3-319-19797-5_2

relation on states), and *Post* is the postcondition (either a set of states or a binary relation on states, depending on the presentation).

Many semantic models of rely-guarantee have appeared in the literature. This paper proposes a new classification of the approaches into two broad groups, which essentially differ in their treatment of guarantee conditions. The main goals of the paper are to capture these differences in a single abstract setting and to investigate their consequences.

The first group stipulates that each step of the program must satisfy the guarantee relation when the initial state satisfies the precondition and the environment satisfies the rely condition. Most mainstream models of rely-guarantee use this interpretation, for example [2,3,11,13,15,18].

The second group stipulates that each step of the program must satisfy the guarantee relation, irrespective of the initial state and the environment. Recent models of rely-guarantee that are based on refinement use this interpretation, for example [1,5,6]. The proofs in these papers suggest that the decoupling of the guarantee condition from the precondition and the rely enables refinement-style reasoning that is much more algebraic in flavour.

The models in previous work often differ in detail (e.g. programming constructs, operational semantics, etc.) which make it hard to study their merits and differences. To avoid this problem, the current paper uses a unified setting of traces as a semantic foundation. It then constructs two models of rely-guarantee that represent the two groups mentioned above.

In order to concentrate only on essential aspects, the semantic setting abstracts from many details. The presentation might therefore seem a bit unconventional to some readers. For example, it makes no assumptions about the (abstract) syntax of programs, it treats computational states abstractly, and it assumes no operational semantics. The idea is that such constraints can be added independently if and when needed. For example, we derive an operational calculus later in the paper to investigate whether the models are sound (i.e. correct) with respect to familiar small-step execution.

The judgements of the two models also make minimal demands. For example, the pre- and postconditions are not required to be 'stable' with respect to the rely condition, and the rely and/or guarantee relations need not be reflexive and/or transitive. Moreover, there is no fixed language for describing interference, and the same holds for assertions.

The models contribute several insights about the two semantic approaches:

- The judgement of the first model is weaker than the one of the second model. Despite this, the models validate mostly the same inference rules, but the weaker judgement supports more general rules for atomic commands and disjunction.
- Only the second model allows for the elegant decomposition of the rely-guarantee quintuple into *rely* and *guar* constructs due to Hayes et al. [5]. The separation of the rely from the guarantee permits refinement-style proofs which humans might find easier to construct.

- The models use postconditions that are single-state predicates, but they generalise nicely to models where postconditions are binary relations on states as is often the case in the rely-guarantee literature [2,5,8]. Interestingly, the attempt to generalise both models in a naive way fails because of their differences.
- Both models are sound with respect to big-step and small-step execution. The soundness proofs are not 'structural', but can be viewed as a simplification of the proof by Coleman and Jones [2]. The soundness results are decoupled from many operational concerns, such as the particular choice of execution rules, and the decision of which atomic operations are easy to implement in a computer.

This paper tries to present rely-guarantee incrementally from first principles, so it might be a good point to start learning about the main ideas. Working in a minimalistic setting also means that many of the proofs are shorter than their counterparts in other literature. All the proofs have been mechanised in Isabelle/HOL and are available online [7] to encourage further exploration.

Outline. Section 2 describes the first model. Section 3 presents the second model which decouples the guarantee from the precondition and the rely. Section 4 generalises the models to rely-guarantee calculi where postconditions are binary relations and not sets of states. Section 5 derives operational calculi to show the soundness of the models with respect to execution. Section 6 discusses related work and Sect. 7 concludes.

2 The First Model

2.1 Formalising the Judgement

Many mainstream treatments of rely-guarantee (e.g. [11,13,17,18]) give the following informal meaning to the quintuple judgement $S\ R\ \{P\}\ G\ S'$:

if

1. **program P is executed in a state which satisfies S, and** ***(precondition)***
2. **every environment step satisfies R,** ***(rely)***

then

1. **every step of P satisfies G, and** ***(guarantee)***
2. **if the execution terminates, then the final state satisfies S'.** ***(postcondition)***

Here, and in the rest of the paper, P will be a program, S and S' are sets of states, and R and G are binary relations on states.

Instead of treating programs as syntactic objects that are generated by a particular (abstract) syntax, we model them generically as sets of traces. Each trace is a sequence of state pairs, called steps, that describe the program's ability to transform states. We use σ to range over states and t to range over traces. The empty trace is denoted by $[]$ and the infix operator $:$ prepends a step to a trace.

Consider the trace $t = (\sigma_1, \sigma_1') : (\sigma_2, \sigma_2') : \cdots : (\sigma_n, \sigma_n')$. Step i transforms state σ_i into σ_i' before step $i + 1$ can be executed. However, step $i + 1$ does not have to be executed immediately after step i, because the model allows the (concurrent) environment to interfere between steps. If $\sigma'_i \neq \sigma_{i+1}$, for example, then step $i+1$ can only be executed when the environment interferes upon the completion of step i to transform state σ'_i into σ_{i+1}. A program's traces therefore describe its potential behaviour and allow for interference by concurrently executing programs. A typical program will have many traces that can never be observed in isolation. These "dormant" behaviours make concurrent programming especially tricky, and it is important to record them in the semantic model.

To formalise the informal interpretation of the rely-guarantee quintuple, it is helpful to consider how each trace of P must behave to satisfy the specification. This is the purpose of the auxiliary judgement *rg-trace*:

Definition 1. $\textit{rg-trace}\ S\ R\ []\ G\ S' \overset{def}{=} R^*(S) \subseteq S'$
$\textit{rg-trace}\ S\ R\ ((\sigma, \sigma') : t)\ G\ S' \overset{def}{=} \sigma \in R^*(S) \Rightarrow (\sigma, \sigma') \in G \wedge \textit{rg-trace}\ \left\{\sigma'\right\}\ R\ t\ G\ S'$.

The base case describes what should happen when the trace is empty. As the trace then contains no steps, it holds vacuously that every step is contained in G. Since the empty trace has no ability to alter the state, the environment must, irrespective of how many steps it performs, transform states satisfying the precondition into ones satisfying the postcondition. In formal terms, the image of S under the relation R^* must be contained in S', where R^* is the reflexive transitive closure of R.

The inductive case describes the situation for a non-empty trace whose first step is (σ, σ'). The first step can become enabled from precondition S and interference R whenever $\sigma \in R^*(S)$. If this is possible, then the step should be in G and the remainder of the trace must fulfill the specification where the new precondition $\left\{\sigma'\right\}$ is the result of the step.

The judgement for a program requires the corresponding auxiliary judgement for all its traces:

Definition 2. $S\ R\ \{P\}\ G\ S' \overset{def}{=} \forall t \in P : \textit{rg-trace}\ S\ R\ t\ G\ S'$.

2.2 Inference Rules

The definition allows a formal investigation of how the judgement can help us to reason about programs. Figures 1 and 2 show a collection of theorems in the form of inference rules[1]. (The reason for separating the rules in two figures is that only the ones of Fig. 1 will also hold as theorems in the second model of Sect. 3.) Roughly speaking, there is one rule for reasoning about each programming operator, and there are additional rules for adapting the specification parts of a judgement.

The programming operators are the familiar ones from formal language theory and are summarised here for reference:

[1] The prefix 'J' in the names of inference rules stands for 'Jones' in tribute to [8].

(Jskip)	$S\,R\,\{skip\}\,G\,(R^*(S))$
(Jseq)	$S\,R\,\{P\}\,G\,S' \;\wedge\; S'\,R\,\{Q\}\,G\,S'' \;\Rightarrow\; S\,R\,\{P\,;Q\}\,G\,S''$
(Jconc)	$S_1\,R_1\,\{P\}\,G_1\,S_1' \;\wedge\; S_2\,R_2\,\{Q\}\,G_2\,S_2' \;\wedge\; G_1 \subseteq R_2 \;\wedge\; G_2 \subseteq R_1 \;\Rightarrow$ $(S_1 \cap S_2)\,(R_1 \cap R_2)\,\{P \parallel Q\}\,(G_1 \cup G_2)\,(S_1' \cap S_2')$
(Jchoice)	$(\forall P \in X : S\,R\,\{P\}\,G\,S') \;\Rightarrow\; S\,R\,\{\bigcup X\}\,G\,S'$
(Jiter)	$S\,R\,\{P\}\,G\,S \;\wedge\; R(S) \subseteq S \;\Rightarrow\; S\,R\,\{P^*\}\,G\,S$
(Jrec)	$(\forall P : S\,R\,\{P\}\,G\,S' \;\Rightarrow\; S\,R\,\{f(P)\}\,G\,S') \;\Rightarrow\; S\,R\,\{lfp\,f\}\,G\,S'$
(Jweak)	$S_1\,R_1\,\{P\}\,G_1\,S_1' \;\wedge\; S_2 \subseteq S_1 \;\wedge\; R_2 \subseteq R_1 \;\wedge\; G_1 \subseteq G_2 \;\wedge\; S_1' \subseteq S_2' \;\Rightarrow$ $S_2\,R_2\,\{P\}\,G_2\,S_2'$
(Jstren)	$S\,R\,\{P\}\,G\,S' \;\Rightarrow\; (R^*(S))\,R^*\,\{P\}\,(G \cap steps(P))\,(S' \cap (R \cup G)^*(S))$
(Jconj)	$S_1\,R_1\,\{P\}\,G_1\,S_1' \;\wedge\; S_2\,R_2\,\{Q\}\,G_2\,S_2' \;\Rightarrow$ $(S_1 \cap S_2)\,(R_1 \cap R_2)\,\{P \cap Q\}\,(G_1 \cap G_2)\,(S_1' \cap S_2')$

Fig. 1. Common rely-guarantee inference rules

- *skip* is the language $\{[]\}$. It does nothing, because its only trace is empty and cannot transform the state.
- a stands for an atom, i.e. a program whose traces all have length one. Every such trace models a single step, so the atoms model (possibly nondeterministic) atomic operations. Atoms are isomorphic to binary relations on states. The binary relation that corresponds to atom a is given by $rel(a)$.
- ; is language concatenation. It corresponds to sequential composition of programs.
- $\parallel$ is language interleaving, also known as shuffle. It corresponds to concurrent composition.
- $\bigcup X$ is the union of all languages in X, i.e. the nondeterministic choice between programs in X. Its binary variant is $\cup$.
- * is the Kleene star, which iterates its operand zero or more times in sequence.
- $lfp\,f$ is the least fixpoint of a monotone function f on languages. It is the meaning of a program P that is defined by recursion as $P = f(P)$. For example, P^* can be defined as the least fixpoint of the monotone function $(\lambda x \,.\, skip \cup (P\,;x))$.

Most of the rules in Fig. 1 are self-explanatory, so a few observations should be sufficient to see how they operate:

- (Jskip): Any guarantee, including the empty relation, is acceptable because *skip* performs no steps. The postcondition takes into account that the environment might still transform states that satisfy the precondition.
- (Jseq): The postcondition of the first program must be the precondition of the second one.
- (Jconc): The guarantee of each program must be compatible with what its concurrent partner relies on. In the consequent of the rule, the environment can

(J1atom) $\quad rel(a) \cap (R^*(S)) \times \Sigma \subseteq G \ \wedge \ (R^*; rel(a); R^*)(S) \subseteq S' \ \Rightarrow \ S\, R\, \{a\}\, G\, S'$

(J1disj) $\quad (\forall S \in Y : S\, R\, \{P\}\, G\, S') \ \Rightarrow \ (\bigcup Y)\, R\, \{P\}\, G\, S'$

Fig. 2. Rules that are specific to the first model

only do what both components have relied on, while the guarantee condition accommodates steps of both components.

- (Jchoice): When every program in a collection satisfies a specification, then the nondeterministic choice between them will also satisfy it.
- (Jiter): Interference from the environment should not invalidate S, which functions as the loop invariant.
- (Jrec): To verify a recursive program, it suffices to check that unfolding the definition once meets the specification when all recursive occurrences meet it.
- (Jweak): This rule can weaken the specification of a judgement and is sometimes known as the 'rule of consequence'.
- (Jstren): This rule can stengthen a specification, i.e. enlarge the precondition and rely, and shrink the guarantee and postcondition. This is not magic – it simply exploits redundancy that exists in specifications. It is simplest to understand this rule in a piecewise fashion as follows. If a judgement holds with precondition S and rely condition R, then the precondition $R^*(S)$ will also work. Moreover, the environment can safely do steps described by R^*. The program can at most perform the steps mentioned in its traces, so the guarantee condition can always be restricted to them ($steps(P)$ relates a pair of states iff the pair is a step in some trace of P). And finally, since the rely and guarantee conditions over-approximate what the environment and program can respectively do, their combination cannot yield final states that are outside $(R \cup G)^*(S)$.
- (Jconj): This rule conjoins the specifications of two judgements. It allows the programs to be different and intersects them in the resulting judgement.

The rules in Fig. 2 should not present difficulties either:

- (J1atom) requires the guarantee to include the relation on states that is isomorphic to the atom, but allows restricting the domain of this relation to those states that the environment can reach from the precondition in zero or more steps. (Σ is the set of all states and $\times$ the Cartesian product operator.) The restriction captures the fact that the atom will never have to perform its step from any other state. Furthermore, the postcondition must at least include those states that can be reached from the precondition under $R^*; rel(a); R^*$, which describes the effect of interleaving the atomic operation with environment steps from R. (When applied to binary relations, ; is the familiar composition operator. Another common symbol for it is $\circ$.)
- (J1disj) says that if a program can meet a specification from each precondition in a set, then it must also be able to meet the specification from their union.

As in most calculi, there is some degree of flexibility in the presentation of the rules. For example, instead of (Jskip) one can prefer the rule:

$$R^*(S) \subseteq S' \;\Rightarrow\; S\,R\,\{skip\}\;G\;S'$$

and it is possible to collapse (J1atom) to:

$$S\,R\,\{a\}\;(rel(a) \cap (R^*(S)){\times}\Sigma)\;((R^*; rel(a); R^*)(S))$$

In both cases one can justify the change and regain the original rule with the weakening rule (Jweak), so nothing is really gained or lost. It is also easy to build aspects of the strengthening rule (Jstren) into other rules, because (Jstren) and (Jweak) imply equivalences such as:

$$\begin{aligned} S\,R\,\{P\}\,G\,S' &\Leftrightarrow (R^*(S))\,R\,\{P\}\,G\,S' \\ S\,R\,\{P\}\,G\,S' &\Leftrightarrow S\,R^*\,\{P\}\,G\,S' \\ S\,R\,\{P\}\,G\,S' &\Leftrightarrow S\,R\,\{P\}\,G\,(S' \cap (R{\cup}G)^*(S)) \end{aligned}$$

These equivalences will of course hold in *any* model that validates (Jstren) and (Jweak).

2.3 Proofs

The formal justification of the inference rules requires rigorous proofs. There are two main reasons why such proofs can be instructive. Firstly, they describe in detail why each rule must work. Secondly, the proofs can collectively communicate the general style of reasoning that the model promotes. It will become clear in Sect. 3 that both aspects can differ considerably between models.

The rely-guarantee judgement in the current model is defined in terms of the auxiliary judgement *rg-trace*. Many rules will therefore directly follow from similar ones about *rg-trace*. Induction on traces is the main mathematical tool of this model, as *rg-trace* was defined by recursion on traces.

For example, the rule (Jconc) follows directly from the following lemma about *rg-trace*, where $t_1 \otimes t_2$ denotes the set of all interleavings of traces t_1 and t_2. Unsurprisingly, the proof proceeds by induction on t:

Lemma 1. *$rg\text{-}trace\; S_1\; R_1\; t_1\; G_1\; S_1' \,\wedge\, rg\text{-}trace\; S_2\; R_2\; t_2\; G_2\; S_2' \,\wedge\, G_1 \subseteq R_2 \,\wedge\, G_2 \subseteq R_1 \,\wedge\, t \in t_1 \otimes t_2 \;\Rightarrow\; rg\text{-}trace\; (S_1 \cap S_2)\; (R_1 \cap R_2)\; t\; (G_1 \cup G_2)\; (S_1' \cap S_2')$.*

Proof. By induction on the structure of t:

- *Base case. We must show that it holds for $t = []$. Since $t \in t_1 \otimes t_2$, we know $t_1 = []$ and $t_2 = []$. Expanding the two rg-trace assumptions gives ${R_1}^*(S_1) \subseteq S_1'$ and ${R_2}^*(S_2) \subseteq S_2'$, which in turn imply $(R_1 \cap R_2)^*(S_1 \cap S_2) \subseteq S_1' \cap S_2'$ because the Kleene star and relational image operators are both monotone.*
- *Step case. Suppose the property holds for a trace t for all t_1, t_2, S_1, S_2. We must show that it will also hold for $(\sigma, \sigma'){:}t$. So assume the rule's antecedents $rg\text{-}trace\; S_1\; R_1\; t_1\; G_1\; S_1'$ and $rg\text{-}trace\; S_2\; R_2\; t_2\; G_2\; S_2'$ and $G_1 \subseteq R_2$ and*

$G_2 \subseteq R_1$ and $(\sigma,\sigma') : t \in t_1 \otimes t_2$. The last assumption implies $\exists t_1' : t_1 = (\sigma,\sigma'){:}t_1' \wedge t \in t_1' \otimes t_2$ or $\exists t_2' : t_2 = (\sigma,\sigma'){:}t_2' \wedge t \in t_1 \otimes t_2'$. The two cases are symmetric, so we will only show the reasoning for the first one. The goal is to show that $\sigma \in (R_1 \cap R_2)^*(S_1 \cap S_2)$ implies both $(\sigma,\sigma') \in G_1 \cup G_2$ and rg-trace $\left\{\sigma'\right\}$ $(R_1 \cap R_2)$ t $(G_1 \cup G_2)$ $(S_1' \cap S_2')$.

Suppose $\sigma \in (R_1 \cap R_2)^*(S_1 \cap S_2)$. Then $\sigma \in {R_1}^*(S_1)$ and $\sigma \in {R_2}^*(S_2)$. Expanding the rg-trace assumption for t_1 now gives rg-trace $\left\{\sigma'\right\}$ R_1 t_1' G_1 S_1' and $(\sigma,\sigma') \in G_1$. So clearly $(\sigma,\sigma') \in G_1 \cup G_2$. Moreover, $(\sigma,\sigma') \in G_1$ and $G_1 \subseteq R_2$ and $\sigma \in {R_2}^*(S_2)$ imply that $\sigma' \in {R_2}^*(S_2)$. So we can enlarge the precondition in the rg-trace assumption for t_2 to ${R_2}^*(S_2)$ and then shrink the precondition of the result to $\left\{\sigma'\right\}$. Applying the induction hypothesis to rg-trace $\left\{\sigma'\right\}$ R_1 t_1' G_1 S_1' and rg-trace $\left\{\sigma'\right\}$ R_2 t_2 G_2 S_2' concludes the proof.

This proof and proofs for all the other rules have been mechanised in Isabelle/HOL. The proof script is available online [7]. The level of detail in formal proofs is typically greater than in pen-and-paper ones, so they can help to clarify gaps in the reasoning. They can also help to formulate different proofs of existing results, or to explore variations on the rules showed here, or even to establish the validity of entirely new ones.

Another use of the Isabelle/HOL formalisation is the discovery of counterexamples. It is straightforward to show, for example, that neither the precondition nor the postcondition have to be 'stable' with respect to the rely condition in this model. In other words, there are examples where $S\,R\,\{P\}\,G\,S'$ holds, yet $R(S) \not\subseteq S$ and $R(S') \not\subseteq S'$. Similarly, the guarantee condition need not be reflexive. Section 6 will discuss why such constraints can be useful in more concrete settings, but there was no need to impose them here.

2.4 The Bigger Picture

The introduction mentioned that rely-guarantee is a generalisation of Hoare logic that augments judgements with rely and guarantee conditions. This intuition can be formalised in a theorem that relates the rely-guarantee quintuple to the Hoare triple (an intuitive understanding of the Hoare triple suffices here; its formal treatment is postponed to Sect. 5):

Theorem 1. $(\exists R\, G : S\, R\, \{P\}\, G\, S') \Leftrightarrow S\,\{P\}\,S'$.

The theorem says that one can establish the Hoare triple $S\,\{P\}\,S'$ by finding some rely R and guarantee G and establishing the rely-guarantee judgement $S\,R\,\{P\}\,G\,S'$ instead. Moreover, if the Hoare triple holds, then it will always be possible to find appropriate rely and guarantee conditions.

One way to prove the theorem is to show that the Hoare triple corresponds exactly to the 'interference-free' situation where the environment can do nothing (the rely is empty) but the program can do anything (the guarantee is the universal relation on states):

Lemma 2. $S\ \emptyset\ \{P\}\ (\Sigma{\times}\Sigma)\ S' \ \Leftrightarrow\ S\ \{P\}\ S'$.

Theorem 1 then follows immediately by (Jweak).

Lemma 2 says that the Hoare triple is a special case of the rely-guarantee judgement. Alternatively, one can view it as characterising certain rely-guarantee judgements (those with empty rely and universal guarantee conditions) in terms of Hoare logic. Surprisingly, the next result shows that it is possible to extend this characterisation to account for arbitrary rely conditions:

Lemma 3. $S\ R\ \{P\}\ (\Sigma{\times}\Sigma)\ S' \ \Leftrightarrow\ S\{P \parallel \mathit{traces}(R)\}\ S'$.

In formal language terminology, $\mathit{traces}(R)$ is the set of all words over alphabet R. Now $\mathit{traces}(\emptyset) = \mathit{skip}$, and skip is the unit of $\parallel$, so Lemma 2 is a straightforward consequence.

One can prove Lemma 3 by first showing:

$$\mathit{rg\text{-}trace}\ S\ R\ t\ (\Sigma{\times}\Sigma)\ S' \ \Leftrightarrow\ S\{\left\{t\right\} \parallel \mathit{traces}(R)\}\,S'$$

This lemma holds by induction on t and [7] contains the full proof.

Unfortunately, there appears to be no straightforward way to extend the characterisation of Lemma 3 to cover arbitrary guarantee conditions. A guarantee condition in this model is quite complicated: its fulfillment depends not only on the program, but also on the precondition and the behaviour of the environment. This is also the case in most mainstream treatments of rely-guarantee. In [2], for example, the auxiliary judgement $\{S, R\} \models P$ **within** G makes the dependency very clear.

3 The Second Model

The quintuple of the second model combines the insight of Lemma 3 with a simple treatment of guarantee conditions:

Definition 3. $S\ R\ \{P\}\ G\ S' \ \stackrel{def}{=}\ S\{P \parallel \mathit{traces}(R)\}\,S' \ \wedge\ P \subseteq \mathit{traces}(G)$.

It demands that all the steps[2] of P must be in G, irrespective of the precondition and the rely. The informal meaning of the quintuple is therefore:

every step of program P satisfies G, and (*guarantee*)
if
1. P is executed in a state which satisfies S, and (*precondition*)
2. every environment step satisfies R, (*rely*)

then
1. if the execution terminates, then the final state satisfies S'. (*postcondition*)

It is easy to show that the new judgement is stronger than the previous one.

3.1 Inference Rules

The fact that the judgement is stronger means that we must again determine which inference rules are theorems. Fortunately, it turns out that all the rules in Fig. 1 remain valid in this model. The ones in Fig. 2 are now invalid, but one can use the variants that appear in Fig. 3:

[2] Note the Galois connection $\mathit{steps}(P) \subseteq G \ \Leftrightarrow\ P \subseteq \mathit{traces}(G)$.

(J2atom)	$rel(a) \subseteq G \ \wedge\ (R^*; rel(a); R^*)(S) \subseteq S' \ \Rightarrow\ S\,R\,\{a\}\,G\,S'$
(J2disj)	$Y \neq \emptyset \ \wedge\ (\forall S \in Y : S\,R\,\{P\}\,G\,S') \ \Rightarrow\ (\bigcup Y)\,R\,\{P\}\,G\,S'$

Fig. 3. Rules that are specific to the second model

- (J2atom): Note that $rel(a)$ must now be fully included in G – there is no possibility to restrict its domain to $R^*(S)$ before checking the inclusion.
- (J2disj): The additional restriction that Y should not be empty reflects the fact that $\emptyset\ R\ \{P\}\ G\ S'$ is not a theorem in this model: the empty precondition does *not* ensure that the steps of P are included in G.

The additional restriction in the rule of disjunction might be a cause for concern. However, by and large the judgement is well-behaved. The next two subsections show that it supports elegant proofs and an interesting decomposition of the judgement.

3.2 Proofs

The seemingly minor act of decoupling the satisfaction of the guarantee condition from the precondition and the rely has a significant impact on the style of the proofs. It becomes possible to justify the inference rules with algebraic reasoning that leverages program refinement. This is similar in spirit to more recent formulations of rely-guarantee [1,5,6].

In our simple setting, we say that P refines P' if and only if $P \subseteq P'$. Here is a refinement-style proof of (Jconc), for example:

Lemma 4. $S_1\,R_1\,\{P\}\,G_1\,S_1' \ \wedge\ S_2\,R_2\,\{Q\}\,G_2\,S_2' \ \wedge\ G_1 \subseteq R_2 \ \wedge\ G_2 \subseteq R_1 \ \Rightarrow$ $(S_1 \cap S_2)\,(R_1 \cap R_2)\,\{P \parallel Q\}\,(G_1 \cup G_2)\,(S_1' \cap S_2')$.

Proof. Assume the antecedents. The first one gives $S_1\,\{P \parallel traces(R_1)\}\,S_1'$ *and* $P \subseteq traces(G_1)$. *The second gives* $S_2\,\{Q \parallel traces(R_2)\}\,S_2'$ *and* $Q \subseteq traces(G_2)$. *Clearly* $P \parallel Q \subseteq traces(G_1 \cup G_2)$, *so it remains to show the validity of the Hoare triple* $S_1 \cap S_2\,\{P \parallel Q \parallel traces(R_1 \cap R_2)\}\,S_1' \cap S_2'$. *Consider the refinement development:*

$$
\begin{array}{ll}
 & P \parallel traces(R_1) \\
= & \textit{// Since } traces(R_1) = traces(R_1) \parallel traces(R_1). \\
 & P \parallel traces(R_1) \parallel traces(R_1) \\
\supseteq & \textit{// Because } G_2 \subseteq R_1. \\
 & P \parallel traces(G_2) \parallel traces(R_1) \\
\supseteq & \textit{// By assumption, } Q \subseteq traces(G_2). \\
 & P \parallel Q \parallel traces(R_1) \\
\supseteq & \textit{// By simple monotonicity.} \\
 & P \parallel Q \parallel traces(R_1 \cap R_2)
\end{array}
$$

Since Hoare triples remain valid for refined programs, $S_1\,\{P \parallel Q \parallel \mathit{traces}(R_1 \cap R_2)\}\, S_1'$ *holds. By a symmetric argument, we obtain* $S_2\,\{P \parallel Q \parallel \mathit{traces}(R_1 \cap R_2)\}\, S_2'$. *Applying the Hoare rule of conjunction to these triples completes the proof.*

3.3 Decomposition of the Quintuple

Another interesting consequence of the judgement's definition is that it can be decomposed into 'rely' and 'guar' constructs along the lines of [5].

Let $P \dashv\!\!\parallel P' \stackrel{\text{def}}{=} \bigcup \left\{ P'' \mid P'' \parallel P \subseteq P' \right\}$. Consider the following definitions:

Definition 4. $\mathit{rely}\ R\ P \stackrel{\mathit{def}}{=} \mathit{traces}(R) \dashv\!\!\parallel P$.

Definition 5. $\mathit{guar}\ G\ P \stackrel{\mathit{def}}{=} \mathit{traces}(G) \cap P$.

The Galois connection $P \parallel P' \subseteq P'' \Leftrightarrow P \subseteq P' \dashv\!\!\parallel P''$ and Definition 4 imply that *rely* R P is the largest program (i.e. the least refined or the most nondeterministic one) that, when placed in an environment R, will refine P:

Lemma 5. $P' \parallel \mathit{traces}(R) \subseteq P \Leftrightarrow P' \subseteq \mathit{rely}\ R\ P$.

Definition 5 is simpler: *guar* G P is the largest program that refines P whose steps are all in G.

Let $[S, S']$ denote the 'specification statement' [10], i.e. the largest program that satisfies the Hoare triple with precondition S and postcondition S'. Formally, $[S, S']$ can be defined as $\bigcup \left\{ P \mid S\,\{P\}\, S' \right\}$. Then $S\,\{P\}\, S' \Leftrightarrow P \subseteq [S, S']$. Lemma 5 implies $S\,\{P \parallel \mathit{traces}(R)\}\, S' \Leftrightarrow P \subseteq \mathit{rely}\ R\ [S, S']$, so the *rely* and *guar* constructs elegantly factor the judgement into smaller parts:

Lemma 6. $S\ R\ \{P\}\ G\ S' \Leftrightarrow P \subseteq \mathit{guar}\ G\ (\mathit{rely}\ R\ [S, S'])$.

Instead of using the inference rules in Figs. 1 and 3, one can use refinement and the algebraic properties of *rely* and *guar* to the same effect. This alternative way to reason about programs allows for a more general presentation, as there is no obligation to restrict attention to constructs of the form $\mathit{guar}\ G\ (\mathit{rely}\ R\ [S, S'])$. The work by Hayes et al. [5] offers an excellent example of this approach.

3.4 Bigger Picture

It is not hard to see that Theorem 1 and Lemmas 2 and 3 remain valid in this model. So once again the quintuple represents a conservative extension of the Hoare triple. However, the two models provide different extensions! This raises the exciting possibility that new extensions with pleasant properties might still await future discovery.

From a practical point of view, one might prefer the rules of the first model, because (J1atom) and (J1disj) are more powerful than (J2atom) and (J2disj). The second model will generally use larger guarantee conditions, so in order to apply the concurrency rule (Jconc), the rely conditions must also be larger.

Whether this will create problems during the verification of concrete programs remains to be seen. The examples in [5], which uses *rely* and *guar* constructs, and [1], which uses a quintuple judgement, suggest that this is perhaps not a serious drawback.

4 Using Binary Relations for Postconditions

Many treatments of rely-guarantee (e.g. [2,5,8]) do not use postconditions that are sets of states. Instead, they use predicates that relate the pre and the post state, i.e. binary relations on states. Some treatments of Hoare logic also follow this convention. The Hoare triple with a relation T in the postcondition can be defined in terms of the usual one as follows[3]:

Definition 6. $S\,\{P\}\,T \overset{def}{=} \forall\sigma \in S : \left\{\sigma\right\}\{P\}\,T(\left\{\sigma\right\})$.

This suggests a similar definition for the rely-guarantee quintuple where post-conditions are relations:

Definition 7. $S\;R\;\{P\}\;G\;T \overset{def}{=} \forall\sigma \in S : \left\{\sigma\right\}\,R\;\{P\}\;G\;(T(\left\{\sigma\right\}))$.

(Rskip) $S\,R\,\{skip\}\,G\,R^*$

(Rseq) $S\,R\,\{P\}\,G\,(T \cap \Sigma{\times}S') \wedge S'\,R\,\{Q\}\,G\,T' \Rightarrow S\,R\,\{P;Q\}\,G\,(T;T')$

(Rconc) $S\,(R{\cup}G_2)\,\{P\}\,G_1\,T_1 \wedge S\,(R{\cup}G_1)\,\{Q\}\,G_2\,T_2 \Rightarrow$
$S\,R\,\{P\,\|\,Q\}\,(G_1{\cup}G_2)\,(T_1{\cap}T_2{\cap}(R{\cup}G_1{\cup}G_2)^*)$

(Rchoice) $(\forall P \in X : S\,R\,\{P\}\,G\,T) \Rightarrow S\,R\,\{\bigcup X\}\,G\,T$

(Riter) $S\,R\,\{P\}\,G\,(T' \cap \Sigma{\times}S) \wedge R^* \cap S{\times}\Sigma \subseteq T \wedge T';T \subseteq T \Rightarrow S\,R\,\{P^*\}\,G\,T$

(Rrec) $(\forall P : S\,R\,\{P\}\,G\,T \Rightarrow S\,R\,\{f(P)\}\,G\,T) \Rightarrow S\,R\,\{\mathit{lfp}\,f\}\,G\,T$

(Rweak) $S_1\,R_1\,\{P\}\,G_1\,T_1 \wedge S_2 \subseteq S_1 \wedge R_2 \subseteq R_1 \wedge G_1 \subseteq G_2 \wedge T_1 \subseteq T_2 \Rightarrow$
$S_2\,R_2\,\{P\}\,G_2\,T_2$

(Rstren) $S\,R\,\{P\}\,G\,T \Rightarrow S\,R^*\,\{P\}\,(G{\cap}\mathit{steps}(P))\,(T \cap (R{\cup}G)^* \cap S{\times}\Sigma)$

(Rconj) $S_1\,R_1\,\{P\}\,G_1\,T_1 \wedge S_2\,R_2\,\{Q\}\,G_2\,T_2 \Rightarrow$
$(S_1{\cap}S_2)\,(R_1{\cap}R_2)\,\{P \cap Q\}\,(G_1{\cap}G_2)\,(T_1{\cap}T_2)$

Fig. 4. Rely-guarantee rules with relations for postconditions

It is now possible to explore the conditions under which this definition validates familiar inference rules. For example, the rule[4] (Rweak) in Fig. 4 follows directly from (Jweak). Similarly, (Rconc) holds by (Jconc), (Jweak) and (Jstren). One can also show that (Rseq) follows from (Jseq), (Jweak) and (J1disj). But (J1disj) is not valid in the second model! Indeed, (Rseq) is not a theorem when Definition 7

[3] Because of the equivalence $S\,\{P\}\,S' \Leftrightarrow S\,\{P\}\,S{\times}S'$, it is also possible to go in the opposite direction, i.e., one can define the usual triple in terms of the triple where postconditions are relations.

[4] The prefix 'R' in the names of inference rules stands for 'relation'.

is applied to its quintuple. So although the naive approach of Definition 7 successfully generalises the first model to the setting where postconditions are relations, it fails to generalise the second model whose quintuple behaves slightly differently.

Nevertheless, it is possible to mirror Definition 3 using the Hoare triple of Definition 6:

Definition 8. $S\,R\,\{P\}\,G\,T \overset{def}{=} S\,\{P \parallel \mathit{traces}(R)\}\,T \;\wedge\; P \subseteq \mathit{traces}(G)$.

This definition facilitates algebraic proofs of the inference rules in Fig. 4, so it successfully generalises the second model. It is also straightforward to see that the *rely* and *guar* constructs of Sect. 3.3 need no adaptation: if $[S,T]$ denotes the specification statement with relation T as postcondition, i.e. $[S,T] \overset{\text{def}}{=} \bigcup\left\{P \mid S\,\{P\}\,T\right\}$, then the judgement of Definition 8 satisfies $S\,R\,\{P\}\,G\,T \Leftrightarrow P \subseteq \mathit{guar}\;G\;(\mathit{rely}\;R\;[S,T])$.

Definition 8 validates the following equivalence where the second model's quintuple appears in the right-hand side:

$$S\,R\,\{P\}\,G\,T \;\Leftrightarrow\; (\forall \sigma \in S : \left\{\sigma\right\} R\,\{P\}\,G\,(T(\left\{\sigma\right\}))) \;\wedge\; \emptyset\,R\,\{P\}\,G\,\emptyset$$

This equivalence shows that Definition 8 strengthens Definition 7 with an additional conjunct that caters specifically for the case where the precondition is false.

The apparent differences between Definitions 7 and 8 can now be resolved by noticing their underlying unity – the generalised judgements of both models satisfy:

Lemma 7. $S\,R\,\{P\}\,G\,T \;\Leftrightarrow\; \forall S' \subseteq S : S'\,R\,\{P\}\,G\,(T(S'))$.

Of course this does not imply that the two generalised models will support the same rules. Figure 4 contains some common rules[5], while Figs. 5 and 6 show rules that are specific to each generalisation. Notice that Figs. 5 and 6 mirror the differences that were present in Figs. 2 and 3.

In retrospect, one can see that since $S\,\{P\}\,T \;\Leftrightarrow\; \forall S' \subseteq S : S'\,\{P\}\,T(S')$ is also valid, it would have been easier to start the generalisation from this characterisation instead of the one in Definition 6. In general, defining judgements

(R1atom) $\quad \mathit{rel}(a) \cap (R^*(S)) {\times} \Sigma \subseteq G \;\wedge\; (R^*; \mathit{rel}(a); R^*) \subseteq T \;\Rightarrow\; S\,R\,\{a\}\,G\,T$

(R1disj) $\quad (\forall S \in Y : S\,R\,\{P\}\,G\,T) \;\Rightarrow\; (\bigcup Y)\,R\,\{P\}\,G\,T$

Fig. 5. Rules specific to the generalisation of the first model

[5] When postconditions are sets of states, one can always change a precondition S into $R^*(S)$ where R is the rely condition. This is invalid when postconditions are relations between input and output states, so (Rstren) does not change the precondition.

(R2atom)	$rel(a) \subseteq G \ \wedge \ (R^*; rel(a); R^*) \subseteq T \ \Rightarrow \ S\,R\,\{a\}\,G\,T$
(R2disj)	$Y \neq \emptyset \ \wedge \ (\forall S \in Y : S\,R\,\{P\}\,G\,T) \ \Rightarrow \ (\bigcup Y)\,R\,\{P\}\,G\,T$

Fig. 6. Rules specific to the generalisation of the second model

in terms of others is a powerful technique to construct sophisticated notions and inference rules in a stepwise fashion, but a little experimentation is often necessary to find suitable definitions for derived judgements.

5 Soundness

This paper presents the two models of rely-guarantee independently of operational calculi. The presentation is fairly self-contained – only the reference to Hoare logic involves another judgement. In fact the Hoare triple also has a direct definition that does not presuppose an operational judgement or calculus:

Definition 9. $S\,\{P\}\,S' \ \stackrel{def}{=} \ \textit{IF-traces-ending-in}(S)\,;P \subseteq \textit{IF-traces-ending-in}(S') \ \cup \ \textit{WithInterference}$.

Here, $\textit{IF-traces-ending-in}(S)$ denotes the set of all traces that are interference-free and end in a state that is also in S. A trace is interference-free when, for each step in the trace, the second state of the step is the same as the first state of the next step if such a step exists. The set of all traces that are not interference-free is denoted by $\textit{WithInterference}$. Definition 9 says that all the interference-free traces of P that start in a state in S must end in a state in S'. Moreover, if P contains the empty trace (which is trivially interference-free), then it must be the case that $S \subseteq S'$.

Despite the independence from operational calculi, the expected soundness relationships nonetheless hold. To show this, we first give direct definitions of familiar operational judgements and then prove that the soundness relationships are theorems. The same technique was used in [14] to demonstrate the soundness of the Views program logic.

The big-step operational judgement is defined as follows:

Definition 10. $\langle P, \sigma\rangle \longrightarrow \sigma' \ \stackrel{def}{=} \ \exists t \in \textit{IF-traces-ending-in}(\sigma) : \exists t' \in \textit{IF-traces-ending-in}(\sigma') : \Big\{t\Big\}\,; P \supseteq \Big\{t'\Big\}$.

It says that P has an interference-free trace that can transform the initial state σ into the final state σ'. The familiar soundness relationship holds between the Hoare triple and the big-step judgement ([7] contains a short and simple formal proof):

Lemma 8. $S\,\{P\}\,S' \ \Leftrightarrow \ (\forall \sigma \in S : \forall \sigma' : \langle P, \sigma\rangle \longrightarrow \sigma' \ \Rightarrow \ \sigma' \in S')$.

This result and Theorem 1 imply that both models of rely-guarantee are sound with respect to big-step rules:

Theorem 2. $(\exists R\ G : S\, R\,\{P\}\, G\, S') \Leftrightarrow (\forall \sigma \in S : \forall \sigma' : \langle P, \sigma\rangle \longrightarrow \sigma' \Rightarrow \sigma' \in S')$.

However, it is much more common to establish the soundness of rely-guarantee with respect to a small-step judgement in the style of Plotkin [12]. The reason is that operational judgements are conventionally *defined* in terms of syntax-directed rules, and the fine-grained interleaving of concurrent composition cannot be expressed by considering only the big steps of each operand.

Although such considerations are not problematic in this more semantic treatment where judgements are not defined by sets of inference rules, it is quite easy to accommodate small-step calculi. The small-step judgement can be defined in terms of a set *Actions* that contains the 'small' operations or actions that are easy to implement in a computer:

Definition 11. $\langle P, \sigma\rangle \longrightarrow \langle P', \sigma'\rangle \stackrel{def}{=} \exists Q \in \mathit{Actions} : P \supseteq Q\,; P' \wedge \langle Q, \sigma\rangle \longrightarrow \sigma'$.

It says that one way of executing P is to execute some action followed by P'. The action itself is hidden – only its effect on the state is explicit in the judgement.

There is a simple relationship between the reflexive transitive closure of the small-step judgement and the big-step one:

Lemma 9. $\langle P, \sigma\rangle \longrightarrow^{*} \langle \mathit{skip}, \sigma'\rangle \Rightarrow \langle P, \sigma\rangle \longrightarrow \sigma'$.

Whether or not the converse holds depends on the choice of *Actions*, but Lemma 9 is sufficient to prove the soundness of both rely-guarantee models with respect to small-step execution:

Theorem 3. $(\exists R\ G : S\, R\,\{P\}\, G\, S') \Rightarrow (\forall \sigma \in S : \langle P, \sigma\rangle \longrightarrow^{*} \langle \mathit{skip}, \sigma'\rangle \Rightarrow \sigma' \in S')$.

It is remarkable that this result is independent of the choice of machine-executable actions. The soundness also remains valid regardless of the choice of operational rules: any rule that is a theorem can be used to discover executions. For example, when *Actions* includes *skip*, then the familiar operational rules in Fig. 7 are all acceptable[6], and one can also (or alternatively) adopt the following rule for nondeterministic choice:

$$P \in X \wedge \langle P, \sigma\rangle \longrightarrow \langle P', \sigma'\rangle \Rightarrow \langle \textstyle\bigcup X, \sigma\rangle \longrightarrow \langle P', \sigma'\rangle$$

None of these decisions or changes can jeopardise the validity of Theorem 3. The formalisation effectively decouples deductive (i.e. program logic) and operational concerns, yet it enforces soundness at the same time.

6 Related Work

Basic Setup and Definitions of the Judgement. Compared to this paper, the formalisations in most conventional presentations of rely-guarantee (e.g.

[6] The prefix 'P' in the names of inference rules stands for 'Plotkin' in tribute to [12].

(Patom)	$a \in \mathit{Actions} \wedge (\sigma,\sigma') \in \mathit{rel}(a) \Rightarrow \langle a, \sigma\rangle \longrightarrow \langle \mathit{skip}, \sigma'\rangle$
(Pseq1)	$\langle \mathit{skip}\,; P, \sigma\rangle \longrightarrow \langle P, \sigma\rangle$
(Pseq2)	$\langle P, \sigma\rangle \longrightarrow \langle P', \sigma'\rangle \Rightarrow \langle P\,; P'', \sigma\rangle \longrightarrow \langle P'\,; P'', \sigma'\rangle$
(Pchoice)	$P \in X \Rightarrow \langle \bigcup X, \sigma\rangle \longrightarrow \langle P, \sigma\rangle$
(Piter1)	$\langle P^*, \sigma\rangle \longrightarrow \langle \mathit{skip}, \sigma\rangle$
(Piter2)	$\langle P^*, \sigma\rangle \longrightarrow \langle P\,; P^*, \sigma\rangle$
(Pconc1)	$\langle \mathit{skip} \parallel P, \sigma\rangle \longrightarrow \langle P, \sigma\rangle$
(Pconc2)	$\langle P \parallel \mathit{skip}, \sigma\rangle \longrightarrow \langle P, \sigma\rangle$
(Pconc3)	$\langle P, \sigma\rangle \longrightarrow \langle P', \sigma'\rangle \Rightarrow \langle P \parallel P'', \sigma\rangle \longrightarrow \langle P' \parallel P'', \sigma'\rangle$
(Pconc4)	$\langle P, \sigma\rangle \longrightarrow \langle P', \sigma'\rangle \Rightarrow \langle P'' \parallel P, \sigma\rangle \longrightarrow \langle P'' \parallel P', \sigma'\rangle$
(Prec)	$\langle f(\mathit{lfp}\, f), \sigma\rangle \longrightarrow \langle P, \sigma'\rangle \Rightarrow \langle \mathit{lfp}\, f, \sigma\rangle \longrightarrow \langle P, \sigma'\rangle$

Fig. 7. Small-step operational rules

[2,3,5,11,15,18]) proceed in a rather different way. They start by giving a grammar that fixes the abstract syntax of programs. Next, they usually give a representation for states (e.g. a state is a function from identifiers to integers). Programs are then equipped with a small-step operational semantics by choosing a set of inference rules similar to the ones in Fig. 7. However, the rules are postulated (i.e. not derived as theorems) and serve to define the small-step judgement. Next, a new small-step judgement is introduced to allow interference by the environment. It uses explicit labels to track whether the program or the environment is responsible for a transition. A popular form of the new judgement is defined by the two inference rules:

$$\langle P, \sigma\rangle \xrightarrow{e} \langle P, \sigma'\rangle$$
$$\langle P, \sigma\rangle \longrightarrow \langle P', \sigma'\rangle \Rightarrow \langle P, \sigma\rangle \xrightarrow{p} \langle P', \sigma'\rangle$$

A program P is then associated with its execution traces, which are finite or infinite sequences of the form:

$$\langle P_0, \sigma_0\rangle \xrightarrow{l_0} \langle P_1, \sigma_1\rangle \xrightarrow{l_1} \langle P_2, \sigma_2\rangle \ldots$$

where $P_0 = P$, each $l_i \in \left\{e, p\right\}$, and each transition in the sequence must be a valid labelled judgement. The rely-guarantee quintuple is then defined in terms of these execution traces of a program. The traces are sometimes summarised by so-called *Aczel traces*, which discard the program components. For example, the Aczel trace of the above execution trace would begin as follows:

$$[(\sigma_0, l_0, \sigma_1), (\sigma_1, l_1, \sigma_2), \ldots$$

Notice that environment steps are incorporated into Aczel traces, and that the labelling helps to determine whether the rely and the guarantee conditions are fulfilled. Alternatively, execution traces can be summarised by so-called *transition traces*. These traces are obtained by dropping all environment transitions from Aczel traces and removing the (now redundant) p-labels. Transition traces look very similar to our traces, but they can be finite or infinite as a result of the operational rules.

This paper proposes a classification of rely-guarantee models into two main groups, but there also exist minor variations in the formal definition of the rely-guarantee judgement within each group. For example:

- Many formalisations place restrictions on rely and guarantee relations such as reflexivity and/or transitivity [2,8,11,18]. Jones originally argued in [8, Chap. 4] that interference should be reflexive and transitive. Subsequent works [11,18] discussed the difficulty of finding transitive conditions in practical examples, and require only reflexivity so that the evaluation of Boolean conditions will automatically satisfy guarantee conditions. Other treatments [3,15] use sets of single-state predicates for rely and guarantee conditions. Dingel [3] showed that such an interference condition corresponds to a binary relation that is both reflexive and transitive.
- Some treatments [2,4,5] require that the pre- and postcondition must be 'stable' with respect to the rely condition in all judgements. An assertion S is stable with respect to interference R iff $R(S) \subseteq S$. The utility of the idea is that it implies $R^*(S) = S$. This means, for example, that a precondition S will still hold when the program takes its first step. If this step does not change the state (e.g. it evaluates a Boolean condition), then S will still hold after the test. One can also assume the test condition if it is stable under the rely. Likewise, stability can ensure that the environment will preserve the assertion that was established by the last step of the program, thereby turning it into a valid postcondition despite interference.

In contrast to the work discussed before, references [1,6] propose general definitions of the rely-guarantee judgement in algebraic terms. The development in [6] augments a Concurrent Kleene Algebra (CKA) with a set of elements called *invariants* to obtain a rely/guarantee-CKA. It is well known that the set of formal languages with interleaving is a model of CKA (the Isabelle formalisation of this paper also contains a proof). It is hence also the case for trace sets under interleaving. Moreover, by considering each trace set of the form $traces(R)$ for some R to be an invariant, our second model can be viewed as an instance of the abstract model in [6]. It is also an instance of the abstract model in [1], because the sets of traces and invariants also satisfy the laws of *rely-guarantee algebra* proposed there. Equipping rely-guarantee algebra with residuals is briefly considered in [1], and their result (6) can be viewed as an abstract version of our Lemma 6. The trace model that is used to verify examples in [1] is similar to ours, but in order to use laws for Boolean tests such as $test(P); test(Q) = test(P \cap Q)$, it additionally demands that trace sets must be closed under stuttering and mumbling.

Inference Rules, their Proofs and Soundness. Figures 1, 2, 3, 4, 5 and 6 include inference rules such as Conjunction and Disjunction that seldom appear in other presentations of rely-guarantee, but which turn out to be useful here for validating other rules (Conjunction helps to strengthen guarantee conditions in the strengthening rules, and Disjunction is used in Sect. 4 to generalise the rule for sequential composition). Most presentations include weakening rules such as (Jweak) or (Rweak), but they almost never include explicit strengthening rules like (Jstren) or (Rstren). A notable exception is the rule RG-AdjustPost in [16, p. 20], which strengthens the postcondition. This strengthening is often built into the rule for concurrent composition, for example in the rule **Par-I** in [2] and the rule $\|$-I of [9]. In Sect. 4, the rule (Rconc) strengthens the postcondition of the resulting judgement in a similar way.

As mentioned before, models that are based on operational semantics define the rely-guarantee judgement in terms of execution traces. The proof of the validity of an inference rule such as (Jconc) then directly or indirectly involves the operational rules for concurrent composition such as (Pconc1) through (Pconc4). The resulting proofs can become quite lengthy and involved (see e.g. [2]), but because the definition of the judgement already captures the intended soundness relationship with the operational semantics, there is no need for a separate soundness proof.

Treatments that propose general algebraic definitions for the rely-guarantee judgement prove that inference rules are valid by assuming certain algebraic laws. All models of e.g. a rely/guarantee-CKA must satisfy these laws, and by doing so they automatically gain the rules. The laws thus factor the proofs of the rules into two parts. The proofs that the rules follow from the laws can be surprisingly elegant. Moreover, the laws can have many concrete models (they can also rule out potentially useful models, such as our first one). Each model should explain why the abstract definition of the judgement is meaningful or appropriate in its context. Although [1,6] do not investigate soundness with repect to operational calculi, each model could also consider it independently. The soundness result in Sect. 5 demonstrates this for a model (our second one) of both rely/guarantee-CKA and rely-guarantee algebra.

Formalisation in Proof Assistants. Previous formalisations of rely-guarantee in Isabelle/HOL include one by Nieto [11] and one by Armstrong et al. [1]. Nieto's treatment uses a while-language with non-nested concurrent composition and deterministic atomic commands. The language is equipped with an operational semantics, and the rely-guarantee judgement is defined such that the satisfaction of the guarantee condition depends on the precondition and the rely (similar to our first model). Armstrong et al. focus on deriving rely-guarantee rules from algebraic laws. They use an abstract definition of the judgement where the guarantee condition is independent of the precondition and the rely (similar to our second model), and they demonstrate that the algebraic principles can be used to verify while-programs with concurrency.

7 Conclusion

This paper proposes a new classification of semantic models for rely-guarantee into two groups that differ in their treatment of guarantee conditions. To compare them, it constructs an abstract model for each group in a unified setting. The first model supports more powerful inference rules. However, by decoupling the satisfaction of the guarantee from the precondition and the rely, the second model allows algebraic reasoning and an elegant decomposition of the judgement. Both models successfully generalise to the setting where postconditions are binary relations. Both are also sound with respect to operational calculi. Perhaps our classification will have to be extended in the future, but efforts to unify rely-guarantee techniques should at least be flexible enough to accommodate models from both groups described here.

Acknowledgements. This work was supported by the SNSF. Comments by Tony Hoare, Georg Struth and the anonymous referees helped to improve the presentation significantly.

References

1. Armstrong, A., Gomes, V.B.F., Struth, G.: Algebraic principles for rely-guarantee style concurrency verification tools. In: Jones, C., Pihlajasaari, P., Sun, J. (eds.) FM 2014. LNCS, vol. 8442, pp. 78–93. Springer, Heidelberg (2014)
2. Coleman, J.W., Jones, C.B.: A structural proof of the soundness of rely/guarantee rules. J. Log. Comput. **17**(4), 807–841 (2007). Comments refer to the revised version which appeared as technical report CS-TR-1029, University of Newcastle, June 2007
3. Dingel, J.: A refinement calculus for shared-variable parallel and distributed programming. Formal Asp. Comput. **14**(2), 123–197 (2002)
4. Dinsdale-Young, T., Birkedal, L., Gardner, P., Parkinson, M., Yang, H.: Views: compositional reasoning for concurrent programs. In: Proceedings of the 40th Annual ACM SIGPLAN-SIGACT Symposium on Principles of Programming Languages, POPL 2013, pp. 287–300. ACM, New York (2013)
5. Hayes, I.J., Jones, C.B., Colvin, R.J.: Refining rely-guarantee thinking. Technical report CS-TR-1334, School of Computing Science, Newcastle University, May 2012
6. Hoare, C.A.R., Möller, B., Struth, G., Wehrman, I.: Concurrent Kleene Algebra. In: Bravetti, M., Zavattaro, G. (eds.) CONCUR 2009. LNCS, vol. 5710, pp. 399–414. Springer, Heidelberg (2009)
7. Isabelle/HOL proofs (2014). http://www0.cs.ucl.ac.uk/staff/s.vanstaden/proofs/RG.tgz
8. Jones, C.B.: Development Methods for Computer Programs including a Notion of Interference. Ph.D. thesis, Oxford University, June 1981. printed as: Programming Research Group, Technical Monograph 25
9. Jones, C.B., Hayes, I.J., Colvin, Rj: Balancing expressiveness in formal approaches to concurrency. Formal Aspects Comput. **27**(3), 475–497 (2015)
10. Morgan, C.: The specification statement. ACM Trans. Program. Lang. Syst. **10**, 403–419 (1988)
11. Prensa Nieto, L.: The rely-guarantee method in Isabelle/HOL. In: Degano, P. (ed.) ESOP 2003. LNCS, vol. 2618, pp. 348–362. Springer, Heidelberg (2003)
12. Plotkin, G.D.: A structural approach to operational semantics. Technical report DAIMI FN-19, Computer Science Department, Aarhus University, Aarhus, Denmark, September 1981

13. de Roever, W.P., de Boer, F.S., Hannemann, U., Hooman, J., Lakhnech, Y., Poel, M., Zwiers, J.: Concurrency Verification: Introduction to Compositional and Non-compositional Methods. Cambridge Tracts in Theoretical Computer Science, vol. 54. Cambridge University Press, Cambridge (2001)
14. van Staden, S.: Constructing the views framework. In: Naumann, D. (ed.) UTP 2014. LNCS, vol. 8963, pp. 62–83. Springer, Heidelberg (2015)
15. Stirling, C.: A generalization of Owicki-Gries's Hoare logic for a concurrent while language. Theor. Comput. Sci. **58**, 347–359 (1988)
16. Vafeiadis, V.: Modular fine-grained concurrency verification. Technical report UCAM-CL-TR-726, University of Cambridge, Computer Laboratory, July 2008
17. Wickerson, J., Dodds, M., Parkinson, M.: Explicit stabilisation for modular rely-guarantee reasoning. In: Gordon, A.D. (ed.) ESOP 2010. LNCS, vol. 6012, pp. 610–629. Springer, Heidelberg (2010)
18. Xu, Q., de Roever, W.P., He, J.: The rely-guarantee method for verifying shared variable concurrent programs. Formal Asp. Comput. **9**(2), 149–174 (1997)

A Relation-Algebraic Approach to Multirelations and Predicate Transformers

Rudolf Berghammer[1] and Walter Guttmann[2](✉)

[1] Institut für Informatik,
Christian-Albrechts-Universität zu Kiel, Kiel, Germany
[2] Department of Computer Science and Software Engineering,
University of Canterbury, Christchurch, New Zealand
walter.guttmann@canterbury.ac.nz

Abstract. The correspondence between up-closed multirelations and isotone predicate transformers is well known. Less known is that multirelations have also been used for modelling topological contact, not only computations. We investigate how properties from these two lines of research translate to predicate transformers. To this end, we express the correspondence of multirelations and predicate transformers using relation algebras. It turns out to be similar to the correspondence between contact relations and closure operations. Many results generalise from up-closed to arbitrary multirelations.

1 Introduction

Predicate transformers have been used for defining the semantics of programs and constructing correct programs for a long time; see, for example, [9,12]. A multirelational representation of isotone predicate transformers has been given in [25]. Multirelations – relations between a set and a powerset – have previously been used for defining the semantics of game-based computations and logics [22] as well as to model contact and related notions from topology [1,2].

In the companion paper [5] we have started to bring together the computational and topological lines of research on multirelations. In particular, we consider various properties of multirelations that have been used in these lines of research. We investigate how these properties are related and under which operations they are closed by introducing general algebras. An observation from this work is that being up-closed is just one among many useful properties.

Research about multirelations in program semantics commonly makes the restriction to up-closed multirelations, perhaps due to the corresponding assumption that predicate transformers are isotone. Motivated by our previous work we raise the question whether the investigation can be and should be liberated from the restriction to up-closed multirelations. That it is possible is indicated by both this paper and the companion paper: many results do not require this restriction. That it is desirable is indicated by some applications in programming, notably concurrent dynamic logic [23], which cannot be restricted to up-closed multirelations.

R. Hinze and J. Voigtländer (Eds.): MPC 2015, LNCS 9129, pp. 50–70, 2015.
DOI:10.1007/978-3-319-19797-5_3

The present paper continues the programme started in the companion paper. Our aim is to investigate the correspondence between multirelations and predicate transformers generally and to use it for translating multirelational operations and properties to predicate transformers. Our contributions are mainly presented in Sect. 4 after a basic discussion of relations, multirelations, contact relations and predicate transformers in Sects. 2 and 3. They are (1) a relation-algebraic description of an order isomorphism between multirelations and predicate transformers in Sect. 4.1, (2) a relation-algebraic translation of multirelational operations to predicate transformers in Sect. 4.2, (3) an alternative composition of multirelations in Sect. 4.3, and (4) relation-algebraic and logical translations of multirelational properties to predicate transformers in Sect. 4.4. Some properties from the topological line of research might be less well understood on the computational side. We show an example that uses such a property to weaken the assumptions for a program transformation, though the focus of this paper is a foundational investigation.

Among related work we chiefly mention [18], which uses power allegories for a categorical approach to up-closed multirelations. The article [20] studies the relationship of up-closed multirelations, predicate transformers and other models for representing higher-order functions. The present paper uses relation algebras and investigates arbitrary multirelations. Other related work is discussed throughout the paper.

2 Preliminaries

In this section we present facts of relation algebras needed in the remainder of the paper. For more details on relations and relation algebras, see [28,30,31].

2.1 Relation Algebras

Following the Z notation, we write $R : A \leftrightarrow B$ if R is a (typed, binary) relation with source A and target B, that is, a subset of the Cartesian product $A \times B$. If the sets A and B of the *type* $A \leftrightarrow B$ of R are finite we may consider R as a Boolean matrix with $|A|$ rows and $|B|$ columns. This interpretation is well suited for many purposes. Therefore, we use matrix notation in this paper and write $R_{x,y}$ instead of $(x, y) \in R$ or $x\,R\,y$.

We assume the reader is familiar with the basic operations on relations, namely R^{c} (transposition, converse), $\overline{R}$ (complement, negation), $R \cup S$ (union, join), $R \cap S$ (intersection, meet), RS (composition, product), the predicates indicating $R \subseteq S$ (inclusion) and $R = S$ (equality), and the special relations O (empty relation), T (universal relation) and I (identity relation).

We use $^{-}$, $\cup$, $\cap$ and $\subseteq$ for arbitrary sets, not just relations. With these set operations, the subset order and the constants $\mathsf{O} : A \leftrightarrow B$ and $\mathsf{T} : A \leftrightarrow B$, the set of relations of a type $A \leftrightarrow B$ forms a complete Boolean lattice. Well-known rules involving transposition and composition are, for instance, $R^{\mathsf{cc}} = R$, $\overline{R^{\mathsf{c}}} = \overline{R}^{\mathsf{c}}$, $(R \cup S)^{\mathsf{c}} = R^{\mathsf{c}} \cup S^{\mathsf{c}}$, $(R \cap S)^{\mathsf{c}} = R^{\mathsf{c}} \cap S^{\mathsf{c}}$, $(RS)^{\mathsf{c}} = S^{\mathsf{c}}R^{\mathsf{c}}$, $Q(R \cup S) = QR \cup QS$

and $Q(R \cap S) \subseteq QR \cap QS$. Moreover, transposition is $\subseteq$-isotone and union, intersection and composition are $\subseteq$-isotone in both arguments.

The theoretical framework for these rules and many others is that of an (axiomatic, heterogeneous) relation algebra in the sense of [28,30], which generalises the original homogeneous approach of [31]. As constants and operations of this algebraic structure we have those of concrete (that is, set-theoretic) relations. The axioms of a relation algebra are those of a complete Boolean lattice for the set operations, the associativity of composition, the neutrality of identity relations for composition, and the equivalences

$$QR \subseteq S \iff Q^{\mathsf{c}}\overline{S} \subseteq \overline{R} \iff \overline{S}R^{\mathsf{c}} \subseteq \overline{Q}$$

for all relations $Q : A \leftrightarrow B$, $R : B \leftrightarrow C$ and $S : A \leftrightarrow C$ or – equivalently – the so-called Dedekind rule

$$QR \cap S \subseteq (Q \cap SR^{\mathsf{c}})(R \cap Q^{\mathsf{c}}S).$$

We assume all relation-algebraic expressions and formulae to be well-typed and suppress type information if appropriate. Many relation-algebraic notions are also available in the settings of categories and allegories [10]. In particular, related category-theoretic axiomatisations are used in [8] for program development.

Residuals are the greatest solutions of certain inclusions. They appear as weakest prespecifications in [15] and as factors in [4]. The *left residual* of the relation $S : A \leftrightarrow C$ over $R : B \leftrightarrow C$, in symbols $S/R : A \leftrightarrow B$, is the greatest relation $X : A \leftrightarrow B$ such that $XR \subseteq S$. So, we have the Galois connection $XR \subseteq S$ if and only if $X \subseteq S/R$, for all relations $X : A \leftrightarrow B$. Similarly, the *right residual* of $S : A \leftrightarrow C$ over $R : A \leftrightarrow B$, in symbols $R \setminus S : B \leftrightarrow C$, is the greatest relation $X : B \leftrightarrow C$ such that $RX \subseteq S$. This implies that $RX \subseteq S$ if and only if $X \subseteq R \setminus S$, for all relations $X : B \leftrightarrow C$.

We will also need relations which are left and right residuals simultaneously. The *symmetric quotient* $\mathsf{syq}(R, S) : B \leftrightarrow C$ of two relations $R : A \leftrightarrow B$ and $S : A \leftrightarrow C$ is defined as the greatest relation $X : B \leftrightarrow C$ such that $RX \subseteq S$ and $XS^{\mathsf{c}} \subseteq R^{\mathsf{c}}$. In terms of the basic operations we have for all relations R and S of appropriate type the following descriptions:

$$S/R = \overline{\overline{S}R^{\mathsf{c}}} \qquad R \setminus S = \overline{R^{\mathsf{c}}\overline{S}} \qquad \mathsf{syq}(R, S) = (R \setminus S) \cap (R^{\mathsf{c}}/S^{\mathsf{c}})$$

Further properties of the two residuals and the symmetric quotient we will use in this paper can be found in [28,30]; for example, $\mathsf{syq}(R, S)^{\mathsf{c}} = \mathsf{syq}(S, R)$. We assume that the unary relation-algebraic operations have the highest precedence, then composition follows and its precedence is higher than that of union, intersection and the residuals, all of which have the same precedence.

2.2 Mappings and Predicate Transformers

The basic operations and constants mentioned in Sect. 2.1 can be used for defining specific classes of relations in a purely algebraic way. In the following we introduce the classes that will be used in the remainder of this paper. For more details we refer again to [28,30].

A relation $R : A \leftrightarrow B$ is *univalent* if $R^{\mathsf{c}}R \subseteq \mathsf{I}$, and *total* if $R\mathsf{T} = \mathsf{T}$. The latter equation is equivalent to the inclusion $\mathsf{I} \subseteq RR^{\mathsf{c}}$. We call R a *mapping* (from A to B) if it is univalent and total. In the case of mappings we use small letters and the common type annotation. Hence, if we write $f : A \rightarrow B$, then f is a relation of type $A \leftrightarrow B$ that is univalent and total. In pointwise arguments we also use $f(x)$ for function application; for example, $f(x) = y$ if and only if $f_{x,y}$.

For a univalent relation $R : A \leftrightarrow B$ we have $R\overline{S} \subseteq \overline{RS}$ and for a total relation $R : A \leftrightarrow B$ we have $R\overline{S} \supseteq \overline{RS}$, for all relations $S : B \leftrightarrow C$. Hence, for a mapping $f : A \rightarrow B$ we get $f\overline{S} = \overline{fS}$, for all relations $S : B \leftrightarrow C$. Moreover, the shunting property $Rf \subseteq S$ if and only if $R \subseteq Sf^{\mathsf{c}}$ holds for any mapping $f : B \rightarrow C$ and relations $R : A \leftrightarrow B$ and $S : A \leftrightarrow C$.

The relation $R : A \leftrightarrow B$ is called *injective* if $R^{\mathsf{c}} : B \leftrightarrow A$ is univalent, *surjective* if R^{c} is total and *bijective* if R^{c} is a mapping. Hence $\overline{S}R \subseteq \overline{SR}$ if R is injective, $\overline{S}R \supseteq \overline{SR}$ if R is surjective, and $\overline{S}R = \overline{SR}$ if R is bijective, for all relations $S : C \leftrightarrow A$. The converse shunting property is $fR \subseteq S$ if and only if $R \subseteq f^{\mathsf{c}}S$ for bijective $f : A \rightarrow B$ and relations $R : B \leftrightarrow C$ and $S : A \leftrightarrow C$.

If the relation Q is univalent, the subdistributivity $Q(R \cap S) \subseteq QR \cap QS$ becomes an equality $Q(R \cap S) = QR \cap QS$. A consequence of $f\overline{S} = \overline{fS}$ and $f(R \cap S) = fR \cap fS$ for a mapping f is the following result.

Lemma 1. *For all relations $R : A \leftrightarrow B$ and $S : A \leftrightarrow C$ and all mappings $f : D \rightarrow B$ we have $\mathsf{syq}(Rf^{\mathsf{c}}, S) = f\mathsf{syq}(R, S)$ and for all mappings $g : D \rightarrow C$ we have $R \setminus Sg^{\mathsf{c}} = (R \setminus S)g^{\mathsf{c}}$.*

Proof. The first claim is [6, Theorem 3.3]. We obtain the second claim by

$$R \setminus Sg^{\mathsf{c}} = \overline{R^{\mathsf{c}}\overline{Sg^{\mathsf{c}}}} = \overline{R^{\mathsf{c}}\overline{S}g^{\mathsf{c}}} = \overline{R^{\mathsf{c}}\overline{S}}g^{\mathsf{c}} = (R \setminus S)g^{\mathsf{c}},$$

using that g is a mapping in the second and third steps. □

Following general terminology, a *predicate transformer* (as introduced in [9] for weakest-precondition semantics) is a function that maps a predicate on the state space A of a program to a predicate on the same space. If we consider a predicate on A as a subset of A, that is, as an element of the powerset 2^A, then a predicate transformer is just a function from 2^A to 2^A. Therefore, allowing different state spaces, we call a mapping $f : 2^B \rightarrow 2^A$ in the relational sense a (relational) predicate transformer.

2.3 Relation-Algebraic Specification of Set-Theoretic Constructions

Besides empty relations, universal relations and identity relations, we need further specific relations for fundamental set-theoretic constructions. They are introduced in the following.

Let A be a set. The *membership relation* $\mathsf{E} : A \leftrightarrow 2^A$ is the relation-level equivalent to the set-theoretic predicate "$\in$". Hence, we have $\mathsf{E}_{x,Y}$ if and only if $x \in Y$, for all $x \in A$ and $Y \in 2^A$. With the help of E, a right residual and a symmetric quotient we can introduce two relations on 2^A as follows:

$$\mathsf{S} := \mathsf{E} \setminus \mathsf{E} : 2^A \leftrightarrow 2^A \qquad \mathsf{C} := \mathsf{syq}(\overline{\mathsf{E}}, \mathsf{E}) : 2^A \leftrightarrow 2^A$$

A little pointwise calculation shows that $\mathsf{S}_{X,Y}$ if and only if $X \subseteq Y$ and $\mathsf{C}_{X,Y}$ if and only if $Y = \overline{X}$, for all $X, Y \in 2^A$, where $\overline{X}$ is the complement of the set X relative to its superset A. Therefore, we call S a *subset relation* and C a *set complement relation.* We use $\mathsf{C}_A, \mathsf{S}_A : 2^A \leftrightarrow 2^A$ and $\mathsf{C}_B, \mathsf{S}_B : 2^B \leftrightarrow 2^B$ to clarify the type if necessary.

We next specify the two binary operations of meet and join of sets as two relations $\mathsf{M} : 2^A \times 2^A \leftrightarrow 2^A$ (*meet relation*) and $\mathsf{J} : 2^A \times 2^A \leftrightarrow 2^A$ (*join relation*) such that $\mathsf{M}_{(X,Y),Z}$ if and only if $X \cap Y = Z$ and $\mathsf{J}_{(X,Y),Z}$ if and only if $X \cup Y = Z$, for all sets $X, Y, Z \in 2^A$. To this end, besides the membership relation $\mathsf{E} : A \leftrightarrow 2^A$ we need the two projections of the direct product $2^A \times 2^A$ as relation-algebraic mappings $p : 2^A \times 2^A \to 2^A$ and $r : 2^A \times 2^A \to 2^A$. They satisfy $p_{(X,Y),Z}$ if and only if $X = Z$ and $r_{(X,Y),Z}$ if and only if $Y = Z$, for all sets $X, Y, Z \in 2^A$. This allows us to derive the following relation-algebraic specifications:

$$\mathsf{M} := \mathsf{syq}([\mathsf{E}, \mathsf{E}]\!], \mathsf{E}) : 2^A \times 2^A \leftrightarrow 2^A \qquad \mathsf{J} := \mathsf{syq}(\overline{[\overline{\mathsf{E}}, \overline{\mathsf{E}}]\!]}, \mathsf{E}) : 2^A \times 2^A \leftrightarrow 2^A$$

In these two definitions $[\cdot, \cdot]\!]$ denotes the *right pairing* operation of the direct product $2^A \times 2^A$, also known as fork or tupling operation. Using the projection mappings $p, r : 2^A \times 2^A \to 2^A$, the right pairing of relations $R : B \leftrightarrow 2^A$ and $S : B \leftrightarrow 2^A$ is defined as

$$[R, S]\!] := Rp^{\mathsf{c}} \cap Sr^{\mathsf{c}} : B \leftrightarrow 2^A \times 2^A.$$

As a consequence we have $[\mathsf{E}, \mathsf{E}]\!]_{x,(X,Y)}$ if and only if $x \in X$ and $x \in Y$, for all $x \in A$ and $X, Y \in 2^A$. From this, a little pointwise calculation yields

$$\mathsf{syq}([\mathsf{E}, \mathsf{E}]\!], \mathsf{E})_{(X,Y),Z} \iff (\forall x \in A : x \in X \wedge x \in Y \Leftrightarrow x \in Z) \iff X \cap Y = Z$$

for all sets $X, Y, Z \in 2^A$, that is, $\mathsf{M}_{(X,Y),Z}$ if and only if $X \cap Y = Z$. Similarly it can be verified that the relation J specifies the join of two sets from 2^A.

The specifications of set-theoretic constructions via the relations S, C, M and J are not yet purely relation-algebraic, since they are still based on the pointwise definitions of membership relations and projection mappings. However, both can be specified with purely relation-algebraic means in a monomorphic manner [28,30]. For the membership relation $\mathsf{E} : A \leftrightarrow 2^A$ we have

$$\mathsf{syq}(\mathsf{E}, \mathsf{E}) = \mathsf{I} \qquad \forall R : \mathsf{T}\,\mathsf{syq}(\mathsf{E}, R) = \mathsf{T}$$

as a second-order axiomatisation. From this we get, for example, $\mathsf{E}\,\mathsf{syq}(\mathsf{E}, R) = R$. The property $\mathsf{syq}(\mathsf{E}, \mathsf{E}) = \mathsf{I}$ captures set equality by mutual inclusion given that $\mathsf{syq}(\mathsf{E}, \mathsf{E}) = \mathsf{S} \cap \mathsf{S}^{\mathsf{c}}$. First-order axioms for the projection mappings are

$$p^{\mathsf{c}}p = \mathsf{I} \qquad r^{\mathsf{c}}r = \mathsf{I} \qquad pp^{\mathsf{c}} \cap rr^{\mathsf{c}} = \mathsf{I} \qquad p^{\mathsf{c}}r = \mathsf{T}.$$

From these axioms we immediately get that the relations p and r are in fact mappings and that their transposes p^{c} and r^{c} are total. Hence, p and r are surjective mappings which is characteristic for projections. The axioms for projections work for general products of type $A \times B$, not just for the instance $2^A \times 2^A$ needed above; similar axioms of biproducts have been used in the context of linear algebra [17].

In the following series of three lemmas we collect further properties needed in the remainder of this paper. The following properties of S are shown in [29].

Lemma 2. *Each subset relation* S *is reflexive* ($\mathsf{I} \subseteq \mathsf{S}$), *antisymmetric* ($\mathsf{S} \cap \mathsf{S}^{\mathsf{c}} \subseteq \mathsf{I}$) *and transitive* ($\mathsf{SS} \subseteq \mathsf{S}$), *that is, a partial order relation.*

Basic laws of symmetric quotients [30, Theorems 4.4.1 and 4.4.3] and the second axiom of the membership relation $\mathsf{E} : A \leftrightarrow 2^A$ yield that $\mathsf{syq}(R, \mathsf{E}) : B \leftrightarrow 2^A$ is a mapping for all relations $R : A \leftrightarrow B$. Therefore we may write $\mathsf{C} : 2^A \to 2^A$ and $\mathsf{M}, \mathsf{J} : 2^A \times 2^A \to 2^A$. For each set complement mapping we furthermore have the following result that, in combination with the mapping property, shows that C is bijective. It is proved in [27].

Lemma 3. *For each set complement mapping* C *we have* $\mathsf{C} = \mathsf{C}^{\mathsf{c}}$.

We formulate the properties of the next lemma only for the projection mappings $p, r : 2^A \times 2^A \to 2^A$ of the direct product $2^A \times 2^A$ and two mappings $f, g : 2^A \to 2^A$, but they obviously hold for arbitrary projections and mappings of appropriate type.

Lemma 4. *Let* $f, g : 2^B \to 2^A$ *be mappings. Then*

(1) $[f, g]\!]p = f$ *and* $[f, g]\!]r = g$,
(2) $[f, g]\!]$ *is a mapping,*
(3) $[\mathsf{E}, \mathsf{E}]\!][f, g]\!]^{\mathsf{c}} = \mathsf{E}f^{\mathsf{c}} \cap \mathsf{E}g^{\mathsf{c}}$.

Proof. (1) The first equality holds because of the following calculation, which uses two axioms of the projection mappings and the Dedekind rule; the second equality is proved similarly:

$$f = f \cap g\mathsf{T} = f \cap gr^{\mathsf{c}}p \subseteq (gr^{\mathsf{c}} \cap fp^{\mathsf{c}})p = [f, g]\!]p = (fp^{\mathsf{c}} \cap gr^{\mathsf{c}})p \subseteq fp^{\mathsf{c}}p = f.$$

(2) Using an axiom of the projection mappings, $[f, g]\!]$ is univalent since

$$[f, g]\!]^{\mathsf{c}}[f, g]\!] = (pf^{\mathsf{c}} \cap rg^{\mathsf{c}})(fp^{\mathsf{c}} \cap gr^{\mathsf{c}}) \subseteq pf^{\mathsf{c}}fp^{\mathsf{c}} \cap rg^{\mathsf{c}}gr^{\mathsf{c}} \subseteq pp^{\mathsf{c}} \cap rr^{\mathsf{c}} = \mathsf{I}.$$

Using the first equation of (1) and that the projection p is total, $[f, g]\!]$ is also total because

$$[f, g]\!]\mathsf{T} = [f, g]\!]p\mathsf{T} = f\mathsf{T} = \mathsf{T}.$$

(3) The claim is a special case of [7, Theorem 2.7]. □

3 Relations and Multirelations

In this section we first recall the basic definitions, operations and properties of multirelations. We also present special kinds of multirelations used to model topological contact. Then we show how multirelational composition and the dual can be expressed in terms of relation algebras and present some fundamental properties. For more details we refer to [5,13].

3.1 Multirelations

A *multirelation* (as introduced in [22,25]) is a relation in the sense of Sect. 2.1 with the additional property that the target is a powerset. So, for sets A and B a multirelation $R : A \leftrightarrow 2^B$ relates an element of A with a subset of B. The Boolean operations *union* $R \cup S$, *intersection* $R \cap S$ and *complement* $\overline{R}$ apply to multirelations R and S (of the same type) as to general relations. Particular multirelations are the *empty multirelations* $\mathsf{O} : A \leftrightarrow 2^B$, the *universal multirelations* $\mathsf{T} : A \leftrightarrow 2^B$ and the *membership multirelations* $\mathsf{E} : A \leftrightarrow 2^A$. Hence $\mathsf{O} = \emptyset$ and $\mathsf{T} = A \times 2^B$. The *composition* of the multirelations $Q : A \leftrightarrow 2^B$ and $R : B \leftrightarrow 2^C$ is the multirelation $Q;R : A \leftrightarrow 2^C$ pointwise defined by

$$(Q;R)_{x,Z} \iff \exists Y \in 2^B : Q_{x,Y} \wedge \forall y \in Y : R_{y,Z},$$

for all $x \in A$ and $Z \in 2^C$. Being relations, multirelations also can be transposed, but the result is not a multirelation whence this plays no role. Instead a dual operation is used. The *dual* of a multirelation $R : A \leftrightarrow 2^B$ is the multirelation $R^{\mathsf{d}} : A \leftrightarrow 2^B$ pointwise defined by

$$R^{\mathsf{d}}{}_{x,Y} \iff \neg R_{x,\overline{Y}},$$

for all $x \in A$ and $Y \in 2^B$, where $\overline{Y}$ denotes the complement of the set Y relative to its superset B. We assume that the unary operations complement and dual have the highest precedence, then composition follows which has higher precedence than union and intersection. Finally, a multirelation $R : A \leftrightarrow 2^B$ is *up-closed* if

$$R_{x,Y} \wedge Y \subseteq Z \implies R_{x,Z}$$

for all $x \in A$ and $Y, Z \in 2^B$. This means that if an element of A is related to a set Y it also has to be related to all supersets of Y.

3.2 Contact Relations

Already before applications in program semantics, multirelations of type $A \leftrightarrow 2^A$ were used in [1] for modelling contact in order to introduce topological concepts. In particular, the following axioms for $R : A \leftrightarrow 2^A$ were considered:

(K_0) $\neg \exists x \in A : R_{x,\emptyset}$
(K_1) $\forall x \in A : R_{x,\{x\}}$
(K_2) $\forall x \in A : \forall Y, Z \in 2^A : R_{x,Y} \wedge Y \subseteq Z \Rightarrow R_{x,Z}$
(K_3) $\forall x \in A : \forall Y, Z \in 2^A : R_{x,Y} \wedge (\forall y \in Y : R_{y,Z}) \Rightarrow R_{x,Z}$
(K_4) $\forall x \in A : \forall Y, Z \in 2^A : R_{x,Y \cup Z} \Leftrightarrow R_{x,Y} \vee R_{x,Z}$

Multirelations satisfying axioms (K_1) to (K_3) are called "contact relations" and those satisfying axioms (K_0) to (K_4) are called "topological contact relations" in [1]. Algebraic characterisations of these logical formulas are derived in [5] and will be discussed in Sect. 4.4. Axiom (K_2) states that R is up-closed.

3.3 Modelling Multirelational Composition and the Dual

Since multirelations are relations, with respect to the Boolean operations the set of multirelations of a given type $A \leftrightarrow 2^B$ forms a complete Boolean lattice with $\mathsf{O} : A \leftrightarrow 2^B$ as least element and $\mathsf{T} : A \leftrightarrow 2^B$ as greatest element of the subset order. To get algebraic laws for multirelational composition and dual we express these operations and the property of being up-closed in terms of relation-algebraic operations and constants, namely composition, complement, right residual, membership relation E, set complement relation C and subset relation S.

Theorem 5. *Let $Q : A \leftrightarrow 2^B$ and $R : B \leftrightarrow 2^C$ be multirelations. Then*

$$\text{(1) } Q\,;R = Q(\mathsf{E} \setminus R) \qquad\qquad \text{(2) } Q^{\mathsf{d}} = \overline{Q}\mathsf{C} = \overline{Q\mathsf{C}}.$$

Moreover, Q is up-closed if and only if $Q = Q\mathsf{S}$, which is equivalent to $Q\mathsf{S} \subseteq Q$.

The algebraic laws of the multirelational operations become more diversified if multirelational composition is taken into account. The familiar relation-algebraic laws that composition distributes over union and has the empty relation as a zero only hold from one side. Other laws of relation algebras, namely, that composition is associative and has the identity relation as a neutral element, hold for up-closed multirelations (with I replaced by E) but need to be weakened in the general case. On the other hand, composition remains $\subseteq$-isotone. These and related properties are summarised in the following result.

Theorem 6. *For all multirelations P, Q and R we have*

$$\text{(1) } \mathsf{O}\,;R = \mathsf{O} \qquad \text{(2) } \mathsf{E}\,;R = R \qquad \text{(3) } \mathsf{T}\,;R = \mathsf{T} \qquad \text{(4) } R \subseteq R\,;\mathsf{E},$$

where in (4) equality holds if and only if R is up-closed, and also

$$\text{(5) } (P \cup Q)\,;R = P\,;R \cup Q\,;R, \qquad \text{(6) } (P \cap Q)\,;R \subseteq P\,;R \cap Q\,;R,$$

where in (6) equality holds if P and Q are up-closed, and also

$$\text{(7) } (P\,;Q)\,;R \subseteq P\,;(Q\,;R),$$

where in (7) equality holds if Q is up-closed, and finally

$$\text{(8) } P\,;Q \cup P\,;R \subseteq P\,;(Q \cup R) \qquad \text{(9) } P\,;(Q \cap R) \subseteq P\,;Q \cap P\,;R.$$

Finally, we consider algebraic laws of the dual. This operation reverses the lattice order and distributes over composition of up-closed multirelations. Again this needs to be weakened in the general case. These and further properties are summarised in the following result.

Theorem 7. *For all multirelations Q and R we have*

(1) $\mathsf{O}^\mathsf{d} = \mathsf{T}$	*(4)* $R^{\mathsf{dd}} = R$	
(2) $\mathsf{E}^\mathsf{d} = \mathsf{E}$	*(5)* $(Q \cup R)^\mathsf{d} = Q^\mathsf{d} \cap R^\mathsf{d}$	*(7)* $(Q;R)^\mathsf{d} \subseteq Q^\mathsf{d};R^\mathsf{d}$
(3) $\mathsf{T}^\mathsf{d} = \mathsf{O}$	*(6)* $(Q \cap R)^\mathsf{d} = Q^\mathsf{d} \cup R^\mathsf{d}$	*(8)* $(Q;R)^\mathsf{d} = (Q;\mathsf{E})^\mathsf{d};R^\mathsf{d}$,

where in (7) equality holds if Q is up-closed.

For proofs of Theorems 5–7 we refer to [5,13,18,28].

4 Multirelations and Predicate Transformers

A one-to-one correspondence between contact relations and closure operations is given in [1] and treated relation-algebraically in [29]. In this section we first generalise this correspondence to one between multirelations and predicate transformers. Based on this, we show how to translate operations and properties between these structures. We come back to contact relations in Sect. 4.4.

4.1 Connecting Multirelations and Predicate Transformers

Let $R : A \leftrightarrow 2^B$ be a multirelation. By forming the symmetric quotient with the membership multirelation $\mathsf{E} : A \leftrightarrow 2^A$ we obtain

$$\Psi(R) := \mathsf{syq}(R, \mathsf{E}) : 2^B \leftrightarrow 2^A.$$

From Sect. 2.3 we know that the specific symmetric quotient construction $\mathsf{syq}(R, \mathsf{E})$ is always a mapping in the sense of Sect. 2.2. As a consequence, we are allowed to write $\Psi(R) : 2^B \to 2^A$ for typing the relation $\Psi(R)$ and the mapping $\Psi(R)$ becomes a predicate transformer in the sense of Sect. 2.2. Conversely, let $f : 2^B \to 2^A$ be a predicate transformer. Using the membership multirelation $\mathsf{E} : A \leftrightarrow 2^A$, by typing reasons we get

$$\Phi(f) := \mathsf{E}f^\mathsf{c} : A \leftrightarrow 2^B,$$

whence the relation $\Phi(f)$ becomes a multirelation. Set-theoretic definitions of the two functions (in the usual mathematical sense) Ψ and Φ between the set of relations of type $A \leftrightarrow 2^B$ and the set of functions (again in the usual mathematical sense) from 2^B to 2^A have already been given in [1]. The above relation-algebraic specifications of Ψ and Φ come from [29]. For up-closed multirelations and isotone predicate transformers they are expressed using the power transpose of power allegories in [18]. In [8] the power transpose Λ is characterised by the universal property $\Lambda R = f$ if and only if $R = f\mathsf{E}^\mathsf{c}$ and it is shown that $\Lambda R = \mathsf{syq}(R^\mathsf{c}, \mathsf{E})$. Hence $\Psi(R) = \Lambda(R^\mathsf{c})$ holds, and $\Phi(f) = \mathsf{E}f^\mathsf{c}$ also follows from the universal property. The following result appears as [29, Corollary 4.2 and Theorem 4.4] and, for up-closed multirelations, as [18, Lemma 6.4].

Theorem 8. *The functions Ψ and Φ are mutually inverse and, thus, constitute a one-to-one correspondence between the multirelations of type $A \leftrightarrow 2^B$ and the mappings of type $2^B \to 2^A$. They are order isomorphisms with respect to the inclusion of multirelations and the pointwise order of mappings.*

4.2 Translating Operations

By means of the one-to-one correspondence of Sect. 4.1 we are able to translate the operations of union, intersection, composition, dual and complement from multirelations to predicate transformers and to do the same for the three constant multirelations and the inclusion order. The corresponding definitions are as follows; we use the notation of [24].

Definition 9. *Given predicate transformers $f, g : 2^B \to 2^A$ and $h : 2^C \to 2^B$ we define the operations and constants*

$$\begin{array}{rlll} \textit{union} & f \sqcup g := \Psi(\Phi(f) \cup \Phi(g)) : 2^B \to 2^A & & \bot := \Psi(\mathsf{O}) : 2^B \to 2^A \\ \textit{intersection} & f \sqcap g := \Psi(\Phi(f) \cap \Phi(g)) : 2^B \to 2^A & & \top := \Psi(\mathsf{T}) : 2^B \to 2^A \\ \textit{composition} & f \circ h := \Psi(\Phi(f) ; \Phi(h)) & : 2^C \to 2^A & 1 := \Psi(\mathsf{E}) \ : 2^A \to 2^A \\ \textit{dual} & f^\circ := \Psi(\Phi(f)^{\mathsf{d}}) & : 2^B \to 2^A & \\ \textit{complement} & f^- := \Psi(\overline{\Phi(f)}) & : 2^B \to 2^A . & \end{array}$$

Moreover, we define the relation $\sqsubseteq$ on the set of predicate transformers of type $2^B \to 2^A$ by $f \sqsubseteq g$ if and only if $\Phi(f) \subseteq \Phi(g)$, for all $f, g : 2^B \to 2^A$.

The following result elaborates these operations, constants and the relation $\sqsubseteq$ in terms of relation algebras. It expresses the following:

- Multirelational union (intersection, complement) translates to the pointwise union (intersection, complement) of predicate transformers.
- Multirelational dual translates to a mapping that, in the usual mathematical sense, is written as the function $f^\circ(X) = \overline{f(\overline{X})}$.
- The empty (universal) multirelation translates to the predicate transformer which maps everything to the empty (full) set and the membership multirelation translates to the identity predicate transformer.
- The lattice order on the set of multirelations of type $A \leftrightarrow 2^B$ translates to the pointwise order on the set of predicate transformers of type $2^B \to 2^A$, which is induced by the inclusion order of the target 2^A.

The pointwise interpretations are as expected from [24]. Note that a restriction to isotone predicate transformers is not required.

Theorem 10. *Let $f, g : 2^B \to 2^A$ be predicate transformers. Then*

(1) $f \sqcup g = [f, g]\mathsf{J}$, *where* $\mathsf{J} : 2^A \times 2^A \to 2^A$ *is the join mapping,*
(2) $f \sqcap g = [f, g]\mathsf{M}$, *where* $\mathsf{M} : 2^A \times 2^A \to 2^A$ *is the meet mapping,*
(3) $f^\circ = \mathsf{C}_B f \mathsf{C}_A$ *and* $f^- = f\mathsf{C}_A$, *where* $\mathsf{C}_A : 2^A \to 2^A$ *and* $\mathsf{C}_B : 2^B \to 2^B$ *are set complement mappings,*
(4) $\bot = \overline{\mathsf{T}\overline{\mathsf{E}}}$ *and* $\top = \overline{\mathsf{T}\overline{\overline{\mathsf{E}}}}$, *where* $\mathsf{E} : A \leftrightarrow 2^A$ *is the membership relation,*
(5) $1 = \mathsf{I}$,
(6) $f \sqsubseteq g$ *if and only if* $f \subseteq g\mathsf{S}^{\mathsf{c}}$, *where* $\mathsf{S} : 2^A \leftrightarrow 2^A$ *is the subset relation.*

Proof. (1) Using the projection mappings $p, r : 2^A{\times}2^A \to 2^A$ of the direct product $2^A{\times}2^A$ for the right pairings we obtain:

$$
\begin{aligned}
f \sqcup g &= \Psi(\Phi(f) \cup \Phi(g)) \\
&= \Psi(\mathsf{E}f^{\mathsf{c}} \cup \mathsf{E}g^{\mathsf{c}}) \\
&= \mathsf{syq}(\mathsf{E}(f^{\mathsf{c}} \cup g^{\mathsf{c}}), \mathsf{E}) \\
&= \mathsf{syq}(\mathsf{E}(p^{\mathsf{c}}[f,g]^{\mathsf{c}} \cup r^{\mathsf{c}}[f,g]^{\mathsf{c}}), \mathsf{E}) && \text{Lemma 4 (1)} \\
&= \mathsf{syq}(\mathsf{E}(p^{\mathsf{c}} \cup r^{\mathsf{c}})[f,g]^{\mathsf{c}}, \mathsf{E}) \\
&= [f,g]\,\mathsf{syq}((\mathsf{E}p^{\mathsf{c}} \cup \mathsf{E}r^{\mathsf{c}}), \mathsf{E}) && \text{Lemma 4 (2) and Lemma 1} \\
&= [f,g]\,\mathsf{syq}((\overline{\overline{\overline{\mathsf{E}}}p^{\mathsf{c}}} \cup \overline{\overline{\overline{\mathsf{E}}}r^{\mathsf{c}}}), \mathsf{E}) && p^{\mathsf{c}}, r^{\mathsf{c}} \text{ bijective} \\
&= [f,g]\,\mathsf{syq}(\overline{\overline{\overline{\mathsf{E}}}p^{\mathsf{c}} \cap \overline{\overline{\mathsf{E}}}r^{\mathsf{c}}}, \mathsf{E}) \\
&= [f,g]\,\mathsf{syq}(\overline{[\overline{\overline{\mathsf{E}}}, \overline{\overline{\mathsf{E}}}]}, \mathsf{E}) \\
&= [f,g]\mathsf{J}
\end{aligned}
$$

(2) The following calculation shows the claim:

$$
\begin{aligned}
f \sqcap g &= \Psi(\Phi(f) \cap \Phi(g)) \\
&= \Psi(\mathsf{E}f^{\mathsf{c}} \cap \mathsf{E}g^{\mathsf{c}}) \\
&= \Psi([\mathsf{E}, \mathsf{E}][f,g]^{\mathsf{c}}) && \text{Lemma 4 (3)} \\
&= \mathsf{syq}([\mathsf{E}, \mathsf{E}][f,g]^{\mathsf{c}}, \mathsf{E}) \\
&= [f,g]\,\mathsf{syq}([\mathsf{E}, \mathsf{E}], \mathsf{E}) && \text{Lemma 4 (2) and Lemma 1} \\
&= [f,g]\mathsf{M}
\end{aligned}
$$

(3) For the dual we proceed as follows; the complement is treated similarly:

$$
\begin{aligned}
f^{\circ} &= \Psi(\Phi(f)^{\mathsf{d}}) \\
&= \Psi((\mathsf{E}f^{\mathsf{c}})^{\mathsf{d}}) \\
&= \Psi(\overline{\mathsf{E}f^{\mathsf{c}}}\mathsf{C}_B) && \text{Theorem 5 (2)} \\
&= \mathsf{syq}(\overline{\mathsf{E}f^{\mathsf{c}}}\mathsf{C}_B{}^{\mathsf{c}}, \mathsf{E}) && \text{Lemma 3} \\
&= \mathsf{C}_B\,\mathsf{syq}(\overline{\mathsf{E}}f^{\mathsf{c}}, \mathsf{E}) && \mathsf{C}_B \text{ mapping, Lemma 1, } f \text{ mapping} \\
&= \mathsf{C}_B f\,\mathsf{syq}(\overline{\mathsf{E}}, \mathsf{E}) && f \text{ mapping, Lemma 1} \\
&= \mathsf{C}_B f \mathsf{C}_A
\end{aligned}
$$

(4) The two equations are shown by the calculations

$$
\begin{aligned}
\bot &= \Psi(\mathsf{O}) = \mathsf{syq}(\mathsf{O}, \mathsf{E}) = (\mathsf{O} \setminus \mathsf{E}) \cap (\mathsf{O}^{\mathsf{c}}/\mathsf{E}^{\mathsf{c}}) = \mathsf{T} \cap \overline{\overline{\mathsf{O}}\mathsf{E}} = \overline{\mathsf{T}\mathsf{E}} \\
\top &= \Psi(\mathsf{T}) = \mathsf{syq}(\mathsf{T}, \mathsf{E}) = (\mathsf{T} \setminus \mathsf{E}) \cap (\mathsf{T}^{\mathsf{c}}/\mathsf{E}^{\mathsf{c}}) = \overline{\mathsf{T}^{\mathsf{c}}\overline{\overline{\mathsf{E}}}} \cap \mathsf{T} = \overline{\mathsf{T}\overline{\overline{\mathsf{E}}}}.
\end{aligned}
$$

(5) Using the first axiom of membership relations, a proof of the claim is:

$$1 = \Psi(\mathsf{E}) = \mathsf{syq}(\mathsf{E}, \mathsf{E}) = \mathsf{I}$$

(6) To verify this fact we calculate as follows:

$$
\begin{aligned}
f \sqsubseteq g &\iff \Phi(f) \subseteq \Phi(g) \\
&\iff \mathsf{E}f^{\mathsf{c}} \subseteq \mathsf{E}g^{\mathsf{c}} \\
&\iff \mathsf{E}f^{\mathsf{c}}g \subseteq \mathsf{E} && g \text{ mapping, shunting} \\
&\iff f^{\mathsf{c}}g \subseteq \mathsf{E} \setminus \mathsf{E} && \text{Galois right residual} \\
&\iff g^{\mathsf{c}}f \subseteq \mathsf{S}^{\mathsf{c}} \\
&\iff f \subseteq g\mathsf{S}^{\mathsf{c}} && g \text{ mapping, shunting}
\end{aligned}
$$

□

The Boolean lattice structure of multirelations is therefore preserved in the pointwise Boolean lattice structure of predicate transformers. As regards composition we have the following result for the restricted set of up-closed multirelations [18, Lemma 6.4]. Their (multirelational) composition amounts to the backward composition of the corresponding isotone predicate transformers.

Theorem 11. *Let $f : 2^B \to 2^A$ and $g : 2^C \to 2^B$ be predicate transformers such that the multirelation $\Phi(f)$ is up-closed. Then $f \circ g = gf$.*

Proof. We calculate as follows:

$$\begin{aligned}
f \circ g &= \Psi(\Phi(f)\,;\Phi(g)) \\
&= \Psi(\Phi(f)\,;\mathsf{E}g^{\mathsf{c}}) \\
&= \Psi(\Phi(f)(\mathsf{E} \setminus \mathsf{E}g^{\mathsf{c}})) && \text{Theorem 5 (1)} \\
&= \Psi(\Phi(f)(\mathsf{E} \setminus \mathsf{E})g^{\mathsf{c}}) && g \text{ mapping, Lemma 1} \\
&= \mathsf{syq}(\Phi(f)\mathsf{S}g^{\mathsf{c}}, \mathsf{E}) \\
&= g\,\mathsf{syq}(\Phi(f)\mathsf{S}, \mathsf{E}) && g \text{ mapping, Lemma 1} \\
&= g\,\mathsf{syq}(\Phi(f), \mathsf{E}) && \Phi(f) \text{ up-closed, Theorem 5} \\
&= g\,\mathsf{syq}(\mathsf{E}f^{\mathsf{c}}, \mathsf{E}) \\
&= gf\,\mathsf{syq}(\mathsf{E}, \mathsf{E}) && f \text{ mapping, Lemma 1} \\
&= gf && \text{axiom of membership relation}
\end{aligned}$$

□

4.3 Alternative Composition of Multirelations

Observe that predicate transformers are mappings, and therefore – when considered as functions in the usual mathematical sense – their composition is associative. The composition of up-closed multirelations is also associative, but the composition of multirelations is not associative in general; see Theorem 6. Similarly, the identity mapping is neutral for composition, but the membership relation is right-neutral only for up-closed multirelations. Hence, the above one-to-one correspondence between predicate transformers and multirelations is not a monoid isomorphism; this is why we restricted attention to up-closed multirelations in Theorem 11.

On the other hand, the one-to-one correspondence induces a monoid isomorphism with a different composition, defined for all multirelations $Q : A \leftrightarrow 2^B$ and $R : B \leftrightarrow 2^C$ by $(Q : R) := \Phi(\Psi(R)\,\Psi(Q)) : A \leftrightarrow 2^C$, which elaborates to

$$\begin{aligned}
(Q : R) &= \Phi(\Psi(R)\,\Psi(Q)) \\
&= \Phi(\mathsf{syq}(R, \mathsf{E})\,\mathsf{syq}(Q, \mathsf{E})) \\
&= \mathsf{E}\,\mathsf{syq}(Q, \mathsf{E})^{\mathsf{c}}\mathsf{syq}(R, \mathsf{E})^{\mathsf{c}} \\
&= \mathsf{E}\,\mathsf{syq}(\mathsf{E}, Q)\,\mathsf{syq}(\mathsf{E}, R) && \text{property of symmetric quotients [30]} \\
&= Q\,\mathsf{syq}(\mathsf{E}, R) && \text{property of symmetric quotients [30]}
\end{aligned}$$

The logical meaning of the new multirelational composition $(Q : R)$ is that for all $x \in A$ and $Z \in 2^C$ we have $(Q : R)_{x,Z}$ if and only if $Q_{x,R^{\mathsf{c}}(Z)}$, where $R^{\mathsf{c}}(Z)$ denotes the set of elements from the set B that R relates with the set Z.

Yet another composition is introduced in [23]. Like the one of Sect. 3.1 it is not associative for general multirelations as shown in [11], but satisfies weaker algebraic properties.

R is ...	if and only if	R is ...	if and only if
up-closed	$R;\mathsf{E} = R$		
total	$R;\mathsf{T} = \mathsf{T}$	co-total	$R;\mathsf{O} = \mathsf{O}$
$\cup$-distributive	$R;(P \cup Q) = R;P \cup R;Q$		
$\cap$-distributive	$R;(P \cap Q) = R;P \cap R;Q$		
reflexive	$\mathsf{E} \subseteq R$	co-reflexive	$R \subseteq \mathsf{E}$
transitive	$R;R \subseteq R$	dense	$R \subseteq R;R$
idempotent	$R;R = R$		
a contact	$R;R \cup \mathsf{E} = R$	a kernel	$R;R \cap \mathsf{E} = R;\mathsf{E}$
a test	$R;\mathsf{T} \cap \mathsf{E} = R$	a co-test	$R;\mathsf{O} \cup \mathsf{E} = R$
a vector	$R;\mathsf{T} = R$		

Fig. 1. Properties of multirelations

4.4 Translating Properties

Theorem 11 is an example of a result which only holds for a restricted set of multirelations. The property of being up-closed is frequently used, but not the only one that is useful. This property and other properties have been discussed in [3,18,19,22,25,26] and – in the context of topology – already in [1].

In the companion paper [5] we have started a systematic investigation of several multirelational properties by giving relational and algebraic definitions, showing how the properties are related and under which operations they are closed. We now discuss how the properties translate to predicate transformers. Specifically, we look at the properties given in Fig. 1 above, where the two distributivity properties universally quantify over the multirelations P and Q. We proceed from top to bottom.

Up-closed Multirelations. We start with the well-known connection of up-closed multirelations and isotone predicate transformers; see [3,14,18,22,25]. The following result expresses it in terms of relation algebras. Note that being up-closed is precisely axiom (K_2) of contact relations [1].

Theorem 12. *Let $f : 2^B \to 2^A$ be a predicate transformer. Then $\Phi(f)$ is up-closed if and only if $\mathsf{S}_B f \subseteq f \mathsf{S}_A$.*

Proof. We calculate as follows:

$$
\begin{array}{rlll}
\Phi(f) \text{ up-closed} & \iff \Phi(f)\mathsf{S}_B \subseteq \Phi(f) & \text{Theorem 5} \\
& \iff \mathsf{E} f^{\mathsf{c}} \mathsf{S}_B \subseteq \mathsf{E} f^{\mathsf{c}} & \\
& \iff \mathsf{E} f^{\mathsf{c}} \mathsf{S}_B f \subseteq \mathsf{E} & f \text{ mapping, shunting} \\
& \iff f^{\mathsf{c}} \mathsf{S}_B f \subseteq \mathsf{E} \backslash \mathsf{E} & \text{Galois right residual} \\
& \iff \mathsf{S}_B f \subseteq f \mathsf{S}_A & f \text{ mapping, shunting} & \square
\end{array}
$$

The inclusion $\mathsf{S}_B f \subseteq f \mathsf{S}_A$ of Theorem 12 expresses relation-algebraically that $X \subseteq Y$ implies $f(X) \subseteq f(Y)$ for all $X, Y \in 2^B$ [21,27].

Total and Co-total Multirelations. The multirelation $R : A \leftrightarrow 2^B$ is called co-total if the empty set is not in its image, that is, $\neg R_{x,\emptyset}$, for all $x \in A$.

This is precisely axiom (K_0) of [1]. The empty multirelation O is not a right-zero of composition, but it is so for co-total multirelations (see [26], where such multirelations are called "total"). As shown in [5], being co-total amounts to $R;\mathsf{O} = \mathsf{O}$ and this is equivalent to $R \subseteq \mathsf{TE}$.

Neither is the universal multirelation T a right-zero of composition, but it is so for total multirelations (see [25], where such multirelations are called "proper"). This is because $R;\mathsf{T} = \mathsf{T}$ is equivalent to $R\mathsf{T} = \mathsf{T}$ as shown in [5]. How these properties translate to isotone predicate transformers is shown in [25, Theorem 10]. The following result generalises this to arbitrary predicate transformers. It follows that the restriction to co-total multirelations corresponds to Dijkstra's Law of the Excluded Miracle [9].

Theorem 13. *Let $f : 2^B \to 2^A$ be a predicate transformer. Then $\Phi(f)$ is co-total if and only if $f(\emptyset) = \emptyset$ and $\Phi(f)$ is total if and only if $f(B) = A$.*

Proof. The first claim follows by

$$\begin{aligned}
\Phi(f);\mathsf{O} = \mathsf{O} &\iff \Phi(f) \subseteq \mathsf{TE} && \text{shown in [5]} \\
&\iff \mathsf{E}f^{\mathsf{c}} \subseteq \mathsf{TE} \\
&\iff \forall x \in A : \forall Y \in 2^B : (\mathsf{E}f^{\mathsf{c}})_{x,Y} \Rightarrow (\mathsf{TE})_{x,Y} \\
&\iff \forall x \in A : \forall Y \in 2^B : x \in f(Y) \Rightarrow Y \neq \emptyset \\
&\iff \forall Y \in 2^B : (\exists x \in A : x \in f(Y)) \Rightarrow Y \neq \emptyset \\
&\iff \forall Y \in 2^B : f(Y) \neq \emptyset \Rightarrow Y \neq \emptyset \\
&\iff \forall Y \in 2^B : Y = \emptyset \Rightarrow f(Y) = \emptyset \\
&\iff f(\emptyset) = \emptyset.
\end{aligned}$$

To prove the second claim, we start with $\Phi(f);\mathsf{T} = \Phi(f)\mathsf{T}$. Hence, $\Phi(f);\mathsf{T} = \mathsf{T}$ is equivalent to $\mathsf{T} \subseteq \mathsf{E}f^{\mathsf{c}}\mathsf{T}$. That $\mathsf{T} \subseteq \mathsf{E}f^{\mathsf{c}}\mathsf{T}$ is equivalent to $f(B) = A$ can again be verified using pointwise reasoning; we omit the details. □

Distributive Multirelations. Next, we investigate $\cup$-distributivity. Assume that $f : 2^B \to 2^A$ is a predicate transformer. Consider for each element $x \in A$ the subset $I_x := \{Y \in 2^B \mid \forall Z \in 2^Y : x \notin f(Z)\}$ of 2^B. The following result relates the sets I_x to $\cup$-distributivity of the multirelation corresponding to f. Recall that a subset I of 2^B is an *ideal* in the lattice $(2^B, \cup, \cap)$ if it is a lower set and closed under binary unions and that a subset F of 2^B is a *filter* in this lattice if it is an upper set and closed under binary intersections.

Theorem 14. *Let $f : 2^B \to 2^A$ be a predicate transformer. Then $\Phi(f)$ is $\cup$-distributive if and only if for all $x \in A$ the set I_x is an ideal in $(2^B, \cup, \cap)$.*

Proof. The multirelation $\Phi(f)$ is $\cup$-distributive if and only if for all predicate transformers $g, h : 2^C \to 2^B$ we have

$$\Phi(f);(\Phi(g) \cup \Phi(h)) = \Phi(f);\Phi(g) \cup \Phi(f);\Phi(h). \tag{1}$$

In the first part of the proof we transform (1) as follows:

$$\begin{aligned}
&\Phi(f);(\Phi(g) \cup \Phi(h)) = \Phi(f);\Phi(g) \cup \Phi(f);\Phi(h) \\
\iff &\Phi(f);(\Phi(g) \cup \Phi(h)) \subseteq \Phi(f);\Phi(g) \cup \Phi(f);\Phi(h) && \text{Theorem 6 (8)} \\
\iff &\mathsf{E}f^{\mathsf{c}}(\mathsf{E} \setminus \mathsf{E}(g \cup h)^{\mathsf{c}}) \subseteq \mathsf{E}f^{\mathsf{c}}(\mathsf{E} \setminus \mathsf{E}g^{\mathsf{c}}) \cup \mathsf{E}f^{\mathsf{c}}(\mathsf{E} \setminus \mathsf{E}h^{\mathsf{c}}) && \text{Theorem 5 (1)} \\
\iff &\mathsf{E}f^{\mathsf{c}}(\mathsf{E} \setminus \mathsf{E}(g \cup h)^{\mathsf{c}}) \subseteq \mathsf{E}f^{\mathsf{c}}\mathsf{S}g^{\mathsf{c}} \cup \mathsf{E}f^{\mathsf{c}}\mathsf{S}h^{\mathsf{c}} && \text{Lemma 1}
\end{aligned}$$

Furthermore, for all sets $Y \in 2^C$ and $Z \in 2^B$ we obtain

$$\begin{aligned}(\mathsf{E} \setminus \mathsf{E}(g \cup h)^{\mathsf{c}})_{Z,Y} &\iff \forall w \in B : w \in Z \Rightarrow w \in g(Y) \cup h(Y) \\ &\iff Z \subseteq g(Y) \cup h(Y).\end{aligned}$$

Hence, (1) holds if and only if for all $x \in A$ and $Y \in 2^C$ the following holds:

$$\begin{aligned}&(\exists Z \in 2^B : x \in f(Z) \wedge Z \subseteq g(Y) \cup h(Y)) \\ &\Rightarrow \\ &(\exists Z \in 2^B : x \in f(Z) \wedge Z \subseteq g(Y)) \vee (\exists Z \in 2^B : x \in f(Z) \wedge Z \subseteq h(Y))\end{aligned} \tag{2}$$

Next, we define for each $x \in A$ the set $M_x := \{Z \in 2^B \mid x \in f(Z)\}$. Substituting $G = g(Y)$ and $H = h(Y)$ in (2) we get the condition

$$(\exists Z \in M_x : Z \subseteq G \cup H) \;\Rightarrow\; (\exists Z \in M_x : Z \subseteq G) \vee (\exists Z \in M_x : Z \subseteq H). \tag{3}$$

This works because for each g and Y in (2), the set G of (3) can be instantiated to $g(Y)$, and for each G in (3), the mapping g of (2) can be instantiated to the constant mapping $\lambda Y.G : 2^C \to 2^B$, and similarly for h and H.

A consequence is that (2) holds for all predicate transformers $g, h : 2^C \to 2^B$, $x \in A$ and $Y \in 2^C$ if and only if (3) holds for all $x \in A$ and sets $G, H \in 2^B$. Summing up, in the first part we have shown that $\Phi(f)$ is $\cup$-distributive if and only if (3) holds for all $x \in A$ and $G, H \in 2^B$.

In the second part of the proof, we define for all $x \in A$ and sets $Y \in 2^B$ the set $M_{x,Y} := M_x \cap \{Z \in 2^B \mid Z \subseteq Y\}$. Then (3) simplifies to

$$M_{x,G \cup H} \neq \emptyset \;\Rightarrow\; M_{x,G} \neq \emptyset \vee M_{x,H} \neq \emptyset$$

which, in turn, is equivalent to

$$M_{x,G} = \emptyset \wedge M_{x,H} = \emptyset \;\Rightarrow\; M_{x,G \cup H} = \emptyset.$$

Observe furthermore that

$$\begin{aligned}M_{x,G} = \emptyset &\iff M_x \cap \{Z \in 2^B \mid Z \subseteq G\} = \emptyset \\ &\iff \forall Z \in 2^G : Z \notin M_x \\ &\iff \forall Z \in 2^G : x \notin f(Z) \\ &\iff G \in I_x\end{aligned}$$

and similarly for $M_{x,H}$ and $M_{x,G \cup H}$. Hence, (3) is equivalent to

$$G \in I_x \wedge H \in I_x \;\Rightarrow\; G \cup H \in I_x. \tag{4}$$

A consequence is that (3) holds for all $x \in A$ and $G, H \in 2^B$ if and only if (4) holds for all $x \in A$ and $G, H \in 2^B$, but the latter states that I_x is closed under binary union, for all $x \in A$. So, as result of this part we have that $\Phi(f)$ is $\cup$-distributive if and only if for all $x \in A$ the set I_x is closed under binary union.

Finally, each I_x is a lower set since $M_{x,Y}$ is $\subseteq$-isotone in its argument Y. Hence, I_x is closed under binary union if and only if it is an ideal. □

For $\cap$-distributive multirelations we obtain the following dual result, where we consider for each $x \in A$ the set $F_x := \{Y \in 2^B \mid \exists Z \in 2^Y : x \in f(Z)\}$. We omit its proof.

Theorem 15. *Let $f : 2^B \to 2^A$ be a predicate transformer. Then $\Phi(f)$ is $\cap$-distributive if and only if for all $x \in A$ the set F_x is a filter in $(2^B, \cup, \cap)$.*

How distributivity over (arbitrary) union/intersection of up-closed multirelations translates to isotone predicate transformers has been investigated in [25]. In [5] we show that multirelations satisfying axiom (K_4) of [1] are $\cup$-distributive and that this is an equivalence for up-closed multirelations.

Reflexive and Co-reflexive Multirelations. Every reflexive multirelation satisfies axiom (K_1) of [1] and this is an equivalence for up-closed multirelations. The following result shows how reflexivity translates to predicate transformers.

Theorem 16. *Let $f : 2^A \to 2^A$ be a predicate transformer. Then*

(1) $\Phi(f)$ is reflexive if and only if $X \subseteq f(X)$ for each $X \in 2^A$,
(2) $\Phi(f)$ is co-reflexive if and only if $X \supseteq f(X)$ for each $X \in 2^A$.

Proof. (1) Reflexivity translates as follows:

$$
\begin{aligned}
\mathsf{E} \subseteq \Phi(f) &\iff \mathsf{I} \subseteq \mathsf{E} \backslash \mathsf{E} f^{\mathsf{c}} && \text{Galois right residual} \\
&\iff \mathsf{I} \subseteq \mathsf{S} f^{\mathsf{c}} && f \text{ mapping, Lemma 1} \\
&\iff f \subseteq \mathsf{S} && f \text{ mapping, shunting} \\
&\iff \forall X \in 2^A : X \subseteq f(X)
\end{aligned}
$$

(2) Co-reflexivity translates as follows:

$$
\begin{aligned}
\Phi(f) \subseteq \mathsf{E} &\iff \mathsf{E} f^{\mathsf{c}} \subseteq \mathsf{E} \\
&\iff f^{\mathsf{c}} \subseteq \mathsf{S} && \text{Galois right residual} \\
&\iff f \subseteq \mathsf{S}^{\mathsf{c}} \\
&\iff \forall X \in 2^A : f(X) \subseteq X
\end{aligned}
$$

□

Transitive, Dense and Idempotent Multirelations. Transitivity of multirelations is precisely axiom (K_3) of [1]. We first give a result for isotone predicate transformers. The first part shows how transitivity translates to predicate transformers. The remaining parts deal with the properties "dense" and "idempotent".

Theorem 17. *Let $f : 2^A \to 2^A$ be an isotone predicate transformer. Then*

(1) $\Phi(f)$ is transitive if and only if $f(f(X)) \subseteq f(X)$ for each $X \in 2^A$,
(2) $\Phi(f)$ is dense if and only if $f(f(X)) \supseteq f(X)$ for each $X \in 2^A$,
(3) $\Phi(f)$ is idempotent if and only if $f(f(X)) = f(X)$ for each $X \in 2^A$.

Proof. (1) Observe that $\mathsf{E}f^{\mathsf{c}} \subseteq \mathsf{E}f^{\mathsf{c}}\mathsf{S} \subseteq \mathsf{ES}f^{\mathsf{c}} = \mathsf{E}f^{\mathsf{c}}$ by Lemma 2, since f is isotone and E is up-closed. Hence, we get

$$\Phi(f);\Phi(f) = (\mathsf{E}f^{\mathsf{c}});(\mathsf{E}f^{\mathsf{c}}) = \mathsf{E}f^{\mathsf{c}}(\mathsf{E} \setminus \mathsf{E}f^{\mathsf{c}}) = \mathsf{E}f^{\mathsf{c}}(\mathsf{E} \setminus \mathsf{E})f^{\mathsf{c}} = \mathsf{E}f^{\mathsf{c}}\mathsf{S}f^{\mathsf{c}} = \mathsf{E}f^{\mathsf{c}}f^{\mathsf{c}}$$

by Theorem 5 and Lemma 1, since f is a mapping. This implies the claim by

$$\begin{aligned}
\Phi(f);\Phi(f) \subseteq \Phi(f) &\iff \mathsf{E}f^{\mathsf{c}}f^{\mathsf{c}} \subseteq \mathsf{E}f^{\mathsf{c}} \\
&\iff f^{\mathsf{c}}f^{\mathsf{c}} \subseteq \mathsf{E} \setminus \mathsf{E}f^{\mathsf{c}} && \text{Galois right residual} \\
&\iff f^{\mathsf{c}}f^{\mathsf{c}} \subseteq \mathsf{S}f^{\mathsf{c}} && f \text{ mapping, Lemma 1} \\
&\iff ff \subseteq f\mathsf{S}^{\mathsf{c}} \\
&\iff \forall X,Y \in 2^A : (ff)_{X,Y} \Rightarrow (f\mathsf{S}^{\mathsf{c}})_{X,Y} \\
&\iff \forall X,Y \in 2^A : Y = f(f(X)) \Rightarrow Y \subseteq f(X) \\
&\iff \forall X \in 2^A : f(f(X)) \subseteq f(X).
\end{aligned}$$

(2) We use again the equation $\Phi(f);\Phi(f) = \mathsf{E}f^{\mathsf{c}}f^{\mathsf{c}}$ from the proof of (1):

$$\begin{aligned}
\Phi(f) \subseteq \Phi(f);\Phi(f) &\iff \mathsf{E}f^{\mathsf{c}} \subseteq \mathsf{E}f^{\mathsf{c}}f^{\mathsf{c}} \\
&\iff f^{\mathsf{c}} \subseteq \mathsf{E} \setminus \mathsf{E}f^{\mathsf{c}}f^{\mathsf{c}} && \text{Galois right residual} \\
&\iff f^{\mathsf{c}} \subseteq \mathsf{S}f^{\mathsf{c}}f^{\mathsf{c}} && f \text{ mapping, Lemma 1} \\
&\iff f \subseteq ff\mathsf{S}^{\mathsf{c}} \\
&\iff \forall X \in 2^A : f(X) \subseteq f(f(X))
\end{aligned}$$

(3) This follows immediately by combining (1) and (2). □

If f is not required to be isotone, the conditions become more complex; for example, the multirelation $\Phi(f)$ is transitive if and only if $Y \subseteq f(X)$ implies $f(Y) \subseteq f(X)$, for each $X, Y \in 2^A$. We omit further details about this and instead give an example using dense multirelations.

Example 18. We consider a basic program transformation that imports an invariant into the body of a loop. Assume that multirelations P and R represent the invariant and the body, respectively. We wish to prove

$$P;(\nu X.R;X) = P;(\nu X.(P;R);X), \qquad (\star)$$

which uses the $\subseteq$-greatest fixpoint ν to represent the semantics of an endless loop. Such a transformation might be followed by simplifications of the loop body using the invariant P.

The invariant P is frequently represented as a test, which acts as a filter in sequential compositions [16, Lemma 2.3.2]. But transformation $(\star)$ can be proved with much weaker assumptions, namely

(1) P is dense, that is, $P \subseteq P;P$,
(2) P is preserved by R, that is, $P;R \subseteq R;P$, and
(3) $P;(Y;Z) \subseteq (P;Y);Z$ for all multirelations Y and Z.

The latter is required because multirelational composition only satisfies the semi-associativity law given in Theorem 6 (7), namely (4) $(X;Y);Z \subseteq X;(Y;Z)$ for all X, Y, and Z. To show that $(\star)$ follows from assumptions (1) to (3), we start by

$$\begin{aligned} P;(\nu X.R;X) &\subseteq (P;R);(\nu X.R;X) && \text{unfold } \nu \text{ and (3)} \\ &\subseteq (P;(P;R));(\nu X.R;X) && \text{(1) and (4)} \\ &\subseteq (P;R);(P;(\nu X.R;X)) && \text{(2) and (3) and (4).} \end{aligned}$$

The greatest fixpoint property of ν implies (5) $P;(\nu X.R;X) \subseteq \nu X.(P;R);X$. Then one inclusion of $(\star)$ is shown by

$$\begin{aligned} P;(\nu X.R;X) &\subseteq (P;P);(\nu X.R;X) && \text{(1)} \\ &\subseteq P;(\nu X.(P;R);X) && \text{(4) and (5).} \end{aligned}$$

For the other inclusion observe that

$$\begin{aligned} P;(\nu X.(P;R);X) &\subseteq (P;(P;R));(\nu X.(P;R);X) && \text{unfold } \nu \text{ and (3)} \\ &\subseteq ((P;R);P);(\nu X.(P;R);X) && \text{(2) and (3)} \\ &\subseteq (P;R);(P;(\nu X.(P;R);X)) && \text{(4).} \end{aligned}$$

The greatest fixpoint property implies (6) $P;(\nu X.(P;R);X) \subseteq \nu X.(P;R);X$. Hence, we get

$$\begin{aligned} \nu X.(P;R);X &\subseteq (R;P);(\nu X.(P;R);X) && \text{unfold } \nu \text{ and (2)} \\ &\subseteq R;(\nu X.(P;R);X) && \text{(4) and (6).} \end{aligned}$$

Finally, the greatest fixpoint property implies $\nu X.(P;R);X \subseteq \nu X.R;X$, from which the remaining inclusion of $(\star)$ follows by composing P on the left. □

Reasoning of the kind shown in this example can be done algebraically (abstracting from concrete multirelations) and is well supported by automated theorem provers; see [5, 11], for example.

Contacts and Kernels. In [5] we show algebraically that a multirelation is a contact if and only if it is reflexive, transitive and up-closed (hence also idempotent). As a consequence, the corresponding isotone predicate transformer $f : 2^A \to 2^A$ satisfies $X \subseteq f(X) = f(f(X))$, for all $X \in 2^A$. In other words, contact multirelations (in the sense of [1]) correspond to predicate transformers that are closure operations – as already shown in [1].

Dually, a multirelation is a kernel if and only if it is co-reflexive, dense and up-closed (hence also idempotent). This implies that the corresponding isotone predicate transformer $f : 2^A \to 2^A$ satisfies $X \supseteq f(X) = f(f(X))$ for all $X \in 2^A$; that is, it is a kernel operator in the sense of lattice theory.

Tests and Co-tests. The following result shows how tests translate from multirelations to predicate transformers.

Theorem 19. *Let $f : 2^A \to 2^A$ be a predicate transformer. Then $\Phi(f)$ is a test if and only if $f(X) = X \cap f(A)$ for each $X \in 2^A$.*

Proof. In terms of relation algebras, we have

$$\begin{aligned}\Phi(f) \text{ is a test } &\iff \Phi(f) = \Phi(f)\mathsf{T} \cap \mathsf{E} \\ &\iff \mathsf{E}f^{\mathsf{c}} = \mathsf{E}f^{\mathsf{c}}\mathsf{T} \cap \mathsf{E} \\ &\iff \mathsf{E}f^{\mathsf{c}} \subseteq \mathsf{E} \ \wedge \ \mathsf{E}f^{\mathsf{c}}\mathsf{T} \cap \mathsf{E} \subseteq \mathsf{E}f^{\mathsf{c}}.\end{aligned}$$

The first expression amounts to co-reflexivity; see Theorem 16 (2):

$$\mathsf{E}f^{\mathsf{c}} \subseteq \mathsf{E} \iff \forall X \in 2^A : f(X) \subseteq X$$

The second expression can be transformed as follows:

$$\begin{aligned}&\mathsf{E}f^{\mathsf{c}}\mathsf{T} \cap \mathsf{E} \subseteq \mathsf{E}f^{\mathsf{c}} \\ \iff &\forall x \in A : \forall X \in 2^A : (\mathsf{E}f^{\mathsf{c}}\mathsf{T} \cap \mathsf{E})_{x,X} \Rightarrow (\mathsf{E}f^{\mathsf{c}})_{x,X} \\ \iff &\forall x \in A : \forall X \in 2^A : (\mathsf{E}f^{\mathsf{c}}\mathsf{T})_{x,X} \wedge \mathsf{E}_{x,X} \Rightarrow x \in f(X) \\ \iff &\forall x \in A : \forall X \in 2^A : (\exists Y, Z \in 2^A : \mathsf{E}_{x,Y} \wedge f^{\mathsf{c}}{}_{Y,Z}) \wedge x \in X \Rightarrow x \in f(X) \\ \iff &\forall x \in A : \forall X \in 2^A : (\exists Z \in 2^A : x \in f(Z)) \wedge x \in X \Rightarrow x \in f(X) \\ \iff &\forall X, Z \in 2^A : \forall x \in A : x \in f(Z) \wedge x \in X \Rightarrow x \in f(X) \\ \iff &\forall X, Z \in 2^A : f(Z) \cap X \subseteq f(X)\end{aligned}$$

Both expressions together imply that f is isotone, since from $X \subseteq Y$ we get

$$f(X) = f(X) \cap X \subseteq f(X) \cap Y \subseteq f(Y),$$

for all $X, Y \in 2^A$. Hence, we have $f(X) \subseteq f(A)$ for all $X \in 2^A$ and, thus,

$$\begin{aligned}\Phi(f) \text{ is a test } &\iff f \text{ is isotone } \wedge \ \forall X, Z \in 2^A : f(Z) \cap X \subseteq f(X) \subseteq X \\ &\iff f \text{ is isotone } \wedge \ \forall X \in 2^A : f(A) \cap X \subseteq f(X) \subseteq X \\ &\iff f \text{ is isotone } \wedge \ \forall X \in 2^A : f(A) \cap X = f(X) \\ &\iff \forall X \in 2^A : f(A) \cap X = f(X)\end{aligned}$$

since the latter also implies that f is isotone. □

For co-tests we obtain the following dual result; we omit its proof.

Theorem 20. *Let $f : 2^A \to 2^A$ be a predicate transformer. Then $\Phi(f)$ is a co-test if and only if $f(X) = X \cup f(\emptyset)$ for each $X \in 2^A$.*

Vectors. A multirelation R is a vector if and only if $R \mathbin{;} \mathsf{T} = R$, which is equivalent to $R\mathsf{T} = R$. The final result in our series of translations shows how the vector property translates to predicate transformers.

Theorem 21. *Let $f : 2^A \to 2^A$ be a predicate transformer. Then $\Phi(f)$ is a vector if and only if f is a constant function.*

Proof. The claim is proved by the following calculation:

$$
\begin{aligned}
\Phi(f) \text{ is a vector} &\iff \Phi(f)\mathsf{T} = \Phi(f) \\
&\iff \Phi(f)\mathsf{T} \subseteq \Phi(f) \\
&\iff \mathsf{E}f^{\mathsf{c}}\mathsf{T} \subseteq \mathsf{E}f^{\mathsf{c}} \\
&\iff \mathsf{E}f^{\mathsf{c}}\mathsf{T}f \subseteq \mathsf{E} && f \text{ mapping, shunting} \\
&\iff f^{\mathsf{c}}\mathsf{T}f \subseteq \mathsf{S} && \text{Galois right residual} \\
&\iff (\mathsf{T}f)^{\mathsf{c}}(\mathsf{T}f) \subseteq \mathsf{S} && \mathsf{TT} = \mathsf{T} \\
&\iff (\mathsf{T}f)^{\mathsf{c}}(\mathsf{T}f) \subseteq \mathsf{S} \cap \mathsf{S}^{\mathsf{c}} && \text{LHS is symmetric} \\
&\iff (\mathsf{T}f)^{\mathsf{c}}(\mathsf{T}f) \subseteq \mathsf{I} && \text{Lemma 2} \\
&\iff \mathsf{T}f \text{ univalent} \\
&\iff f \text{ is constant} && f \text{ mapping}
\end{aligned}
$$

□

5 Conclusion

In the present paper we have investigated how properties from research on multirelations and contact relations translate to predicate transformers. Similar to the correspondence between contact relations and closure operations in [29] we have expressed the correspondence of multirelations and predicate transformers within the language of relation algebras. This approach allowed us to generalise many results from up-closed to arbitrary multirelations. Looking beyond up-closed multirelations is a necessary step to cover approaches to program semantics such as concurrent dynamic logic [23]. We thus hope to establish new links between program semantics and other areas (for example, topology and games) which use multirelations as a conceptual and methodological basis.

Further exploration of the new composition operator for multirelations and a comparison with the other composition operators mentioned in Sect. 4.3 should be carried out in the future. Another topic for further work are probabilistic multirelations, which are relations of the type $A \leftrightarrow D(A)$ where $D(A)$ is the set of all probabilistic sub-distributions over A; see [32] for details.

Acknowledgments. We thank Hitoshi Furusawa and Georg Struth for pointing out related work and the anonymous referees for making helpful comments.

References

1. Aumann, G.: Kontakt-Relationen. Sitzungsberichte der Bayerischen Akademie der Wissenschaften, Mathematisch-Naturwissenschaftliche Klasse, pp. 67–77 (1970)
2. Aumann, G.: AD ARTEM ULTIMAM - Eine Einführung in die Gedankenwelt der Mathematik. R. Oldenbourg Verlag (1974)
3. Back, R.J., von Wright, J.: Refinement Calculus. Springer, New York (1998)
4. Backhouse, R.C., van der Woude, J.: Demonic operators and monotype factors. Math. Struct. Comput. Sci. **3**(4), 417–433 (1993)
5. Berghammer, R., Guttmann, W.: Closure, properties and closure properties of multirelations (submitted 2015)

6. Berghammer, R., Schmidt, G., Zierer, H.: Symmetric quotients and domain constructions. Inf. Process. Lett. **33**(3), 163–168 (1989)
7. Berghammer, R., Zierer, H.: Relational algebraic semantics of deterministic and nondeterministic programs. Theor. Comput. Sci. **43**, 123–147 (1986)
8. Bird, R., de Moor, O.: Algebra of Programming. Prentice Hall (1997)
9. Dijkstra, E.W.: Guarded commands, nondeterminacy and formal derivation of programs. Commun. ACM **18**(8), 453–457 (1975)
10. Freyd, P.J., Ščedrov, A.: Categories, Allegories. North-Holland Mathematical Library, vol. 39. Elsevier Science Publishers (1990)
11. Furusawa, H., Struth, G.: Concurrent dynamic algebra (2014). CoRR http://arxiv.org/abs/1407.5819
12. Gries, D.: The Science of Programming. Springer, New York (1981)
13. Guttmann, W.: Multirelations with infinite computations. J. Log. Algebr. Meth. Program. **83**(2), 194–211 (2014)
14. Hesselink, W.H.: Multirelations are predicate transformers (2004). Available from http://www.cs.rug.nl/~wim/pub/whh318.pdf
15. Hoare, C.A.R., He, J.: The weakest prespecification. Inf. Process. Lett. **24**(2), 127–132 (1987)
16. Kozen, D.: Kleene algebra with tests. ACM Trans. Progr. Lang. Syst. **19**(3), 427–443 (1997)
17. Macedo, H.D., Oliveira, J.N.: Typing linear algebra: a biproduct-oriented approach. Sci. Comput. Program. **78**(11), 2160–2191 (2013)
18. Martin, C.E., Curtis, S.A.: The algebra of multirelations. Math. Struct. Comput. Sci. **23**(3), 635–674 (2013)
19. Martin, C.E., Curtis, S.A., Rewitzky, I.: Modelling angelic and demonic nondeterminism with multirelations. Sci. Comput. Program. **65**(2), 140–158 (2007)
20. Morris, J.M., Tyrrell, M.: Modelling higher-order dual nondeterminacy. Acta Inf. **45**(6), 441–465 (2008)
21. Naumann, D.A.: A categorical model for higher order imperative programming. Math. Struct. Comput. Sci. **8**(4), 351–399 (1998)
22. Parikh, R.: Propositional logics of programs: new directions. In: Karpinski, M. (ed.) FCT 1983. LNCS, vol. 158, pp. 347–359. Springer, Heidelberg (1983)
23. Peleg, D.: Concurrent dynamic logic. J. ACM **34**(2), 450–479 (1987)
24. Preoteasa, V.: Algebra of monotonic Boolean transformers. In: Simao, A., Morgan, C. (eds.) SBMF 2011. LNCS, vol. 7021, pp. 140–155. Springer, Heidelberg (2011)
25. Rewitzky, I.: Binary multirelations. In: de Swart, H., Orłowska, E., Schmidt, G., Roubens, M. (eds.) Theory and Applications of Relational Structures as Knowledge Instruments. LNCS, vol. 2929, pp. 256–271. Springer, Heidelberg (2003)
26. Rewitzky, I., Brink, C.: Monotone predicate transformers as up-closed multirelations. In: Schmidt, R.A. (ed.) RelMiCS/AKA 2006. LNCS, vol. 4136, pp. 311–327. Springer, Heidelberg (2006)
27. Schmidt, G.: Partiality I: embedding relation algebras. J. Log. Algebr. Program. **66**(2), 212–238 (2006)
28. Schmidt, G.: Relational Mathematics. Cambridge University Press, Cambridge (2011)
29. Schmidt, G., Berghammer, R.: Contact, closure, topology, and the linking of row and column types of relations. J. Log. Algebr. Program. **80**(6), 339–361 (2011)
30. Schmidt, G., Ströhlein, T.: Relationen und Graphen. Springer, Heidelberg (1989)
31. Tarski, A.: On the calculus of relations. J. Symb. Log. **6**(3), 73–89 (1941)
32. Tsumagari, N.: Probabilistic relational models of complete IL-semirings. Bull. Inf. Cybern. **44**, 87–109 (2012)

Preference Decomposition and the Expressiveness of Preference Query Languages

Patrick Roocks(✉)

Institut für Informatik, Universität Augsburg, 86135 Augsburg, Germany
roocks@informatik.uni-augsburg.de

Abstract. Preferences in the scope of relational databases allow modeling user wishes by queries with soft constraints. There are different frameworks for database preferences including commercially available systems. They slightly vary in semantics and expressiveness but have in common that preferences induce strict partial orders on a given data set. In the present paper we study the expressiveness of preference operators in the available implementations. Particularly, we search for decompositions of strict partial orders into fundamental preference constructs. We study which preference operators and operands are necessary to express any strict partial order. Finally, we present two decomposition algorithms and show their correctness.

Keywords: Relational algebra · Preferences · Expressiveness · Query languages

1 Motivation

In the field of relational databases, the concept of *preferences* was introduced to realize queries with *soft constraints*. They allow selecting highly relevant tuples that are optimal compromises for the user. For example, assume that one is interested in cars having low fuel consumption and high power. We formulate the *Pareto preference* to optimize both dimensions simultaneously, i.e., a car is considered to be better than another one if it is not worse in both dimensions (fuel consumption and power) and strictly better in at least one of these dimensions, cf. Fig. 1. In the literature this kind of data selection is also called the *Skyline operator* [1]. Typically such techniques are used to find optimal tuples w.r.t. potentially conflicting goals.

Implementations of such preferences are available in the *rPref* package [12], formerly R-Pref [13], and the commercial database *Exasolution 5* with the *Skyline* feature [3,7]. In these tools, user wishes such as *"high power and low fuel consumption"* can be modeled and the optimal tuples can be retrieved.

In the present work we investigate the expressiveness of these preference frameworks. We try to construct any strict partial order by a minimal set of fundamental building blocks and call this process *preference decomposition*. These

R. Hinze and J. Voigtländer (Eds.): MPC 2015, LNCS 9129, pp. 71–92, 2015.
DOI:10.1007/978-3-319-19797-5_4

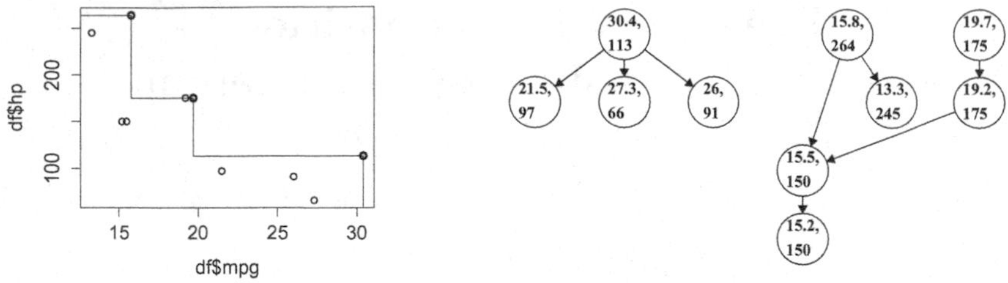

Fig. 1. Pareto optima of cars (`mtcars[21:30,]` in R) with low fuel consumption (i.e., high miles per gallon *(mpg)* value) and high horsepower *(hp)*. The corresponding Better-Than-Graph (Usual visualization in rPref; corresponds to the Hasse diagram of the converse preference order, i.e. arrows point from better to worse tuples), containing the values for mpg and horsepower as vertex labels, is depicted alongside.

building blocks are called *preference constructors* and their formal definition in this paper is strongly connected to the implementations mentioned above.

In the preference framework [6] an *"Explicit(E)"-constructor* is suggested, which constructs a preference from a given set of edges E. The transitive hull of E is the *Better-Than-Graph.* This solves our problem in a very pragmatic way. Such generic constructors lack syntactic neatness and are error-prone, e.g., the user may accidentally specify a non-strict-order within E. In the preference implementations considered in this work, such a constructor is not contained.

Instead, we use more fundamental preference operators like *Pareto composition* (exemplified above) and *Prioritisation* (lexicographic order) together with *Boolean preferences.* The latter are the simplest preference constructs, both contained in the *rPref* and *Exasolution* implementations. This motivated us to search for decompositions of any strict partial order using just these simple constructs. The answer to this problem precisely characterizes the expressiveness of the preference query languages in the mentioned implementations.

For a better reproducibility of the examples in the paper the R source code (based on *rPref*) for all (counter)-examples, their visualizations and the algorithms in this paper is provided on the web [11]. A small code snippet implementing a model finder for a certain kind of decomposition is given in the Appendix of the paper.

The remainder of the paper is structured as follows: In Sect. 2 we recapitulate the formal foundations. Next, we give an intuition of the problem and prove some lower and upper bounds for the expressiveness of preferences under some restrictions. In Sect. 4, as our main contribution, the decomposition algorithms are specified and their correctness is proved. We end with a conclusion and outlook.

2 Preference Algebra Background

In the following we briefly recapitulate some results and definitions of the algebraic calculus of database preferences from [8–10]. This calculus allows point-free derivation of optimization laws for preference queries and correctness proofs of preference evaluation algorithms. The present paper is founded on this calculus and keeps the same notation. As it is more appropriate for this work, we assume a concrete relational instance and not an abstract relation algebra.

2.1 Relation Algebra and Fundamental Definitions

Assume a domain D. We define a concrete relation algebra of binary homogeneous relations on $D^2 = D \times D$ in the same notation as in [10] using the following conventions. First we define the special relations

$$
\begin{aligned}
0 &=_{df} \emptyset && \text{(empty relation)},\\
1\!\!1 &=_{df} \{(t,t) \mid t \in D\} && \text{(identity relation)},\\
\top &=_{df} D^2 && \text{(universal relation)}.
\end{aligned}
$$

The fundamental operations for relations u, v are the relational union $u + v$, the composition $u \cdot v$, the intersection $u \sqcap v$ and the inclusion order $u \leq v$. We use the following conventions regarding literals in formulas:

- a, b, c, d are binary homogeneous relations.
- p, q, r, s are subidentities, i.e. $p \leq 1\!\!1$, each representing a set $M_p \subseteq D$ which is related to p via $p = \{(t,t) \mid t \in M_p\}$. They are also called *tests*.
- x, y, z are subidentities which are atomic w.r.t. $+$, also called *atomic tests*. They represent singletons, i.e., $x = \{(t,t)\}$ for some $t \in D$.

According to the database scenario, which is our field of application, a test is a *data set* or a *set of tuples*, i.e. a (partial) table of a database, and an atomic test is a *tuple* (a row of a data set). Hence we use the term *(atomic) test* synonymously to *data set/tuple*. For tests, composition and intersection coincide, i.e., we have $p \cdot q = p \sqcap q$.

To express that tuples x, y are in relation via a we use the notation

$$x \, a \, y \quad \Leftrightarrow_{df} \quad (t,t') \in a \qquad \text{where } x = (t,t) \wedge y = (t',t').$$

Based on the above conventions we define the following operations:

- The converse relation $a^{-1} =_{df} \{(t',t) \in \top \mid (t,t') \in a\}$,
- the general complement $\overline{a} =_{df} \{(t,t') \in \top \mid (t,t') \notin a\}$,
- the complement of tests $\neg p =_{df} \{(t,t) \in 1\!\!1 \mid (t,t) \notin p\}$,
- the difference of tests $p - q =_{df} p \cdot \neg q$,
- the (inverse) image of a relation a w.r.t. a test p:

$$
\begin{aligned}
\langle a | p &=_{df} \{(t,t) \in 1\!\!1 \mid \exists t' \in p : (t',t) \in a\} && \text{(image)},\\
| a \rangle p &=_{df} \{(t,t) \in 1\!\!1 \mid \exists t' \in p : (t,t') \in a\} && \text{(inverse image)}.
\end{aligned}
$$

The finite iteration a^+ is defined by $a^+ =_{df} \sum_{i=1}^{\infty} a^i$ and the Kleene star by $a^* =_{df} 1\!\!1 + a^+$. Powers are given by $a^0 =_{df} 1\!\!1$ and $a^i =_{df} a \cdot a^{i-1}$.

An algebraic axiomatisation of these operations based on abstract relation algebra is given in [8].

We will also need the *Hasse diagram* of a relation, i.e. the transitive reduction. For a transitive relation a this is given by $a \sqcap \overline{a^2}$, cf. [14], Proposition 3.2.5.

In the following we will use the convention that $x \in p$ for a data set p picks a tuple x from p, formally $x = \{(t,t)\}$ with $(t,t) \in p$.

2.2 The Preference Framework

Definition 2.1 (Preferences with SV). A relation a is a *preference* if and only if it is irreflexive and transitive, i.e., is a *strict partial order*. In relation algebra, this means $a \sqcap \mathbb{1} = 0$ and $a^2 \leq a$.

Every preference a will be associated with an SV-relation s_a. This has to be an equivalence relation on the domain of a, where the equivalence classes contain "equally good" objects (SV is short for "substitutable values"). It must fulfill the compatibility conditions

1. $\mathsf{s}_a \cdot a \leq a$,
2. $a \cdot \mathsf{s}_a \leq a$.

A preference a is the special case of a *layered preference*, also known as *strict weak order*, if and only if additionally *negative transitivity* $(\overline{a})^2 \leq \overline{a}$ holds. □

The intuition behind SV relations is that in a predicate $(x\,a\,y)$ the tuple x can be substituted by x', if $(x\,\mathsf{s}_a\,x')$ holds (analogous for y), which follows from the compatibility conditions. Compatibility for an SV relation implies $\mathsf{s}_a \sqcap a = 0$. For a layered preference a the relation $\mathsf{s}_a = \overline{a + a^{-1}}$ (equivalently we have $a + \mathsf{s}_a = \overline{a^{-1}}$) is always an SV-relation, and this also the most intuitive way to define which tuples are equivalent w.r.t. a preference. In general (i.e. for non-layered preferences) this construction does not fulfill the compatibility conditions. Proofs and counterexample can be found in [9].

If a tuple x is a-related to y, formally $(x\,a\,y)$ we say that *y is better than x w.r.t. to the preference a*. The compatibility conditions in Definition 2.1 ensure for a tuple y better than x that all tuples from the SV-equivalence class of y are better than those equivalent to x. By convention, the empty preference 0 has the SV-relation $\mathsf{s}_0 =_{df} \top$, i.e., all tuples are equivalent.

Corollary 2.2. *Assume a countable domain D. A relation a is a layered preference if and only if there exists a measure function $f : \mathbb{1} \to \mathbb{N}$ such that*

$$x\,a\,y \Leftrightarrow f(x) < f(y).$$

This corollary is proven in [4], Theorem 2.2. A consequence of this is that $(x\,\mathsf{s}_a\,y) \Leftrightarrow f(x) = f(y)$ holds. This corollary gives an intuitive access to *layered* preferences. The preimages $f^{-1}(k)$ with ascending k (for $k = 1, 2, ...$) form a-related layers, i.e., we have $x\,a\,y \Leftrightarrow \exists k, l \in \mathbb{N}$ with $k < l \wedge x \in f^{-1}(k) \wedge y \in f^{-1}(l)$.

Definition 2.3 (Preference selection). In the scope of database preferences the maximum operator $\rhd$ selecting the a-maximal elements on r is defined by

$$a \rhd r =_{df} r - |a\rangle r,$$

which is also known as *preference selection* on the data set r w.r.t. preference a.

We now recapitulate the definitions of the two most important complex preference operators Pareto composition and prioritisation. Usually, they are used to construct preferences over joins of data sets, i.e. type domains $D_1 \times \ldots \times D_k$ corresponding to different attributes of a database table. In the scope of this paper we restrict our attention to the special case of data sets having a single domain D. Regarding our application in databases, we require that the domain values of D are unique for every tuple, i.e., D is a *primary key* in the database context. Then it is clearly sufficient to define a order on D to characterise an arbitrary order on the data base table.

In the case of a single domain the join operator from the join algebra in [9] coincides with the $\sqcap$ operator. This leads to simplified definitions of the preference operators presented subsequently. Note that $\sqcap$ binds stronger than $+$.

Definition 2.4 (Prioritisation and Pareto on Single Domain). Let a, b be preferences with associated SV-relations $\mathsf{s}_a, \mathsf{s}_b$. The *prioritisation with SV on a single domain* is given by

$$a \,\&\, b =_{df} a + \mathsf{s}_a \sqcap b,$$

whereas the *Pareto composition on a single domain with SV* is defined as

$$a \otimes b =_{df} (a + \mathsf{s}_a) \sqcap b + a \sqcap (b + \mathsf{s}_b)$$

We define the SV-relation of the result by $\mathsf{s}_{a \star b} =_{df} \mathsf{s}_a \sqcap \mathsf{s}_b$ for $\star \in \{\&, \otimes\}$.

The intuition behind the prioritisation is the lexicographic order: $a \,\&\, b$ means "better in a or {equal in a and better in b}". The Pareto composition is used to compose equally important wishes, $a \otimes b$ means "equal or better in {a or b} and strictly better in one of them". Both operators are associative and $\otimes$ is even commutative. They both preserve preferences, i.e., for preferences a, b the object $a \star b$ for $\star \in \{\&, \otimes\}$ is a preference again. Furthermore 0 is a neutral element for both operators (note that $\mathsf{s}_0 = \top$), i.e., we have

$$0 \otimes a = a \otimes 0 = a, \qquad a \,\&\, 0 = 0 \,\&\, a = a.$$

All these are results from [9].

To compare preference relations we introduce a concept of equivalence of preferences w.r.t. a given data set.

Definition 2.5 (r-equivalence). Let a, b be preferences and r a data set. We say that a and b are r-equivalent, if and only if $r \cdot a \cdot r = r \cdot b \cdot r$.

This is equivalent to $(x\,a\,y) \Leftrightarrow (x\,b\,y)$ for all tuples $x, y \in r$. Note that r-equivalence of a and b does *not* imply that $a \star c = b \star c$ for $\star \in \{\&, \otimes\}$ holds in general. We will show a counterexample in Remark 4.2.

Corollary 2.6. *Let r be a finite data set (i.e. with finitely many tuples) and a, b layered preferences. The prioritisation $a \& b$ is r-equivalent to a layered preference.*

Proof. Consider layered preferences a and b with measure functions f_a and f_b (cf. Corollary 2.2). By definition, $c = r \cdot (a \,\&\, b) \cdot r$ and $(a \,\&\, b)$ are r-equivalent. We have to show that c is a layered preference. Consider the measure function $f_c(x) = f_a(x) \cdot (1 + \max\{f_b(y) \mid y \in r\}) + f_b(x)$, where the maximum of f_b over r obviously exists because r is finite. We calculate for $x, y \in r$

$$\begin{aligned} x\, c\, y \;&\Leftrightarrow\; x\, a\, y \;\vee\; (x\, \mathsf{s}_a\, y \;\wedge\; x\, b\, y) \\ &\Leftrightarrow f_a(x) < f_a(y) \;\vee\; (f_a(x) = f_a(y) \;\wedge\; f_b(x) < f_b(y)) \;\Leftrightarrow f_c(x) < f_c(y). \end{aligned}$$

Thus c is a layered preference with measure function f_c. □

Chains of $\otimes$-composition like $a_1 \otimes ... \otimes a_n$ can be interpreted as "better for one i and {better or equal for all i}", formally

$$x\,(a_1 \otimes ... \otimes a_n)\,y \iff (\exists i : x\, a_i\, y) \wedge (\forall i : x\,(a_i + \mathsf{s}_{a_i})\,y). \tag{1}$$

This interpretation plays an important role for the intuition and the arguments in the following sections. We will also need the distributive law for $\otimes$ and & from [10] which can be applied to chains of $\otimes$-compositions, resulting in

$$a \,\&\, (b_1 \otimes ... \otimes b_n) = (a \,\&\, b_1) \otimes ... \otimes (a \,\&\, b_n).$$

3 The Decomposition Problem

Next to the preference operators of Pareto composition and Prioritisation we introduce another fundamental building block for our decomposition approach: Boolean preferences, used as operands connected by &/$\otimes$. Then we will study how a given preference can be constructed using these operators and operands. We call this the *decomposition problem.*

3.1 Boolean Preferences

The following definition is the formal counterpart to Boolean preferences, supported in the preference query languages considered in this paper (*Exasolution Skyline* [3] and *rPref* [12]).

Definition 3.1 (Boolean preference). Let $\rho : D \rightarrow \{\mathsf{true}, \mathsf{false}\}$ be a predicate which can be evaluated over all elements in D. Then $\mathsf{is\_true}(\rho)$ is a *Boolean preference* defined by

$$\mathsf{is\_true}(\rho) =_{df} \rho^{-1}(\mathsf{false}) \times \rho^{-1}(\mathsf{true})$$

We define $\mathsf{s}_{\mathsf{is\_true}(\rho)} =_{df} (\rho^{-1}(\mathsf{false}) \times \rho^{-1}(\mathsf{false})) + (\rho^{-1}(\mathsf{true}) \times \rho^{-1}(\mathsf{true}))$. □

Obviously, a Boolean preference with predicate ρ is a layered preference with the measure function $f(x) = 1$ if $\rho(x) = \mathsf{false}$ and $f(x) = 2$ otherwise.

For our decomposition approach we specialize this concept to preferences on sets and tuples.

Definition 3.2 (Set/Tuple Preference). For a set of tuples $p \leq \mathbb{1}$ we define the *set preference*

$$\mathsf{t}(p) =_{df} \mathsf{is\_true}(\rho_p) \text{ where } \rho_p(t) \Leftrightarrow_{df} \{(t,t)\} \leq p.$$

If $|p| = 1$, i.e., p is a tuple, we also say that $\mathsf{is\_true}(p)$ is a *tuple preference.* □

Hence we have for all $x, y \in r$ that $(x \, \mathsf{t}(p) \, y) \Leftrightarrow (x \notin p \wedge y \in p)$ holds.

Along with our convention that $p, q, ...$ are sets of tuples and $x, y, ...$ are tuples, we use $\mathsf{t}(p)$ for general set preferences, whereas $\mathsf{t}(x)$ refers to a tuple preference. From the definition it is clear that set/tuple preferences are special cases of Boolean preferences and hence they can be constructed in the considered preference query languages.

In the following we will investigate *preference terms* where Boolean preferences are connected by the &/⊗ operators. As a special case we consider terms with *unique tuple preferences*, where each tuple from the data set may occur at most once in the preference term. These preference terms have the pleasant property that the number of operands is restricted by the number of tuples and hence comparisons w.r.t. them are very efficient.

Consider the following example, where the data set $r = x_1 + ... + x_n$ with distinct tuples x_i is given: The tuple preferences in $a = \mathsf{t}(x_1) \,\&\, (\mathsf{t}(x_2) \otimes \mathsf{t}(x_3))$ are unique, because each x_i for $i \in \{1, 2, 3\}$ occurs just once. Instead, the preferences in $a = \mathsf{t}(x_1) \& (\mathsf{t}(x_2) \otimes \mathsf{t}(x_1))$ are not unique because x_1 occurs twice. Subsequently we define uniqueness formally.

Definition 3.3 (Uniqueness). For a data set $r = x_1 + ... + x_n$, operators $\mathsf{op} \subseteq \{\&, \otimes\}$ and a preference a we say that the $\mathsf{t}(x_i)$ are *unique* in a if and only if a is in the set $\mathsf{un_{op}}(r)$ recursively defined by:

$$\mathsf{un_{op}}(s) =_{df} \begin{cases} \{0\} & \text{if } s = 0, \\ \{b \star \mathsf{t}(x) \mid x \in s, \, b \in \mathsf{un_{op}}(s - x), \, \star \in \mathsf{op}\} \cup & \\ \{\mathsf{t}(x) \star b \mid x \in s, \, b \in \mathsf{un_{op}}(s - x), \, \star \in \mathsf{op}\} \cup & \\ \mathsf{un_{op}}(s - x) & \text{otherwise.} \end{cases}$$

We will also call this set *unique set preferences* (over op). □

This set contains the closure of all operators in op and all operands in the set $\{\mathsf{t}(x) \mid x \in r\}$, which may occur just once in each preference.

As in the last step of the recursion $\mathsf{un_{op}}(0)$ is called, we say that all tuple preferences of the empty preference 0 are unique (even if there are none). In the earlier recursion steps $\mathsf{un_{op}}(s)$ with $s \neq 0$ we get $0 \star \mathsf{t}(x) = \mathsf{t}(x) \star 0 = \mathsf{t}(x)$ because of neutrality of 0.

3.2 The Expressiveness Problem

With &/⊗ and Boolean preferences we have defined the fundamental tools which are available in common implementations of preference query languages. We are interested whether any preference (i.e. strict partial order) can be expressed by using these operators and operands. Additionally we consider restrictions of the operands to (unique) tuple preferences. We are also interested which is the minimal (or most restricted) set of preference operators and operands being sufficient to express arbitrary preferences. Table 1 shows the problem. Our aim is to characterize the sets corresponding to each *?*-cell in the table.

Table 1. The expressiveness problem is to characterize which preference orders can be composed from the given operators and operands.

	Unique tuple pref.	Tuple preferences	Set preferences
&	*?*	*?*	*?*
$\otimes$	*?*	*?*	*?*
&/$\otimes$	*?*	*?*	*?*

Note that it is trivial to express any preference by set preferences and $+/\sqcap$ as preference operators, instead of &/$\otimes$: For a data set r and $x, y \in r$ consider that $a_{x,y} = \mathtt{t}(\neg x) \sqcap \mathtt{t}(y)$ constructs a preference where $(z\, a_{x,y}\, z')$ holds if and only if $z = x \wedge z' = y$. Hence a preference a has an r-equivalent decomposition via the union $\sum \{a_{x,y} \mid \exists x, y \in r : (x\, a\, y)\}$. With $+/\sqcap/(\cdot)^{-1}$ as operators even tuple preferences suffice as operands, because $a_{x,y} = \mathtt{t}(x)^{-1} \sqcap \mathtt{t}(y)$ holds.

But in the scope of database preferences the operators &/$\otimes$ have been established as the most common operators for good reasons. They have the pleasant property that they preserve strict partial orders (& preserves even strict weak orders, cf. Corollary 2.6) whereas the union, i.e. $a_1 + a_2$, violates transitivity in general. Its transitive closure $(a_1 + a_2)^+$ does not preserve irreflexivity in general. Hence from the application level it seems reasonable to restrict a preference language to operators like &/$\otimes$, and especially to exclude the union operator.

In the following we will present a first idea how different preferences can be constructed from these building blocks. The intuitive mechanism behind set preferences is to pick some tuples from the data set and declare them to be better than the remainder. Using &-chains we can construct chains in descending order and by $\otimes$-composition we can combine such chains. We illustrate this in the following example.

Example 3.4. Consider the data set $r = x_1 + x_2 + x_3 + x_4$ and the following terms of unique tuple preferences. They are visualized in Fig. 2(1–3), where we use a circled number i in the figure to express the element x_i.

1. $a = \mathtt{t}(x_1)$
2. $b = \mathtt{t}(x_1)\ \&\ \mathtt{t}(x_2)$
3. $c = \mathtt{t}(x_1)\ \&\ (\mathtt{t}(x_2) \otimes \mathtt{t}(x_3))$

Comparing the preference terms of a, b, c and their Hasse diagrams in Fig. 2, one recognizes that parallel compositions correspond to $\otimes$ and serial compositions to &. This unveils the following interesting link to language theory.

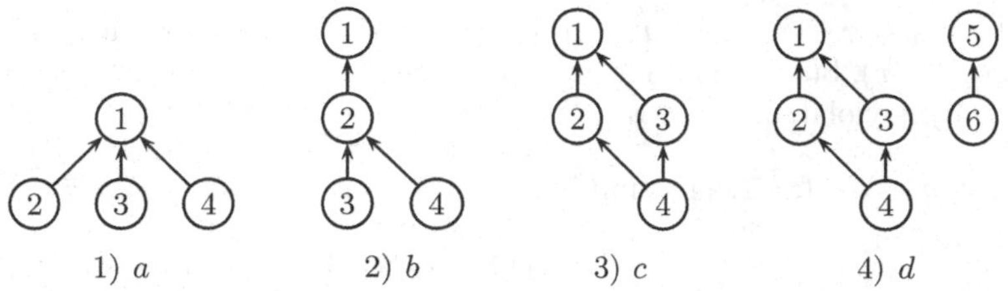

Fig. 2. Hasse diagrams of preference relations in Example 3.4 and Remark 3.5.

Remark 3.5. The paths in the Hasse diagram of a data set r w.r.t. a preference $d \in \mathsf{un}_{\mathsf{op}}(r)$ coincide with the language of a finite automaton without loops where every edge label occurs just once. At the language level, this implies that concatenation and choice are sufficient. For example, the preference

$$d = (\mathsf{t}(x_1) \,\&\, (\mathsf{t}(x_2) \otimes \mathsf{t}(x_3)) \,\&\, \mathsf{t}(x_4)) \otimes (\mathsf{t}(x_5) \,\&\, \mathsf{t}(x_6)),$$

on $r = x_1 \; + \; ... \; + \; x_6$ (see Fig. 2(4)) corresponds to the behavior (cf. [2])

$$\{x_4\}\,\{x_2, x_3\}\,\{x_1\} \,\cup\, \{x_6\}\{x_5\}.$$

There & corresponds to concatenation and $\otimes$ to choice.

Example 3.4 shows that some preferences can be expressed with &/$\otimes$ and unique tuple preferences. This raises the question if this restricted set of operands already suffices to express arbitrary strict partial orders. The following lemma shows that this is not the case.

Lemma 3.6. *Let $r = x_1 + ... + x_n$ be a data set, where x_i are its distinct tuples. Then the following proper inclusions in the sense of r-equivalence hold:*

$$\{ \text{ layered preferences } \} \;\subsetneq\; \mathsf{un}_{\{\&,\otimes\}}(r) \;\subsetneq\; \{ \text{ preferences } \}$$

Proof. First we show the "$\subseteq$" conditions from left to right. Let a be a layered preference and f_a be the measure function of a according to Corollary 2.6. Straightforward verification shows that a is r-equivalent to a' where

$$a' = \mathop{\&}_{i=0}^{N-1} (\mathsf{t}(y_{i,1}) \otimes ... \otimes \mathsf{t}(y_{i,n_i})) \quad \text{with} \quad \{y_{(N-i),1}, ..., y_{(N-i),n_i}\} := f_a^{-1}(i),$$

for $i = 1, ..., N$ with $N := \max\{f_a(x) \mid x \in r\}$.

The preimages $f_a^{-1}(i)$ are disjoint for distinct i, hence the $\mathsf{t}(y_{i,j})$ are unique in a'. This implies $a' \in \mathsf{un}_{\{\&,\otimes\}}(r)$ which shows the first inclusion. The next "$\subseteq$" inclusion is trivially true, as preferences are closed under &/$\otimes$.

Regarding the proper subset conditions consider the following preferences (visualized in Fig. 3(1–2)) that falsify equality.

1. Let $r = x_1 + x_2 + x_3$ and $a = (\mathsf{t}(x_1) \,\&\, \mathsf{t}(x_2)) \otimes \mathsf{t}(x_3)$. Obviously we have $a \in \mathsf{un}_{\{\&,\otimes\}}(r)$, but a is not a layered preference because negative transitivity $((\overline{a})^2 = \overline{a})$ is violated:

$$(x_2 \,\overline{a}\, x_3) \wedge (x_3 \,\overline{a}\, x_1) \quad \text{(implying } x_2 \,(\overline{a})^2\, x_1) \qquad \text{but} \quad \neg(x_2 \,\overline{a}\, x_1).$$

2. Let $r = x_1 + ... + x_4$ and $b = ((\mathsf{t}(x_1) \otimes \mathsf{t}(x_3)) \,\&\, \mathsf{t}(x_2)) \otimes (\mathsf{t}(x_3) \,\&\, \mathsf{t}(x_4))$. We show that there is no r-equivalent preference in $\mathsf{un}_{\{\&,\otimes\}}(r)$ by a model finder checking all possible preference terms, see appendix A. □

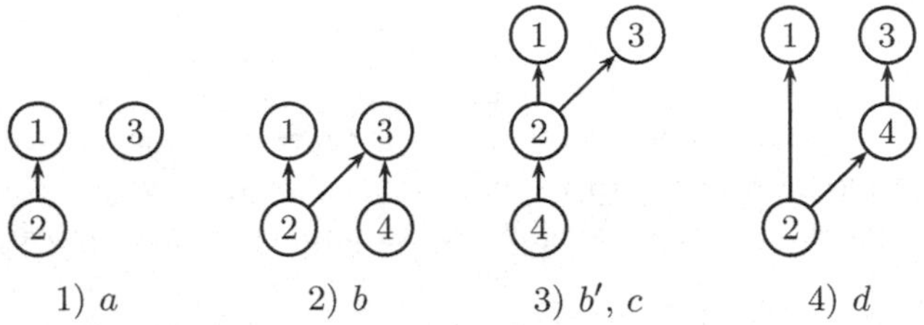

Fig. 3. Hasse diagrams of preferences for Lemma 3.6.

Remark 3.7. In Remark 3.5 we already noticed a connection between the Hasse diagrams of preferences in $\mathsf{un}_{\{\&,\otimes\}}(r)$ and concatenation/choice in formal languages. Their induced orders coincide with series-parallel pomsets as described in [5]. Finite pomsets are series-parallel if and only if they are *N-free* (Theorem 3.1 in [5]), where N-free means that the diagram contains no "N-shaped" subrelation like the preference b ($\notin \mathsf{un}_{\{\&,\otimes\}}(r)$, cf. Fig. 3(2)) from the above proof.

To get an intuitive understanding why this preference b cannot be decomposed within $\mathsf{un}_{\{\&,\otimes\}}(r)$ consider the layered preference $b' = (\mathsf{t}(x_1) \otimes \mathsf{t}(x_3)) \,\&\, \mathsf{t}(x_2))$. Inserting "$\&\, \mathsf{t}(x_4)$" into b' in such a way that $(x_4 \, a \, x_3)$ holds is possible at two positions (the resulting preferences are visualized in Fig. 3(3–4)):

1. Inserting "$\&\, \mathsf{t}(x_4)$" at the end of b', i.e. $c = b' \,\&\, \mathsf{t}(x_4)$. Then b' and c are r-equivalent and $(x_4 \, c \, x_2)$ holds which we do not want.
2. We replace $\mathsf{t}(x_3)$ by $(\mathsf{t}(x_3) \,\&\, \mathsf{t}(x_4))$, i.e., $d = (\mathsf{t}(x_1) \otimes (\mathsf{t}(x_3) \,\&\, \mathsf{t}(x_4))) \,\&\, \mathsf{t}(x_2))$ is the result. This causes the undesired effect of $(x_2 \, d \, x_4)$.

In both cases, the desired incomparability of x_2 and x_4 will be destroyed.

Now we omit the restriction to tuple preferences. For the special case of layered preferences we can simplify the construction from Lemma 3.6.

Lemma 3.8. *Let r be a data set. In the sense of r-equivalence the set of layered preferences over r is equivalent to &-chains of set preferences over r, formally*

$$\{\text{layered preferences}\} = \{\mathsf{t}(p_1) \,\&\, \mathsf{t}(p_2) \;\&\; ... \;\&\; \mathsf{t}(p_k) \mid p_k \leq r\}.$$

Proof. Let a be a layered preference on r. A straightforward verification shows that the following preference b is r-equivalent to a:

$$b = \mathop{\&}_{i=0}^{N-1} \mathsf{t}(y_{i,1} + ... + y_{i,n_i}) \quad \text{where} \quad \{y_{(N-i),1}, ..., y_{(N-i),n_i}\} := f_a^{-1}(i),$$

for $i = 1, ..., N$ with $N := \max\{f_a(x) \mid x \in r\}$.

The reverse inclusion is clear, as set preferences are layered preferences and this property is preserved by &, cf. Corollary 2.6. □

An application of this lemma is the *induced layered preference* from [8], which we can now rewrite via

$$m(a, r) =_{df} \mathop{\&}_{i=1}^{N} \mathsf{t}((r \cdot a)^i \triangleright r) \quad \text{where} \quad N := \max\{i \in \mathbb{N} \mid |(r \cdot a)^i\rangle r \neq 0\}.$$

The intuitive meaning of this transformation is that the maxima are taken away iteratively from the remainder of the data set r and then the maxima of each iteration are assigned to a layer. In some sense this is a "layered approximation" of a, see [8] for more details.

3.3 Decompositions of General Preferences

The results up to now show that unique tuple preferences do not suffice to construct arbitrary preferences. Subsequently, we will search for decompositions under less restrictive conditions. Primarily, we will investigate the following cases:

(i) (Non-unique) tuple preferences and &/⊗,
(ii) set preferences and only ⊗.

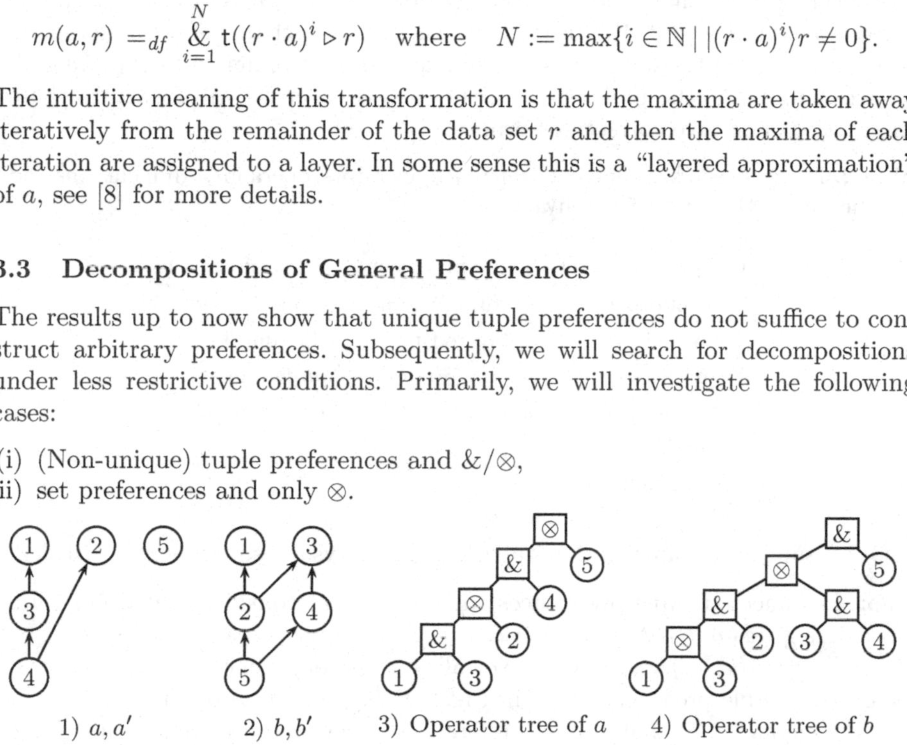

Fig. 4. Hasse diagrams and operator trees of preferences for Example 3.9.

Example 3.9. The preferences depicted in Fig. 4 on the data set $r = \sum_{i=1}^{5} x_i$ can be decomposed in the following ways:

1. Using unique tuple preferences we find the following decomposition of the preference depicted in Fig. 4(1):

$$a = (((\mathsf{t}(x_1) \mathbin{\&} \mathsf{t}(x_3)) \otimes \mathsf{t}(x_2)) \mathbin{\&} \mathsf{t}(x_4)) \otimes \mathsf{t}(x_5)$$

Alternatively we can decompose the preference into a ⊗-composition of set preferences, where each set is closed under $\langle a + \mathbb{1}|(\cdot)$, i.e., upward closed:

$$a' = \mathsf{t}(x_1) \otimes \mathsf{t}(x_2) \otimes \mathsf{t}(x_1 + x_3) \otimes \mathsf{t}(x_1 + x_2 + x_3 + x_4) \otimes \mathsf{t}(x_5)$$

2. For the preference depicted in Fig. 4(2) we get analogously:

$$b = (((\mathtt{t}(x_1) \otimes \mathtt{t}(x_3)) \mathbin{\&} \mathtt{t}(x_2)) \otimes (\mathtt{t}(x_3) \mathbin{\&} \mathtt{t}(x_4))) \mathbin{\&} \mathtt{t}(x_5)$$
$$b' = \mathtt{t}(x_1) \otimes \mathtt{t}(x_3) \otimes \mathtt{t}(x_1 + x_2 + x_3) \otimes \mathtt{t}(x_3 + x_4) \otimes \mathtt{t}(r)$$

The example above leads us to the conjecture that $\otimes$-compositions of set preferences *or* tuple preferences with $\&/\otimes$ are sufficient to express arbitrary preferences up to r-equivalence. The construction principle seems to be strongly connected to the Hasse diagram of the preference. Roughly spoken, we use $\otimes$ to connect parallel chains and & to step down in the Hasse diagram, cf. the operator trees in Fig. 4(3–4). For the construction of $\otimes$-chains of set preferences, every element from the data set is transformed to a set preference containing all its successors w.r.t. the preference. In the next section we will formalize these constructions and give correctness proofs.

Table 2. Expressiveness of different preference operators/operands. All inclusions and equalities are in the sense of r-equivalence.

	Unique tuple pref.	Tuple preferences	Set preferences
&	$\supsetneq$ total orders	$\supsetneq$ total orders	layered preferences
	$\subsetneq$ layered pref.	$\subsetneq$ layered pref.	(Lemma 3.8)
$\otimes$	set preferences	set preferences	*preferences?*
$\&/\otimes$	$\supsetneq$ layered pref.	*preferences?*	*preferences?*
	$\subsetneq$ preferences		
	(Lemma 3.6)		

Our results are summarized in Table 2 and its rows are subsequently justified:

1. For &-connected tuple preferences (first two cells) consider that every total order, e.g. $(x_1 \, a \, x_2) \wedge (x_2 \, a \, x_3) \wedge ... \wedge (x_{n-1} \, a \, x_n)$, can be expressed by $a = \mathtt{t}(x_n) \mathbin{\&} \mathtt{t}(x_{n-1}) \mathbin{\&} ... \mathbin{\&} \mathtt{t}(x_1)$. Not all of these preferences are total orders, e.g., the simple preference $\mathtt{t}(x_1)$ in Fig. 2(1) is not a total order.
   By Lemma 3.8, &-chains of set preferences are layered preferences, justifying the last cell of the first line. As tuple preferences are a special case of set preferences, the proper subset conditions "$\subsetneq$ layered pref." in the first two cells are also justified by Lemma 3.8. To see that &-connected tuple preferences are a proper subset of layered preferences, consider $\mathtt{t}(x_1 + x_2) \mathbin{\&} \mathtt{t}(x_3 + x_4)$. This preference cannot be expressed with &-chains of tuple preferences.
2. All $\otimes$-compositions of (unique) tuple preferences can be expressed by set preferences, as $\mathtt{t}(x_1) \otimes ... \otimes \mathtt{t}(x_n)$ is r-equivalent to $\mathtt{t}(x_1 + ... + x_n)$ for tuples $x_i \in r$. This justifies the first two cells of the second line, while the last cell corresponds to our conjecture above.
3. The first cell of the last line is the result of Lemma 3.6 while the next two cells are again consequences of our conjecture, which will be proved later.

The results summarized in Table 2 show that $\otimes$-compositions of set preferences *or* $\&/\otimes$-compositions of tuple preferences are the simplest hypothetical possibility to decompose general preferences within this operators/operands matrix; the other cells contain proper subsets of preferences. In the following section we will prove the conjecture related to the *"preferences?"* cells in the table.

4 Decomposition Algorithms

In this section we present the main contribution of our work, the decomposition algorithms and proofs of their correctness. This will finally show the conjecture from the last section.

4.1 Pareto Compositions of Set Preferences

The following theorem formalizes the decomposition of arbitrary preferences into $\otimes$-compositions of set preferences.

Theorem 4.1. *Let a be a preference and $r \leq \mathbb{1}$ a finite data set. Then a can be decomposed into a $\otimes$-composition of set preferences where each set is upward closed w.r.t. $a + \mathbb{1}$. Formally,*

$$\text{DECOMP\_PARETO}(a, r) =_{df} \bigotimes_{x \in r} \mathsf{t}(r \cdot \langle a + \mathbb{1} | x),$$

is r-equivalent to a.

Proof. Let $b = \text{DECOMP\_PARETO}(a, r)$ and $r = \sum_{i=1}^{k} x_i$ with distinct tuples x_i. We define the family of sets

$$p_i = r \cdot \langle a + \mathbb{1} | x_i \quad \text{for } i = 1, ..., k,$$

corresponding to the arguments of the $\mathsf{t}(\cdot)$ preferences (i.e. preferred sets) contained in b. For this family it clearly holds that $\sum_{i=1}^{k} p_i = r$. The claim is $u\,a\,v \Leftrightarrow u\,b\,v$ for all $u, v \in r$. We show both implications separately:

"$\Rightarrow$" Assume $u, v \in r$ with $(u\,a\,v)$. We show first that $u \leq p_i \Rightarrow v \leq p_i$ holds for all $i \in \{1, ..., k\}$. From $u, v \in r$ with $(u\,a\,v)$ we get $v \leq r \cdot \langle a | u$. With this, the assumption $u \leq p_i$ and the transitivity of a we calculate:

$$v \leq r \cdot \langle a | u \leq r \cdot \langle a | (r \cdot \langle a + \mathbb{1} | x_i) \leq r \cdot \langle a^2 + a | x_i = r \cdot \langle a | x_i \leq r \cdot \langle a + \mathbb{1} | x_i = p_i.$$

The result above and the definition of $\mathsf{t}(\cdot)$ shows that u is not better than v w.r.t. to all the $\mathsf{t}(\cdot)$ preferences in b, formally $\neg(v\,\mathsf{t}(p_i)\,u)$ for all i. As set preferences are layered preferences, this implies $(u\,(\mathsf{t}(p_i) + \mathsf{s}_{\mathsf{t}(p_i)})\,v)$. To finally prove that $(u\,b\,v)$ holds, according to the definition of $\otimes$ we have to show that there is at least one $\mathsf{t}(\cdot)$ for which v is better than u. For this we consider

$$u\,\mathsf{t}(r \cdot \langle a + \mathbb{1} | v)\,v,$$

where $\mathsf{t}(r \cdot \langle a + \mathbb{1} | v)$ is by definition one of the $\otimes$-operands in b. This predicate is true because:

– It is clear that $v \leq r \cdot \langle a + \mathbb{1}|v$, i.e., v is in the preferred set.
– From $(u\,a\,v)$ it follows that $u \leq |a\rangle v$ and we have $((\langle a + \mathbb{1}|v) \cdot (|a\rangle v) = 0$ because a is a strict partial order. Hence $u \cdot r \cdot \langle a + \mathbb{1}|v = 0$, i.e., u is not in the preferred set.

Therefore, according to the definition of $\mathsf{t}(\cdot)$ we get that v is better than u w.r.t. $\mathsf{t}(r \cdot \langle a + \mathbb{1}|v)$ and not worse w.r.t. the other set preferences in b. By definition of $\otimes$ we conclude $(u\,b\,v)$.

"$\Leftarrow$" We show the contraposition $\neg(u\,a\,v) \Rightarrow \neg(u\,b\,v)$ for all atomic $u, v \in r$. We distinguish the following cases:

1. Assume $(v\,a\,u)$, i.e., u is better than v. Completely analogously to the first part of the proof we get $(v\,b\,u)$ and hence $\neg(u\,b\,v)$ as b is a strict partial order.
2. Otherwise we have $\neg(u\,(a + a^{-1})\,v)$, i.e., u and v are incomparable w.r.t. a. We can assume that $u \neq v$ because for $u = v$ the claim is trivially true as a is irreflexive. From the assumption we get that $u \cdot r \cdot \langle a + \mathbb{1}|v = 0$ and $v \cdot r \cdot \langle a + \mathbb{1}|u = 0$ holds. Hence we obtain:

$$u\,\mathsf{t}(r \cdot \langle a + \mathbb{1}|v)\,v \;\;\wedge\;\; v\,\mathsf{t}(r \cdot \langle a + \mathbb{1}|u)\,u$$

With the definition of b, where these $\mathsf{t}(r \cdot \langle a + \mathbb{1}|x)$ preferences with $x = u$ and $x = v$ are $\otimes$-operands, this implies that u, v are incomparable w.r.t. b. Thus we finally get $\neg(u\,b\,v)$, which shows the claim. □

Remark 4.2. The preference generated according to Theorem 4.1 contains tuple preferences for each tuple from the maximal elements. Formally we have

$$\textsc{Decomp\_Pareto}(a, r) = \mathsf{t}(y_1) \;\otimes\; ... \;\otimes\; \mathsf{t}(y_n) \;\otimes\; ... \quad \text{with } a \triangleright r = \sum_{i=1}^{n} y_i,$$

where $y_i \in r$ are tuples. Even if $\mathsf{t}(y_1) \otimes ... \otimes \mathsf{t}(y_n)$ is r-equivalent to $\mathsf{t}(y_1 + ... + y_n)$ a substitution of $(\mathsf{t}(y_i) \otimes \mathsf{t}(y_j))$ by $\mathsf{t}(y_i + y_j)$ in the generated preference is not allowed in general. Consider the following counterexample, visualized in Fig. 5. Let $r = x_1 + ... + x_4$. The following preferences a and b are not r-equivalent:

$$a = \mathsf{t}(x_1) \otimes \mathsf{t}(x_2) \otimes \mathsf{t}(x_1 + x_3) \otimes \mathsf{t}(x_1 + x_2 + x_3 + x_4),$$
$$b = \mathsf{t}(x_1 + x_2) \otimes \mathsf{t}(x_1 + x_3) \otimes \mathsf{t}(x_1 + x_2 + x_3 + x_4).$$

This means that we need the "induced incomparability" generated by the $\otimes$-connected tuple preferences. In the proof of Theorem 4.1 this is implicitly used in the last step, i.e. in the argument exploiting that the predicates $u\,\mathsf{t}(r \cdot |a + \mathbb{1}\rangle v)\,v$ and $v\,\mathsf{t}(r \cdot |a + \mathbb{1}\rangle u)\,u$ are simultaneously fulfilled. After a substitution of $\mathsf{t}(x) \otimes \mathsf{t}(y)$ by $\mathsf{t}(x + y)$ these predicates may not hold anymore. Hence, such a substitution invalidates the argument in the proof.

4.2 Pareto Compositions and Prioritisations of Tuple Preferences

Now we present an algorithm which decomposes any preference into tuple preferences with &/$\otimes$ as composition operators. First we sketch the basic idea:

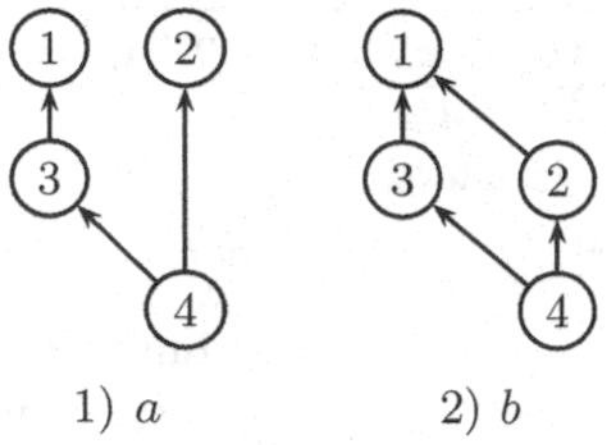

Fig. 5. Hasse diagrams of preferences for Remarks 4.2 and 4.5.

- Initially, the Hasse diagram and the maxima of the given data r w.r.t. a preference a are calculated.
- Starting with the maxima as the *working set*, every node (i.e. tuple) $x \in r$ of the Hasse diagram will be annotated with a preference, expressing a decomposition of a which is correct for the nodes above x in the Hasse diagram (formally an $(r \cdot \langle a + \mathbb{1}|x)$-equivalent decomposition of a). To calculate this preference for each node $x \in r$ we
  1. take the annotations of all successors of x and $\otimes$-compose them,
  2. add ... & $\mathsf{t}(x)$ to this preference.

  This process is iterated downwards the Hasse diagram until the minima in r w.r.t. a are reached. This means that in each step the working set is replaced by the set of its maximal predecessors.
- Before the replacement, the annotations of all those nodes having predecessors are deleted (i.e. set to 0) after they were used to calculate the preferences of the predecessors. This step removes redundancy and leads to shorter preference terms.
- Finally all non-zero annotations are $\otimes$-composed and returned. We claim that the returned preference is r-equivalent to a.

These annotations are stored in the preference-valued array $b[\cdot]$. The assignment $b[p] \leftarrow c$ for non-atomic and non-empty p is used as a shorthand notation for simultaneous assignments $b[x] \leftarrow c$ for all $x \in p$. We also assume that in all assignments the neutrality of 0 is used, implying that $b[p] \leftarrow 0 \star c$ is executed as $b[p] \leftarrow c$ for $\star \in \{\&, \otimes\}$.

In Fig. 6 an example run of the algorithm is visualized where for every step the operator trees of the preferences in $b[\cdot]$ are shown.

Regarding the proof of correctness we will first get rid of line 9 of the algorithm, where the preferences of m-successors are deleted. This line removes redundant preferences, e.g., compare the final preference b_{res} from Fig. 6 with b'_{res}, generated by the algorithm without line 9:

$$\begin{aligned} b_{\text{res}} &= \quad (((\mathsf{t}(x_1) \otimes \mathsf{t}(x_3)) \,\&\, \mathsf{t}(x_2)) \otimes (\mathsf{t}(x_3) \,\&\, \mathsf{t}(x_4))) \,\&\, \mathsf{t}(x_5) \\ b'_{\text{res}} &= b_{\text{res}} \otimes ((\mathsf{t}(x_1) \otimes \mathsf{t}(x_3)) \,\&\, \mathsf{t}(x_2)) \otimes (\mathsf{t}(x_3) \,\&\, \mathsf{t}(x_4)) \otimes \mathsf{t}(x_1) \otimes \mathsf{t}(x_3) \end{aligned} \quad (2)$$

Hence line 9 generates much simpler preference terms. Formally they are equivalent, as we state in the following lemma.

Algorithm 1. Preference decomposition into tuple preferences and &/⊗

Input: Preference to decompose a, data set r
Output: r-equivalent decomposition b_{res}
1: **function** DECOMP_TUPLE(a, r)
2: $\quad a_h \leftarrow a \sqcap \overline{a^2}$ // Hasse diagram
3: $\quad b[r] \leftarrow 0$ // Initialization of array b of preferences
4: $\quad m \leftarrow a \rhd r$ // Start traversing with maxima
5: $\quad$ **while** $m \neq 0$ **do**
6: $\quad\quad$ **for all** $y \in m$ **do** // Pref. for y, collect and ⊗-compose successors,
7: $\quad\quad\quad b[y] \leftarrow \left(\bigotimes_{x \in r \cdot \langle a_h | y} b[x]\right) \mathbin{\&} \mathsf{t}(y)$ // and finally add pref. on y
8: $\quad\quad$ **end for**
9: $\quad\quad b[r \cdot \langle a_h | m] \leftarrow 0$ // Delete preferences of m-successors
10: $\quad\quad m \leftarrow a \rhd (r \cdot | a_h \rangle m)$ // Find a-maximal predecessors of m
11: $\quad$ **end while**
12: $\quad b_{\text{res}} \leftarrow \bigotimes_{x \in r} b[x]$; **return** b_{res} // ⊗-compose final preference and return
13: **end function**

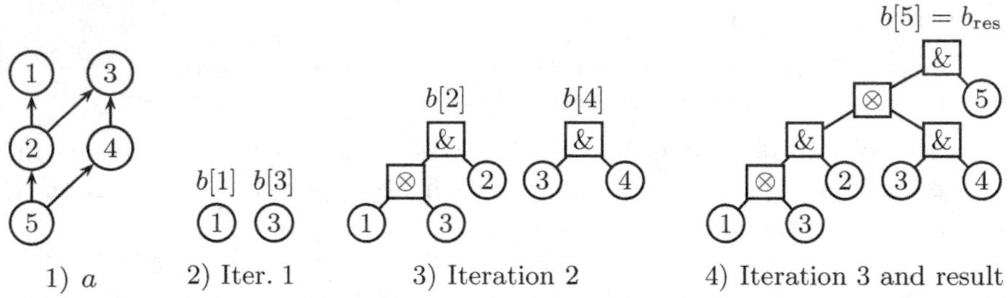

Fig. 6. Example run of Algorithm 1 with $r = x_1 + ... + x_5$ and shown a (Hasse diagram). The $b[x]$ values in every iteration of the while loop are depicted as operator trees where a circled i is short for $\mathsf{t}(x_i)$. We have $b[x] = 0$ for all values of x which are not depicted.

Lemma 4.3. *Let a, b preferences. Then $a \mathbin{\&} b = a \otimes (a \mathbin{\&} b)$ holds (in the sense of strict equivalence, not just r-equivalence).*

Proof. With neutrality of 0 for &/⊗ and the left-distributivity of & over ⊗, cf. [10], we obtain $a \mathbin{\&} b = a \mathbin{\&} (0 \otimes b) = (a \mathbin{\&} 0) \otimes (a \mathbin{\&} b) = a \otimes (a \mathbin{\&} b)$. □

If we apply this iteratively from right to left to b'_{res} in Eq. (2) we immediately see that $b'_{\text{res}} = b_{\text{res}}$ holds. This implies that line 9 from Algorithm 1 can be removed without changing the result formally. We will use this in the proof of the following theorem, stating the correctness of Algorithm 1.

Theorem 4.4. *Let a be a preference and $r \leq \mathbb{1}$ a finite data set. The preference $b =$ DECOMP_TUPLE(a, r) (from Algorithm 1) is r-equivalent to a.*

Proof. We reconsider the algorithm with a helper variable p summing up the traversed nodes of the Hasse diagram, and without the deletions in $b[\cdot]$, as explained above. We formulate a loop invariant (I) stating

(1) the working set m contains the a-maximal elements of the remainder $(r-p)$ and the traversed nodes p are contained in the data set r,
(2) $b[y]$ is $(r \cdot \langle a + \mathbb{1} | y)$-equivalent to a (and also their SV-relations are identical),
(3) y and all tuples above y in the Hasse diagram are preferred w.r.t. $b[y]$ over the remainder $(\neg q)$ and $\neg q$ is one equivalence class w.r.t. $\mathsf{s}_{b[y]}$.

Formally we specify (I) ($\Leftrightarrow$ (I1) $\wedge$ (I2) $\wedge$ (I3)) by

$$\text{(I)} \Leftrightarrow_{df} \quad a \rhd (r-p) = m \ \wedge \ p \leq r \ \wedge \qquad \text{(I1)}$$
$$\forall y \in p : \ (\, b[y] = q \cdot a \cdot q \ \wedge \ \mathsf{s}_{b[y]} = q \cdot \mathsf{s}_a \cdot q \ \wedge \qquad \text{(I2)}$$
$$\mathsf{t}(q) \leq b[y] \ \wedge \ \neg q \cdot \mathsf{s}_{\mathsf{t}(q)} \cdot \neg q \leq \mathsf{s}_{b[y]} \quad \text{where} \ \ q = r \cdot \langle a + \mathbb{1} | y \,) \qquad \text{(I3)}$$

Our proof strategy is to show that (I) is indeed an invariant and then conclude that b_{res} is r-equivalent to a. The modified algorithm reads as follows:

1: $a_h \leftarrow a \sqcap \overline{a^2}$; $b[r] \leftarrow 0$; $m \leftarrow a \rhd r$; $p \leftarrow 0$ // Initialization
2: {{ *(I) is true* }} // The invariant initially
3: **while** $m \neq 0$ **do**
4: $\quad p \leftarrow p + m$ // Helper variable for the invariant
5: $\quad$ **for all** $y \in m$ **do**
6: $\quad\quad b[y] \leftarrow \left(\bigotimes_{x \in r \cdot \langle a_h | y} b[x] \right) \ \& \ \mathsf{t}(y)$
7: $\quad$ **end for**
8: $\quad m \leftarrow a \rhd (r \cdot | a_h \rangle m)$
9: $\quad$ {{ *(I) is true* }} // The invariant after any loop run
10: **end while**
11: $b_{\text{res}} \leftarrow \bigotimes_{x \in r} b[x]$

First, we see that (I) trivially holds initially in line 2. Let (I)' ($\Leftrightarrow$ (I1)' $\wedge$ (I2)' $\wedge$ (I3)') be the invariant for the previous loop run, where $b'[\cdot]$ are the corresponding $b[\cdot]$-values. We have to show that (I)' $\Rightarrow$ (I) holds in line 9.

To see that (I1) holds, consider the p-update $p \leftarrow p + m$ in line 4. Using the strict partial order property of a we get that line 8 of the algorithm is equivalent to $m \leftarrow a \rhd (r-p)$, i.e., the working set is replaced by the maxima of the remainder which shows $a \rhd (r-p) = m$. The p-update and $m \leq r$ imply $p \leq r$.

By definition of $\rhd$ we get $m = a \rhd m$ from (I1) and hence $x + y \leq m$ with $x \in r \cdot \langle a_h | y$ can never occur. This implies that all $b[x]$ with $x \in r \cdot \langle a_h | y$ are calculated before $b[y]$. Thus we have that $b[x] = b'[x]$ holds for the update of $b[\cdot]$ in line 6 and we infer for $b[y]$:

$$\forall y \in m : \ b[y] = \left(\bigotimes_{x \in r \cdot \langle a_h | y} b'[x] \right) \ \& \ \mathsf{t}(y). \qquad \text{(U)}$$

By the definition of $\otimes$ (cf. Eq. (1)) and & we obtain point-wise predicates for $b[y]$ and $\mathsf{s}_{b[y]}$. For all $y \in m$ and $u, v \in r$ we have:

$$u \, b[y] \, v \ \Leftrightarrow \ \left(\forall x \in s : u \, (\mathsf{s}_{b'[x]} + b'[x]) \, v \right) \ \wedge \ \left((\exists x \in s : u \, b'[x] \, v) \ \vee \ u \, \mathsf{t}(y) \, v \right), \qquad \text{(U1)}$$
$$u \, \mathsf{s}_{b[y]} \, v \ \Leftrightarrow \ \left(\forall x \in s : u \, \mathsf{s}_{b'[x]} \, v \right) \ \wedge \ u \, \mathsf{s}_{\mathsf{t}(y)} \, v, \qquad \text{where} \quad s = r \cdot \langle a_h | y. \qquad \text{(U2)}$$

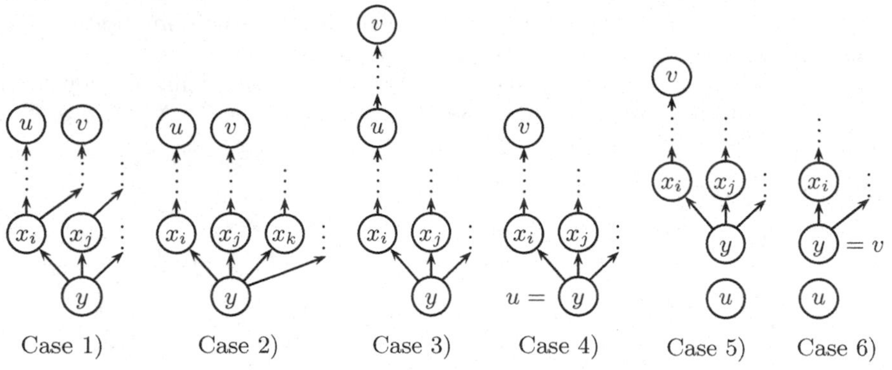

Fig. 7. Partial Hasse diagrams for different positions of u, v with fixed y.

For $y \in (p - m)$ we have $b[y] = b'[y]$ and hence (I)' $\Rightarrow$ (I). Thus we will restrict our attention to all $y \in m$. As (I1) is already clear we just have to check (I2) and (I3). Therefore we show that (I)' $\Rightarrow$ (I2) $\wedge$ (I3) holds by establishing the following point-wise predicates (J2) ($\Leftrightarrow$ (I2)) and (J3) ($\Leftrightarrow$ (I3)),

$$\forall u, v \in r,\ y \in p:\quad u\, b[y]\, v \Leftrightarrow u\, (q \cdot a \cdot q)\, v \ \wedge\ u\, \mathsf{s}_{b[y]}\, v \Leftrightarrow u\, (q \cdot \mathsf{s}_a \cdot q)\, v \ \wedge \qquad \text{(J2)}$$

$$u\, \mathsf{t}(q)\, v \Rightarrow u\, b[y]\, v \ \wedge\ u\, (\neg q \cdot \mathsf{s}_{\mathsf{t}(q)} \cdot \neg q)\, v \Rightarrow u\, \mathsf{s}_{b[y]}\, v. \qquad \text{(J3)}$$

where $q = r \cdot \langle a + \mathbb{1} | y$ as defined in (I).

For $u, v \in r$ we distinguish different cases, visualized in Fig. 7. Each $x_i, x_j, x_k \in r \cdot \langle a_h | y$ in the figure represents a whole class of nodes having the same (depicted) properties. These x are direct successors of y and they are the indices for the $\otimes$-composition in Eq. (U). The values of $b[y]$ and $\mathsf{s}_{b[y]}$ are given by the updates (U1) and (U2).

(1) $u, v \in \langle a | x_i \ \wedge\ \neg(u\, (a + a^{-1})\, v)$ with $x_i \in r \cdot \langle a | y$: This means both u and v are above x_i (which is above y) in the Hasse diagram and $\neg(u\, (q \cdot a \cdot q)\, v) \ \wedge$ $\neg(u\, \mathsf{s}_{q \cdot a \cdot q}\, v)$ holds. We see that $u, v \in r \cdot \langle a + \mathbb{1} | x_i \leq q$ and by (I2)' we get that $\neg(u\, b'[x_i]\, v) \ \wedge\ \neg(u\, \mathsf{s}_{b'[x_i]}\, v)$, and hence (J2) follows.
As $u + v \leq q$, (J3) is clear, which also applies to the cases (2–4).
(2) Otherwise, $u, v \in \langle a | y \ \wedge\ \neg(u\, (a + a^{-1})\, v)$: This means both u and v are above y in the Hasse diagram and $\neg(u\, (q \cdot a \cdot q)\, v) \ \wedge\ \neg(u\, \mathsf{s}_{q \cdot a \cdot q}\, v)$. By (I3)' we get that $\mathsf{t}(r \cdot \langle a + \mathbb{1} | u) \leq b'[x_i]$ and $\mathsf{t}(r \cdot \langle a + \mathbb{1} | v) \leq b'[x_j]$ holds. This yields (J2).
(3) Otherwise, $u, v \in \langle a | y \ \wedge\ (u\, (a + a^{-1})\, v)$: W.l.o.g. we assume $(u\, a\, v)$. This means that v is above u and both are above y. Hence we have $(u\, b'[x_i]\, v)$ and $\neg(u\, \mathsf{s}_{b'[x_i]}\, v)$ from (I2)' and thus we get (J2).
(4) Otherwise, $u, v \in q \ \wedge\ (u\, (a + a^{-1})\, v)$: W.l.o.g. we assume $u = y$ and thus $(u\, a\, v)$. With (I3)' we get $(u\, b'[x_i]\, v)$ and $\neg(u\, \mathsf{s}_{b'[x_i]}\, v)$. We conclude (J2).
(5) Otherwise, $(u + v) \cdot \langle a | y \neq 0 \ \wedge\ (u + v) \cdot \neg q \neq 0$. W.l.o.g. we assume $v \in \langle a | y$ and $u \notin \langle a | y$, hence $v \in q$ and $u \in \neg q$. We have (J2) analogous to case 4. As $u\, b[y]\, v$ is true and $u \notin \neg q$, (J3) also holds.

(6) Otherwise, $(u+v)\cdot y \neq 0 \ \wedge \ (u+v)\cdot \neg q \neq 0$. W.l.o.g. we assume $v = y$ and $u \notin q$. With (I2)' we get $u \, \mathsf{s}_{b'[x_i]} \, v$. With $u \, \mathsf{t}(y) \, v$ (because $y = v$) we retrieve $u \, b[y] \, v$. Thus we have (J2) and by $v \in q \wedge u \notin q$, (J3) also holds.
(7) Otherwise, i.e. $u + v \leq \neg q$ (similar to Fig. 7(6), but $y \neq v$). Then $(u \, \mathsf{s}_{(q \cdot a \cdot q)} \, v)$ holds. By $(u \, \mathsf{s}_{\mathsf{t}(y)} \, v)$ (because $u \neq y \neq v$) we get $(u \, \mathsf{s}_{b[y]} \, v)$, thus (J2) and (J3).

Note that the validity of (I) in the cases (1–5) just follows from the $\otimes$-composition in (U). In the cases (6–7) the &-connected tuple preference ... & $\mathsf{t}(y)$ is also exploited.

When leaving the while loop we have $m = 0$ and hence $0 = a \triangleright (r-p) \wedge p \leq r$. By definition of $\triangleright$ we get $r - p = 0 \ \wedge \ p \leq r$ and thus $p = r$. Hence (I2) can be applied to all $b[x]$ for $x \in r$. Using the definition of $\otimes$ (similar to (U1), but without the & operator) in the assignment of b_{res} (line 11) we get that

$$\forall u, v \in r: \ \ u \, b_{\text{res}} \, v \ \Leftrightarrow \ u \left(\bigotimes\nolimits_{x \in r} b[x] \right) v \ \Leftrightarrow \ u \, (r \cdot a \cdot r) \, v$$

holds, which shows the claim. □

Remark 4.5. Note that the assignment $m \leftarrow a \triangleright (r \cdot |a_h\rangle m)$ ensures that it is impossible to get $x + y \leq m \ \wedge \ (x \, a \, y)$ for any $x, y \in r$ in any run of the while loop. If we omit the maxima calculation in line 10 i.e. alter the assignment to $m = r \cdot |a_h\rangle m$, the algorithm is not correct anymore. In the proof of Theorem 4.4 the requirement that the $b[x]$ with $x \in r \cdot \langle a_h | y$ are calculated before $b[y]$ would by violated. As an example, consider the preference a in Fig. 5. In the altered algorithm we get $m = x_3 + x_4$ in the second loop run, where $x_3 \in r \cdot \langle a_h | x_4$.

Within this section we now have proved all the results from Table 2, i.e., we classified the expressiveness of &, $\otimes$ and tuple/set preferences.

5 Conclusion and Outlook

In this paper we have shown in a constructive way that all strict partial orders on a given data set can be algorithmically constructed in the preference framework. We called this process *preference decomposition.*

5.1 Applicational Relevance

We characterized the expressiveness of database preference frameworks containing constructors for Pareto composition, Prioritisation and Boolean preferences. This functionality is covered by current preference implementations like the *Skyline* feature in the commercial database *Exasolution* [3,7] and the freely available R package *rPref* [12]. Both implementations support low/high/Boolean preferences and &/$\otimes$ in a very similar way.

For example, the preference $a = (\mathsf{t}(x_1 + x_2) \otimes \mathsf{t}(x_3)) \,\&\, \mathsf{t}(x_4)$ can be translated in the SQL dialect of Exasolution as follows, where `id` is a primary key of the data set `R` corresponding to the x_i indices:

```
SELECT * FROM R PREFERRING ((id in (1,2)) PLUS id=3) PRIOR TO id=4
```

Regarding the *rPref* implementation we provide a script [11] containing the implementation of the algorithms in this paper. Hence we showed the applicability of our approach to current implementations of preference query languages.

We have developed the rPref package and we were also involved in the specification of the Exasol Skyline feature due to an academic-industrial cooperation. With this research we want to bridge the gap between theory and application in the scope of database preferences. With rPref we have a tool that supported many steps of this work, especially for evaluating and visualizing preferences, and also for the model finder in the appendix. The expressiveness of preference query languages is our main result and shows why offering such preference operators in database systems is a useful thing to do.

5.2 Minimal Decompositions

For future research, the search for "minimal" decompositions is an interesting question, which we will briefly explain in the following.

Algorithm 1 does not produce minimal preferences in the sense of term length or complexity. For example for the empty preference $a = 0$ on the data set $r = x_1 + ... + x_n$ we get $\text{DECOMP\_TUPLE}(0, r) = \mathsf{t}(x_1) \otimes ... \otimes \mathsf{t}(x_n)$. For $a = \mathsf{t}(x_1 + x_2) \,\&\, \mathsf{t}(x_3 + x_4)$ on $r = x_1 + ... + x_4$ we find an r-equivalent preference $\mathsf{t}(x_1) \otimes \mathsf{t}(x_2)$ where $\text{DECOMP\_TUPLE}(a, r)$ produces the more complicated term $((\mathsf{t}(x_1) \otimes \mathsf{t}(x_2)) \,\&\, \mathsf{t}(x_3)) \otimes ((\mathsf{t}(x_1) \otimes \mathsf{t}(x_2)) \,\&\, \mathsf{t}(x_4))$.

The Pareto decomposition from Theorem 4.1 is not minimal in a very similar way. For example the empty preference is also blown up into a $\otimes$-chain, i.e., we have $\text{DECOMP\_PARETO}(0, x_1 + ... + x_n) = \mathsf{t}(x_1) \otimes ... \otimes \mathsf{t}(x_n)$.

From computational aspects it is cheap to deal with layered preferences (strict weak orders). Chains of prioritisations (lexicographic orders) are such strict weak orders. The costs to evaluate a Pareto composition like $a_1 \otimes ... \otimes a_n$ where all a_i are layered preferences quickly increases with the length n. Hence it is desirable to minimize this length. We know that such decompositions are generally possible, as Theorem 4.1 shows. But they are obviously not optimal as the blow-up of the empty preference illustrates.

Acknowledgement. I am grateful to Bernhard Möller for proofreading many drafts of the paper, plenty of helpful remarks and very fruitful discussions about this topic. I am also grateful to Carla Harth and Alfons Huhn for proofreading and valuable comments, and to the anonymous referees for their helpful remarks.

A Unique Tuple Decompositions

Given the data set $r = x_1 + ... + x_4$ and the preference

$$a = ((\mathsf{t}(x_1) \otimes \mathsf{t}(x_3)) \,\&\, \mathsf{t}(x_2)) \otimes (\mathsf{t}(x_3) \,\&\, \mathsf{t}(x_4)),$$

there is no decomposition into an r-equivalent preference within $\mathsf{un}_{\{\&,\otimes\}}(r)$. We will show this in the following R-Script, which is a snippet from [11].

The model finder function `search` extends the temporary preference term `a_tmp` by ... $\otimes$ $\mathsf{t}(x_i)$ or ... & $\mathsf{t}(x_i)$ in the recursive step. The term extension at the end is sufficient to get all possible terms, as & and $\otimes$ are associative operators. The variable `xs` stores those $x \in r$ which can still be used for $\mathsf{t}(x)$ without violating the uniqueness of the x_i. Hence `xs` is comparable to the parameter s in $\mathsf{un}_{\mathsf{op}}(s)$, Definition 3.3.

The comparison w.r.t. r-equivalency of two preferences (cf. Definition 2.5) is done by comparing the adjacency lists of their Hasse diagrams. Note that we can rely on a predefined sorting (lexicographic) of the adjacency list of a Hasse diagram in the result of `get_hasse_diag`. Hence the equivalency of these adjacency lists imply the r-equivalency of the corresponding preferences.

```
# Include the rPref package
library(rPref)

# Define set preference
t <- function(...) true(id %in% c(...))

# Implement an r-equality check of a and b
is_r_equal <- function(a, b, r)
  identical(get_hasse_diag(r, a), get_hasse_diag(r, b))

# Data set and the given preference to decompose (a_ref)
r <- data.frame(id = 1:4)
a_ref <- ((t(1) * t(3)) & t(2)) * (t(3) & t(4))

# Recursive search for unique tuple decompositions
search <- function(a_tmp, xs) {

  # Check if temporary preference is equivalent to the reference
  if (is_r_equal(a_tmp, a_ref, r)) return(TRUE)

  # Recursively search for other possible terms
  if (length(xs) > 0) {
    for (x in xs) {
      if (search(a_tmp & t(x), setdiff(xs, x))) return(TRUE)
      if (search(a_tmp * t(x), setdiff(xs, x))) return(TRUE)
    }
  }
  return(FALSE)
}

# Do the search (returns FALSE)
search(empty(), 1:4)
```

Note that this is some kind of a brute-force search. It could be optimized by e.g. exploiting the commutativity of $\otimes$ and a more efficient r-equivalence check. We omitted such optimizations to keep the code as simple as possible. The script generates and checks 633 possible terms. On our off-the-shelf computer the execution time of this program is about 8 s. Finally, it returns `FALSE`, i.e., no decomposition is found.

References

1. Borzsony, S., Kossmann, D., Stocker, K.: The skyline operator. In: 17th International Conference on Data Engineering, pp. 421–430 (2001)

2. Eilenberg, S.: Automata, Languages, and Machines, vol. 59. Academic Press, New York (1974)
3. Exasol: Skyline. In: EXASolution User Manual Version 5.0.0, pp. 239–241. http://tinyurl.com/mob2mfm
4. Fishburn, P.C.: Utility theory for decision making. Technical report, New York, NY, USA (1970)
5. Gischer, J.L.: The equational theory of pomsets. Theoret. Comput. Sci. **61**(2), 199–224 (1988)
6. Kießling, W.: Foundations of preferences in database systems. In: VLDB 2002: Proceedings of the 28th International Conference on Very Large Data Bases, pp. 311–322. VLDB, Hong Kong (2002)
7. Mandl, S., Kozachuk, O., Endres, M., Kießling, W.: Preference analytics in EXASolution. In: 16th Conference on Database Systems for Business, Technology, and Web (2015). http://tinyurl.com/pxco8d4
8. Möller, B., Roocks, P.: An algebra of layered complex preferences. In: Kahl, W., Griffin, T.G. (eds.) RAMICS 2012. LNCS, vol. 7560, pp. 294–309. Springer, Heidelberg (2012)
9. Möller, B., Roocks, P.: An algebra of database preferences. J. Logic. Algebraic Methods Program. **84**(3), 456–481 (2015)
10. Möller, B., Roocks, P., Endres, M.: An algebraic calculus of database preferences. In: Gibbons, J., Nogueira, P. (eds.) MPC 2012. LNCS, vol. 7342, pp. 241–262. Springer, Heidelberg (2012)
11. Roocks, P.: R script containing examples and algorithms from the paper (2015). http://www.p-roocks.de/pref-decomp-mpc.r
12. Roocks, P.: The rPref package: preferences and skyline computation in R (2015). http://www.p-roocks.de/rpref
13. Roocks, P., Kießling, W.: R-Pref: rapid prototyping of database preference queries in R. In: DATA 2013, pp. 104–111 (2013)
14. Schmidt, G., Ströhlein, T.: Relations and Graphs: Discrete Mathematics for Computer Scientists. EATCS Monographs on Theoretical Computer Science. Springer, Heidelberg (1993)

Hierarchy in Generic Programming Libraries

José Pedro Magalhães[1](✉) and Andres Löh[2]

[1] Department of Computer Science, University of Oxford, Oxford, UK
jpm@cs.ox.ac.uk
[2] Well-Typed LLP, Regensburg, Germany
andres@well-typed.com

Abstract. Generic programming (GP) is a form of abstraction in programming languages that serves to reduce code duplication by exploiting the regular structure of algebraic datatypes. Several different approaches to GP in Haskell have surfaced, giving rise to the problem of code duplication across GP libraries. Given the original goals of GP, this is a rather unfortunate turn of events. Fortunately, we can convert between the different representations of each approach, which allows us to "borrow" generic functions from different approaches, avoiding the need to reimplement every generic function in every single GP library.

In previous work we have shown how existing GP libraries relate to each other. In this paper we go one step further and advocate "hierarchical GP": through proper design of different GP approaches, each library can fit neatly in a hierarchy, greatly minimizing the amount of supporting infrastructure necessary for each approach, and allowing each library to be specific and concise, while eliminating code duplication overall. We introduce a new library for GP in Haskell intended to sit at the top of the "GP hierarchy". This library contains a lot of structural information, and is not intended to be used directly. Instead, it is a good starting point for generating generic representations for other libraries. This approach is also suitable for being the only library with native compiler support; all other approaches can be obtained from this one by simple conversion of representations in plain Haskell code.

1 Introduction

Generic programs are concise, abstract, and reusable. They allow one single definition to be used for many kinds of data, existing and to come. For example, parsing and pretty-printing, (de-)serialisation, test data generation, and traversals can all be implemented generically, freeing the programmer to implement only datatype-specific functionality.

Given its power, it's no surprise that GP approaches abound: including preprocessors, template-based approaches, language extensions, and libraries, there are well over 15 different approaches to GP in Haskell (Magalhães 2012, Chap. 8). This abundance is partly caused by the lack of a clearly superior approach; each approach has its strengths and weaknesses, uses different implementation mechanisms, a different generic view (Holdermans et al. 2006) (i.e. a different representation of datatypes), or focuses on solving a particular task. Their number

R. Hinze and J. Voigtländer (Eds.): MPC 2015, LNCS 9129, pp. 93–112, 2015.
DOI:10.1007/978-3-319-19797-5_5

and variety makes comparisons difficult, and can make prospective GP users struggle even before actually writing a generic program, since first they have to choose a library that is appropriate for their needs.

We have previously investigated how to model and formally relate some Haskell GP libraries using Agda (Magalhães and Löh 2012), and concluded that some approaches clearly subsume others. Afterwards, we have shown how to reduce code duplication in GP libraries in Haskell by converting between the representations of different approaches, in what we dubbed "generic programming" (Magalhães and Löh 2014).

To help understand the benefits of our work, it is important to distinguish three kinds of users of GP:

Compiler writer. As far as GP goes, the compiler writer is concerned with which approach(es) are natively supported by the compiler. At the moment, in the main Haskell compiler GHC, both `syb` (Lämmel and Peyton Jones 2003, 2004) and `generic-deriving` (Magalhães et al. 2010) are natively supported. This means that the compiler can automatically generate the necessary generic representations to enable using these approaches. A quick analysis reveals that in GHC there are about 226 lines of code for supporting `syb`, and 884 for `generic-deriving`.

GP library author. The library author maintains one or more GP libraries, and possibly creates new ones. Since most approaches are not natively supported, the library author has to deal with generating generic representations for their library (or accept that no end user will use this library, given the amount of boilerplate code they would have to write). Typically, the library author will rely on Template Haskell (TH, Sheard and Peyton Jones 2002) for this task. Given that TH handles code generation at the AST level (thus syntactic), this is not a pleasant task. Furthermore, the TH API changes as frequently as the compiler, so the library author currently has to update its supporting code frequently.

End users. The end users know nothing about compiler support, and ideally not even about library-specific detail. They simply want to use a particular generic function on their data, with minimum overhead.

While Magalhães and Löh (2014) focused mostly on improving the life of the end user, the work we describe in this paper brings more advantages to the compiler writer and GP library author. We elaborate further on the idea of generic generic programming by highlighting the importance of hierarchy in GP approaches. Each new GP library should only be a piece of the puzzle, specialising in one task, or exploring a new generic representation, but obtaining most of its supporting infrastructure (such as example generic functions) from already existing approaches. To facilitate this, we introduce a new GP library, `structured`, which we use as a core to derive representations for other GP libraries. Defining a new library does not mean introducing a lot of new supporting code. In fact, we do not even think many generic functions will ever be defined in our new library, as its representation is verbose (albeit precise). Instead, we use it to guide conversion efforts, as a highly structured approach provides a good foundation to build upon.

From the compiler writer's perspective, this library would be the only one needing compiler support; support for other libraries follows automatically from conversions that are defined in plain Haskell, not through more compiler extensions. Since `structured` has only one representation, as opposed to `generic-deriving`'s two representations, we believe that supporting it in GHC would require fewer lines of code than the existing support for `generic-deriving`. The code for supporting `generic-deriving` and `syb` could then be removed, as those representations can be obtained from `structured`.

Should we ever find that we need more information in `structured` to support converting to other libraries, we can extend it without changing any of the other libraries. This obviates the need to change the compiler for supporting or adapting GP approaches. It also simplifies the life of the GP library author, who no longer needs to rely on meta-programming tools.

Specifically, our contributions are the following:

- A new library for GP, `structured`, which properly encodes the nesting of the different structures within a datatype representation (Sect. 2). We propose this library as a foundation for GP in Haskell, from which many other approaches can be derived. It is designed to be highly expressive and easily extensible, serving as a back-end for more stable and established GP libraries.
- We show how `structured` can provide GP library authors with different views of the nesting of constructors and fields (Sect. 3). Different generic functions prefer different balancings, which we provide through automatic conversion (instead of duplicated encodings differing only in the balancing).
- We position `structured` at the top of GP hierarchy by showing how to derive `generic-deriving` (Magalhães et al. 2010) representations from it (Sect. 4). This also shows how `structured` unifies the two generic representations of `generic-deriving`. Representations for other libraries (`regular` (Van Noort et al. 2008), `multirec` (Rodriguez Yakushev et al. 2009), and `syb` (Lämmel and Peyton Jones 2003, 2004)) can then be obtained from `generic-deriving`.

Figure 1 shows an overview of the hierarchical relationships between different libraries for GP in Haskell. In this paper, we introduce `structured` and its conversion to `generic-deriving`. We refer the reader to Magalhães and Löh (2014) for the conversions from `generic-deriving`.

1.1 Notation

In order to avoid syntactic clutter and to help the reader, we adopt a liberal Haskell notation in this paper. We assume the existence of a **kind** keyword, which allows us to define kinds directly. These kinds behave as if they had arisen from datatype promotion (Yorgey et al. 2012), except that they do not define a datatype and constructors. We omit the keywords **type family** and **type instance** entirely, making type-level functions look like their value-level

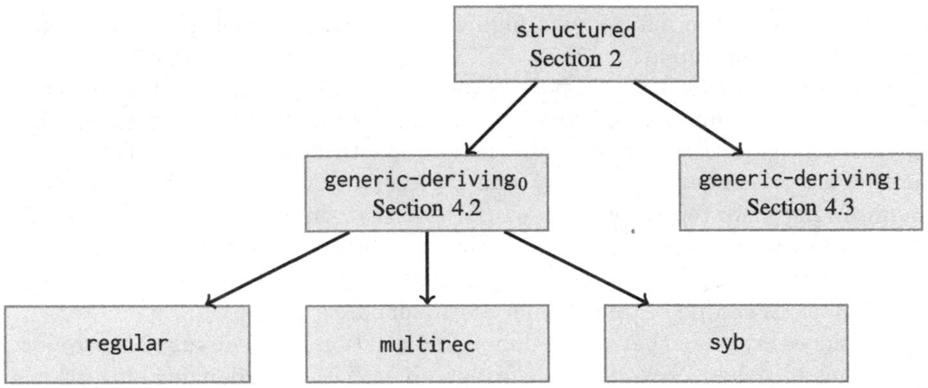

Fig. 1. Hierarchical relationship between GP approaches.

counterparts. We colour constructors in *blue*, types in *red*, and kinds in *green*. In case the colours cannot be seen, the "level" of an expression is clear from the context. Additionally, we use Greek letters for type variables, apart from κ, which is reserved for kind variables.

This syntactic sugar is only for presentation purposes. An executable version of the code, which compiles with GHC 7.8.3, is available at http://dreixel.net/research/code/hgp.zip. We rely on many GHC-specific extensions to Haskell, which are essential for our development. Due to space constraints we cannot explain them all in detail, but we try to point out relevant features as we use them.

1.2 Structure of the Paper

The remainder of this paper is structured as follows. We first introduce the `structured` library (Sect. 2). We then see how to obtain views with different balancings of the constructors and constructor arguments (Sect. 3). Afterwards, we see how to obtain `generic-deriving` from `structured` (Sect. 4). We conclude with a discussion in Sect. 5.

2 A Highly Structured GP Library

Our efforts of modularising a hierarchy of GP libraries stem from a `structured` library intended to sit at the top of the hierarchy. Our goal is to define a library that is highly expressive, without having to worry about convenience of use. Users requiring the level of detail given by `structured` can use it directly, but we expect most to prefer using any of the other, already existing GP libraries. Usability is not our concern here; expressiveness is. Stability is also not guaranteed; we might extend our library as needed to support converting to more approaches. Previous approaches had to find a careful balance between having too little information in the generic representation, resulting in a library with poor expressiveness, and having too much information, resulting in a verbose and hard to use approach. Given our modular approach, we are free from these concerns.

The design of `structured` we give here is preliminary; we plan to extend it in the future in order to support representing more datatypes. In fact, as the type language of GHC grows with new extensions, we expect to keep changing `structured` frequently. However, the simple fact that we introduce `structured`, and show how to use it for decoupling `generic-deriving` from the compiler, improves the current status quo. In particular, if `structured` is supported through automatic deriving in GHC, no more compiler support is required for the other libraries. Using this library also improves modularity; it can be updated or extended more freely, since supporting the other libraries requires only updating the conversions, not the compiler itself (for the automatic derivation of instances).

In our previous work (Magalhães and Löh 2014) we have shown how to obtain the representation of many libraries from `generic-deriving`. Given that `structured`, at the time of writing, serves only to provide a conversion to `generic-deriving`, the reader might think that it is unnecessary. We have several reasons justifying `structured`, however:

- The `generic-deriving` library has been around for some time now, and lots of code using it has been written. Sticking to `generic-deriving` as a foundational approach would force us to break lots of code whenever we would need to update it in order to support new approaches, or to add functionality.
- By introducing `structured`, we can decouple most of `generic-deriving` from the compiler. In particular, the mechanism for deriving `generic-deriving` instances can be simplified, because `generic-deriving` has two representations (which need to be derived separately), while `structured` has only one (from which we can derive both `generic-deriving` representations).
- Being a new approach designed to be an internal representation, `structured` can be changed without worrying too much about breaking existing code; the only code that would need to be adapted is that for the conversion to `generic-deriving`. This is plain Haskell code, not compiler code or TH, so it's easier to update and maintain.

We now proceed to describe the representation types in `structured`, their interpretation as values, and the conversion between user datatypes and their generic representations, together with example encodings.

2.1 Universe

The structure used to encode datatypes in a GP approach is called its *universe* (Morris 2007). The universe of `structured`, for now, is similar to that of `generic-deriving` (Magalhães 2012, Chap. 11), as it supports abstraction over at most one datatype parameter. We choose to restrict this parameter to be the last of the datatype, and only if its kind is $\star$. This is a pragmatic decision: many generic functions, such as *map*, require abstraction over one parameter, but comparatively few require abstraction over more than one parameter. For example, in the type $[\alpha]$, the parameter is α, and in *Either* α β, it is β. The differences to `generic-deriving` lay in the explicit hierarchy of data, constructor, and field, and the absence of two separate ways of encoding constructor arguments. It might seem unsatisfactory that we do not improve on the limitations of `generic-deriving` with regards to datatype parameters, but that is secondary to our goal in this paper (and it would be easy to implement support for multiple parameters in `structured` following the strategy of Magalhães (2014)). Furthermore, `structured` can easily be improved later, keeping the other libraries unchanged, and adapting only the conversions if necessary.

Datatypes are represented as types of *kind Data*. We define new kinds, whose types are not inhabited by values: only types of kind $\star$ are inhabited by values. These kinds can be thought of as datatypes, but their "constructors" will be used as indices of a GADT (Schrijvers et al. 2009) to construct values with a specific structure.

Datatypes have some metadata, such as their name, and contain constructors. Constructors have their own metadata, and contain fields. Finally, each field can have metadata, and contain a value of some structure:

kind *Data* = *Data MetaData* (*Tree Con*)
kind *Con* = *Con MetaCon* (*Tree Field*)
kind *Field* = *Field MetaField Arg*

kind *Tree* κ = *Empty* | *Leaf* κ | *Bin* (*Tree* κ) (*Tree* κ)

We use a binary leaf tree to encode the structure of the constructors in a datatype, and the fields in a constructor. Typically lists are used, but we will see in Sect. 3 that it is convenient to encode the structure as a tree, as we can change the way it is balanced for good effect.

The metadata we store is unsurprising:

```
kind MetaData  = MD Symbol          -- datatype name
                    Symbol          -- datatype module name
                    Bool            -- is it a newtype?
kind MetaCon   = MC Symbol          -- constructor name
                    Fixity          -- constructor fixity
                    Bool            -- does it use record syntax?
kind MetaField = MF (Maybe Symbol)  -- field name
```

kind *Fixity* = *Prefix* | *Infix Associativity Nat*

kind *Associativity* = *LeftAssociative* | *RightAssociative* | *NotAssociative*

kind *Nat* = *Ze* | *Su Nat*
kind *Symbol* -- internal

It is important to note that this metadata is encoded at the type level. In particular, we have type-level strings and natural numbers. We make use of the current (in GHC 7.8.3) implementation of type-level strings, whose kind is *Symbol*.

Finally, *Arg* describes the structure of constructor arguments:

$$\begin{aligned}
&\textbf{kind } \mathit{Arg} = K\ \mathit{KType}\ \star \\
&\qquad\qquad \mid \mathit{Rec}\ \mathit{RecType}\ (\star \to \star) \\
&\qquad\qquad \mid \mathit{Par} \\
&\qquad\qquad \mid (\star \to \star) :\circ: \mathit{Arg} \\
&\textbf{kind } \mathit{KType} \quad = P \mid R\ \mathit{RecType} \mid U \\
&\textbf{kind } \mathit{RecType} = S \mid O
\end{aligned}$$

A field can either be a datatype parameter other than the last ($K\ P$), an occurrence of a different datatype of kind $\star$ ($K\ (R\ O)$), some other type (such as an application of a type variable, encoded with $K\ U$), a datatype of kind (at least) $\star \to \star$ (*Rec*), which can be either the same type we're encoding (S) or a different one (O), the (last) parameter of the datatype (*Par*), or a composition of a type constructor with another argument ($:\circ:$).

The representation is best understood in terms of an example. Consider the following datatype:

$$\textbf{data } D\ \phi\ \alpha\ \beta = D_1\ \mathit{Int}\ (\phi\ \alpha) \mid D_2\ [D\ \phi\ \alpha\ \beta]\ \beta$$

We first show the encoding of each of the four constructor arguments: *Int* is a datatype of kind $\star$, so it's encoded with $K\ (R\ O)\ \mathit{Int}$; $\phi\ \alpha$ depends on the instantiation of ϕ, so it's encoded with $K\ U\ (\phi\ \alpha)$; $[D\ \phi\ \alpha\ \beta]$ is a composition between the list functor and the datatype we're defining, so it's encoded with $[]\ :\circ:\ \mathit{Rec}\ S\ (D\ \phi\ \alpha)$; finally, β is the parameter we abstract over, so it's encoded with *Par*:

$$\begin{aligned}
A_{11} &= K\ (R\ O)\ \mathit{Int} \\
A_{12} &= K\ U\ (\phi\ \alpha) \\
A_{21} &= []\ :\circ:\ \mathit{Rec}\ S\ (D\ \phi\ \alpha) \\
A_{22} &= \mathit{Par}
\end{aligned}$$

The entire representation consists of wrapping of appropriate meta-data around the representation for constructor arguments:

$$\begin{aligned}
&\mathit{Rep}_D\ \phi\ \alpha\ \beta = \\
&\quad \mathit{Data}\ (\mathit{MD}\ \texttt{"D"}\ \texttt{"Module"}\ \mathit{False}) \\
&\qquad (\mathit{Bin}\ (\mathit{Leaf}\ (\mathit{Con}\ (\mathit{MC}\ \texttt{"D1"}\ \mathit{Prefix}\ \mathit{False}) \\
&\qquad\qquad\qquad (\mathit{Bin}\ (\mathit{Leaf}\ (\mathit{Field}\ (\mathit{MF}\ \mathit{Nothing})\ A_{11})) \\
&\qquad\qquad\qquad\qquad (\mathit{Leaf}\ (\mathit{Field}\ (\mathit{MF}\ \mathit{Nothing})\ A_{12}))))) \\
&\qquad\quad (\mathit{Leaf}\ (\mathit{Con}\ (\mathit{MC}\ \texttt{"D2"}\ \mathit{Prefix}\ \mathit{False}) \\
&\qquad\qquad\qquad (\mathit{Bin}\ (\mathit{Leaf}\ (\mathit{Field}\ (\mathit{MF}\ \mathit{Nothing})\ A_{21})) \\
&\qquad\qquad\qquad\qquad (\mathit{Leaf}\ (\mathit{Field}\ (\mathit{MF}\ \mathit{Nothing})\ A_{22}))))))
\end{aligned}$$

2.2 Interpretation

The interpretation of the universe defines the structure of the values that inhabit the datatype representation. Datatype representations will be types of kind *Data*.

We use a data family (Schrijvers et al. 2008) $[\![\_]\!]$ to encode the interpretation of the universe of `structured`:

$$\textbf{data family}\ [\![\_]\!] :: \kappa \to \star \to \star$$

Its first argument is written infix, and the second postfix. Its kind, $\kappa \to \star \to \star$, is overly general in κ; we will only instantiate κ to the types of the universe shown before, and prevent further instantiation by not exporting the family $[\![\_]\!]$ (effectively making it a closed data family). The second argument of $[\![\_]\!]$, of kind $\star$, is the parameter of the datatype which we abstract over.

The top-level inhabitant of a datatype representation is a constructor D_1, which serves only as a proxy to store the datatype metadata in its type:

$$\begin{array}{l}
\textbf{data instance}\ [\![\upsilon :: \mathit{Data}]\!]\ \rho\ \textbf{where} \\
\quad D_1 :: [\![\alpha]\!]\ \rho \to [\![\mathit{Data}\ \iota\ \alpha]\!]\ \rho
\end{array}$$

Constructors, on the other hand, are part of a *Tree* structure, so they can be on the left (L_1) or right (R_1) side of a branch, or be a leaf. As a leaf, they contain the meta-information for the constructor that follows (C_1):

$$\begin{array}{l}
\textbf{data instance}\ [\![\upsilon :: \mathit{Tree}\ \mathit{Con}]\!]\ \rho\ \textbf{where} \\
\quad C_1 :: [\![\alpha]\!]\ \rho \to [\![\mathit{Leaf}\ (\mathit{Con}\ \iota\ \alpha)]\!]\ \rho \\
\quad L_1 :: [\![\alpha]\!]\ \rho \to [\![\mathit{Bin}\ \alpha\ \beta]\!]\ \rho \\
\quad R_1 :: [\![\beta]\!]\ \rho \to [\![\mathit{Bin}\ \alpha\ \beta]\!]\ \rho
\end{array}$$

Constructor fields are similar, except that they might be empty (U_1, as some constructors have no arguments), leaves contain fields (S_1), and branches are inhabited by the arguments of both sides ($:\!\times\!:$):

$$\begin{array}{llll}
\multicolumn{4}{l}{\textbf{data instance}\ [\![\upsilon :: \mathit{Tree}\ \mathit{Field}]\!]\ \rho\ \textbf{where}} \\
\quad U_1 & :: & & [\![\mathit{Empty}]\!]\ \rho \\
\quad S_1 & :: [\![\alpha]\!]\ \rho & \to & [\![\mathit{Leaf}\ (\mathit{Field}\ \iota\ \alpha)]\!]\ \rho \\
\quad (:\!\times\!:) & :: [\![\alpha]\!]\ \rho \to [\![\beta]\!]\ \rho & \to & [\![\mathit{Bin}\ \alpha\ \beta]\!]\ \rho
\end{array}$$

We're left with constructor arguments. We encode base types with K, datatype occurrences with *Rec*, the parameter with *Par*, and composition with *Comp*:

$$\begin{array}{llll}
\multicolumn{4}{l}{\textbf{data instance}\ [\![\upsilon :: \mathit{Arg}]\!]\ \rho\ \textbf{where}} \\
\quad K & :: \{ \mathit{unK_1} & :: \alpha\ \} & \to [\![K\ \iota\ \alpha]\!]\ \rho \\
\quad \mathit{Rec} & :: \{ \mathit{unRec} & :: \phi\ \rho \} & \to [\![\mathit{Rec}\ \iota\ \phi]\!]\ \rho \\
\quad \mathit{Par} & :: \{ \mathit{unPar} & :: \rho\ \} & \to [\![\mathit{Par}]\!]\ \rho \\
\quad \mathit{Comp} & :: \{ \mathit{unComp} & :: \sigma\ ([\![\phi]\!]\ \rho) \} & \to [\![\sigma :\circ: \phi]\!]\ \rho
\end{array}$$

2.3 Conversion to and from User Datatypes

Having seen the generic universe and its interpretation, we need to provide a mechanism to mediate between user datatypes and our generic representation. We use a type class for this purpose:

```
class Generic (α :: ⋆) where
  Rep      α :: Data
  ThePar α :: ⋆
  ThePar α = NoPar
  from     :: α → ⟦ Rep φ ⟧ (ThePar α)
  to       :: ⟦ Rep φ ⟧ (ThePar α) → α

data NoPar  -- empty
```

In the *Generic* class, the type family *Rep* encodes the generic representation associated with user datatype α, and *ThePar* extracts the last parameter from the datatype. In case the datatype is of kind $\star$, we use *NoPar*; a type family default allows us to leave the type instance empty for types of kind $\star$. The conversion functions *from* and *to* perform the conversion between the user datatype values and the interpretation of its generic representation.

2.4 Example Datatype Encodings

We now show two complete examples of how user datatypes are encoded in `structured`. Naturally, users should never have to define these manually; a release version of `structured` would be incorporated in the compiler, allowing automatic derivation of *Generic* instances.

Choice. The first datatype we encode represents a choice between four options:

```
data Choice = A | B | C | D
```

Choice is a datatype of kind $\star$, so we do not need to provide a type instance for *ThePar*. The encoding, albeit verbose, is straightforward:

```
instance Generic Choice where
  Rep Choice =
    Data (MD "Choice" "Module" False)
      (Bin (Bin (Leaf (Con (MC "A" Prefix False) Empty))
                (Leaf (Con (MC "B" Prefix False) Empty)))
           (Bin (Leaf (Con (MC "C" Prefix False) Empty))
                (Leaf (Con (MC "D" Prefix False) Empty))))
  from A = D1 (L1 (L1 (C1 U1)))
  from B = D1 (L1 (R1 (C1 U1)))
  from C = D1 (R1 (L1 (C1 U1)))
  from D = D1 (R1 (R1 (C1 U1)))
  to (D1 (L1 (L1 (C1 U1)))) = A
  ...
```

We use a balanced tree structure for the constructors; in Sect. 3 we will see how this can be changed without any user effort.

Lists. Standard Haskell lists are a type of kind $\star \to \star$. We break down its type representation into smaller fragments using type synonyms, to ease comprehension. The encoding of the metadata of each constructor and the two arguments to (:) follows:

$$
\begin{array}{lll}
MC_{Nil} & = MC\ \text{"[]"}\ Prefix & False \\
MC_{Cons} & = MC\ \text{": "}\ (Infix\ RightAssociative\ 5) & False \\
H & = Leaf\ (Field\ (MF\ Nothing)\ Par) & \\
T & = Leaf\ (Field\ (MF\ Nothing)\ (Rec\ S\ [\,])) &
\end{array}
$$

The encoding of the first argument to (:), H, states that there is no record selector, and that the argument is the parameter *Par*. The encoding of the second argument, T, is a recursive occurrence of the same datatype being defined ($Rec\ S\ [\,]$).

With these synonyms in place, we can show the complete *Generic* instance for lists:

$$
\begin{array}{l}
\mathbf{instance}\ Generic\ [\alpha]\ \mathbf{where} \\
\quad Rep\ [\alpha] = Data\ (MD\ \text{"[]"}\ \texttt{"Prelude"}\ False) \\
\qquad\qquad\qquad (Bin\ (Leaf\ (Con\ MC_{Nil}\ Empty)) \\
\qquad\qquad\qquad\quad (Leaf\ (Con\ MC_{Cons}\ (Bin\ H\ T)))) \\
\quad ThePar\ [\alpha] = \alpha \\
\quad from\ [\,] \qquad = D_1\ (L_1\ (C_1\ U_1)) \\
\quad from\ (h : t) = D_1\ (R_1\ (C_1\ (S_1\ (Par\ h) \times S_1\ (Rec\ t)))) \\
\quad to\ (D_1\ (L_1\ (C_1\ U_1))) \qquad\qquad\qquad\quad = [\,] \\
\quad to\ (D_1\ (R_1\ (C_1\ (S_1\ (Par\ h) \times S_1\ (Rec\ t))))) = h : t
\end{array}
$$

The type function *ThePar* extracts the parameter α from $[\alpha]$; the *from* and *to* conversion functions are unsurprising.

3 Left- and Right-Biased Encodings

The `structured` library uses trees to store the constructors inside a datatype, as well as the fields inside a constructor. So far we have kept these trees balanced, but other choices would be acceptable too. In fact, the balancing choice determines a generic view (Holdermans et al. 2006). Different balancings might be more convenient for certain generic functions. For example, if we are defining a binary encoding function, it is convenient to use the balanced encoding, as then we can easily minimise the number of bits used to encode a constructor. On the other hand, if we are defining a generic function that extracts the first argument to a constructor (if it exists), we would prefer using a right-nested view, as then we can simply pick the first argument on the left. Fortunately, we do not have to provide multiple representations to support this; we can automatically convert between different balancings. As an example, we see in this section how to convert from the (default) balanced encoding to a right-nested one.

This is the first conversion shown in this paper, and as such serves as an introduction to our conversions. Following the style of Magalhães and Löh (2014), we use a type family to adapt the representation, and a type-class to adapt the values. Since this conversion works at the top of the hierarchy (on `structured`), the new balancing persists in future conversions, so a generic function in `generic-deriving` could make use of a right-biased encoding.

3.1 Type Conversion

The essential part of the type conversion is a type function that performs one rotation to the right on a tree:

$$
\begin{array}{l}
\mathit{RotR}\ (\alpha :: \mathit{Tree}\ \kappa) :: \mathit{Tree}\ \kappa \\
\mathit{RotR}\ (\mathit{Bin}\ (\mathit{Bin}\ \alpha\ \beta)\ \gamma) = \mathit{Bin}\ \alpha\ (\mathit{Bin}\ \beta\ \gamma) \\
\mathit{RotR}\ (\mathit{Bin}\ (\mathit{Leaf}\ \alpha)\ \gamma) = \mathit{Bin}\ (\mathit{Leaf}\ \alpha)\ \gamma
\end{array}
$$

We then apply this rotation repeatedly at the top level until the tree contains a *Leaf* on the left subtree, and then proceed to rotate the right subtree:

$$
\begin{array}{l}
S_{\rightarrow}SR_d\ (\alpha :: \mathit{Data}) :: \mathit{Data} \\
S_{\rightarrow}SR_d\ (\mathit{Data}\ \iota\ \alpha) = \mathit{Data}\ \iota\ (S_{\rightarrow}SR_{cs}\ \alpha) \\
S_{\rightarrow}SR_{cs}\ (\alpha :: \mathit{Tree}\ \mathit{Con}) :: \mathit{Tree}\ \mathit{Con} \\
S_{\rightarrow}SR_{cs}\ \mathit{Empty} = \mathit{Empty} \\
S_{\rightarrow}SR_{cs}\ (\mathit{Leaf}\ (\mathit{Con}\ \iota\ \gamma)) = \mathit{Leaf}\ (\mathit{Con}\ \iota\ (S_{\rightarrow}SR_{fs}\ \gamma)) \\
S_{\rightarrow}SR_{cs}\ (\mathit{Bin}\ (\mathit{Bin}\ \alpha\ \beta)\ \gamma) = S_{\rightarrow}SR_{cs}\ (\mathit{RotR}\ (\mathit{Bin}\ (\mathit{Bin}\ \alpha\ \beta)\ \gamma)) \\
S_{\rightarrow}SR_{cs}\ (\mathit{Bin}\ (\mathit{Leaf}\ \alpha)\ \gamma) = \mathit{Bin}\ (S_{\rightarrow}SR_{cs}\ (\mathit{Leaf}\ \alpha))\ (S_{\rightarrow}SR_{cs}\ \gamma) \\
S_{\rightarrow}SR_{fs}\ (\alpha :: \mathit{Tree}\ \mathit{Field}) :: \mathit{Tree}\ \mathit{Field} \\
S_{\rightarrow}SR_{fs}\ \mathit{Empty} = \mathit{Empty} \\
S_{\rightarrow}SR_{fs}\ (\mathit{Leaf}\ \gamma) = \mathit{Leaf}\ \gamma \\
S_{\rightarrow}SR_{fs}\ (\mathit{Bin}\ (\mathit{Bin}\ \alpha\ \beta)\ \gamma) = S_{\rightarrow}SR_{fs}\ (\mathit{RotR}\ (\mathit{Bin}\ (\mathit{Bin}\ \alpha\ \beta)\ \gamma)) \\
S_{\rightarrow}SR_{fs}\ (\mathit{Bin}\ (\mathit{Leaf}\ \alpha)\ \gamma) = \mathit{Bin}\ (\mathit{Leaf}\ \alpha)\ (S_{\rightarrow}SR_{fs}\ \gamma)
\end{array}
$$

The conversion for constructors ($S_{\rightarrow}SR_{cs}$) and selectors ($S_{\rightarrow}SR_{fs}$) differs only in the treatment for leaves, as the leaf of a selector is the stopping point of this transformation.

3.2 Value Conversion

The value-level conversion is witnessed by a type class:

$$
\begin{array}{l}
\mathbf{class}\ \mathit{Convert}_{S_{\rightarrow}SR}\ (\alpha :: \mathit{Data})\ \mathbf{where} \\
\quad s_{\rightarrow}rs :: [\![\alpha]\!]\ \rho \rightarrow [\![S_{\rightarrow}SR_d\ \alpha]\!]\ \rho \\
\quad s_{\leftarrow}rs :: [\![S_{\rightarrow}SR_d\ \alpha]\!]\ \rho \rightarrow [\![\alpha]\!]\ \rho
\end{array}
$$

We skip the definition of the instances, as they are mostly unsurprising and can be found in our code bundle.

3.3 Example

To test the conversion, we define a generic function that computes the depth of the encoding of a constructor:

$$
\begin{array}{l}
\textbf{class } \mathit{CountSums_r}\ \alpha\ \textbf{where} \\
\quad \mathit{countSums_r} :: [\![\, \alpha \,]\!]\ \rho \rightarrow \mathit{Int} \\
\textbf{instance } (\mathit{CountSums_r}\ \alpha) \Rightarrow \mathit{CountSums_r}\ (\mathit{Data}\ \iota\ \alpha)\ \textbf{where} \\
\quad \mathit{countSums_r}\ (D_1\ x) = \mathit{countSums_r}\ x \\
\textbf{instance } \mathit{CountSums_r}\ \mathit{Empty}\quad \textbf{where } \mathit{countSums_r}\ \_ = 0 \\
\textbf{instance } \mathit{CountSums_r}\ (\mathit{Leaf}\ \alpha)\ \textbf{where } \mathit{countSums_r}\ \_ = 0 \\
\textbf{instance } (\mathit{CountSums_r}\ \alpha, \mathit{CountSums_r}\ \alpha) \\
\qquad\qquad \Rightarrow \mathit{CountSums_r}\ (\mathit{Bin}\ \alpha\ \beta :: \mathit{Tree\ Con})\ \textbf{where} \\
\quad \mathit{countSums_r}\ (L_1\ x) = 1 + \mathit{countSums_r}\ x \\
\quad \mathit{countSums_r}\ (R_1\ x) = 1 + \mathit{countSums_r}\ x
\end{array}
$$

We now have two ways of calling this function; one using the standard encoding, and other using the right-nested encoding obtained using $\mathit{Converts}_{S_{\rightarrow}SR}$:

$$
\begin{array}{l}
\mathit{countSumsBal} :: (\mathit{Generic}\ \alpha, \mathit{CountSums_r}\ (\mathit{Rep}\ \alpha)) \Rightarrow \alpha \rightarrow \mathit{Int} \\
\mathit{countSumsBal} = \mathit{countSums_r} \circ \mathit{from} \\
\mathit{countSumsR} :: (\mathit{Generic}\ \alpha, \mathit{Converts}_{S_{\rightarrow}SR}\ (\mathit{Rep}\ \alpha) \\
\qquad\qquad\qquad , \mathit{CountSums_r}\ (S_{\rightarrow}SR_d\ (\mathit{Rep}\ \alpha))) \Rightarrow \alpha \rightarrow \mathit{Int} \\
\mathit{countSumsR} = \mathit{countSums_r} \circ s_{\rightarrow}rs \circ \mathit{from}
\end{array}
$$

Applying these two functions to the constructors of the *Choice* datatype should give different results:

$$
\begin{array}{l}
\mathit{testCountSums} :: ([\mathit{Int}], [\mathit{Int}]) \\
\mathit{testCountSums} = (\mathit{map}\ \mathit{countSumsBal}\ [A, B, C, D] \\
\qquad\qquad\qquad\quad , \mathit{map}\ \mathit{countSumsR}\quad [A, B, C, D])
\end{array}
$$

Indeed, *testCountSums* evaluates to $([2,2,2,2],[1,2,3,3])$ as expected. As we've seen, not only can we obtain a different balancing without having to duplicate the representation, but we can also effortlessly apply the same generic function to differently-balanced encodings. Furthermore, the conversions shown in the coming sections automatically "inherit" the balancing chosen in `structured`, allowing us to provide representations with different balancings to the other GP libraries as well.

4 From `structured` to `generic-deriving`

In this section we show how to obtain `generic-deriving` representations from `structured`.

4.1 Encoding `generic-deriving`

The first step is to define `generic-deriving`. We could use its definition as implemented in the `GHC.Generics` module, but it seems more appropriate to at least make use of proper kinds. We thus redefine `generic-deriving` in this paper to bring it up to date with the most recent compiler functionality[1]. This is not essential for our conversions, and should be seen only as a small improvement. The type representation is similar to a collapsed version of `structured`, where all types inhabit a single kind Un_D:

$$
\begin{aligned}
\textbf{kind}\ Un_D = {} & V_D \\
\mid {} & U_D \\
\mid {} & Par_D \\
\mid {} & K_D \quad KType \quad \star \\
\mid {} & Rec_D\ RecType\ (\star \to \star) \\
\mid {} & M_D\ Meta_D\ Un_D \\
\mid {} & Un_D\ {:}{+}{:}_D\ Un_D \\
\mid {} & Un_D\ {:}{\times}{:}_D\ Un_D \\
\mid {} & (\star \to \star)\ {:}{\circ}{:}_D\ Un_D \\
\textbf{kind}\ Meta_D = {} & D_D\ MetaData \mid C_D\ MetaCon \mid F_D\ MetaField
\end{aligned}
$$

Since many names are the same as those in `structured`, we use the "D" subscript for `generic-deriving` names. V_D, U_D, Par_D, K_D, Rec_D, and $({:}{\circ}{:}_D)$ behave very much like the `structured` *Empty*, *Leaf*, *Par*, *K*, *Rec*, and $({:}{\circ}{:})$, respectively. The binary operators $({:}{+}{:}_D)$ and $({:}{\times}{:}_D)$ are equivalent to *Bin*, and M_D encompasses `structured`'s *Data*, *Con*, and *Field*.

Having seen the interpretation of `structured`, the interpretation of the `generic-deriving` universe is unsurprising:

$$
\begin{aligned}
&\textbf{data}\ [\![\, \alpha :: Un_D \,]\!]_D\ (\rho :: \star) :: \star\ \textbf{where} \\
&\quad U_{1D} && :: [\![\, U_D \,]\!]_D\ \rho \\
&\quad M_{1D} && :: [\![\, \alpha \,]\!]_D\ \rho \to [\![\, M_D\ \iota\ \alpha \,]\!]_D\ \rho \\
&\quad Par_{1D} && :: \rho \to [\![\, Par_D \,]\!]_D\ \rho \\
&\quad K_{1D} && :: \alpha \to [\![\, K_D\ \iota\ \alpha \,]\!]_D\ \rho \\
&\quad Rec_{1D} && :: \phi\ \rho \to [\![\, Rec_D\ \iota\ \phi \,]\!]_D\ \rho \\
&\quad Comp_{1D} && :: \phi\ ([\![\, \alpha \,]\!]_D\ \rho) \to [\![\, \phi\ {:}{\circ}{:}_D\ \alpha \,]\!]_D\ \rho \\
&\quad L_{1D} && :: [\![\, \phi \,]\!]_D\ \rho \to [\![\, \phi\ {:}{+}{:}_D\ \psi \,]\!]_D\ \rho \\
&\quad R_{1D} && :: [\![\, \psi \,]\!]_D\ \rho \to [\![\, \phi\ {:}{+}{:}_D\ \psi \,]\!]_D\ \rho \\
&\quad {:}{\times}{:}_D && :: [\![\, \phi \,]\!]_D\ \rho \to [\![\, \psi \,]\!]_D\ \rho \to [\![\, \phi\ {:}{\times}{:}_D\ \psi \,]\!]_D\ \rho
\end{aligned}
$$

The significant difference from `structured` is the relative lack of structure. The types (and kinds) do not prevent an L_{1D} from showing up under a ${:}{\times}{:}_D$, for example.

---

[1] Along the lines of its proposed kind-polymorphic overhaul described in http://hackage.haskell.org/trac/ghc/wiki/Commentary/Compiler/GenericDeriving#Kind-polymorphicoverhaul.

User datatypes are converted to the generic representation using two type classes:

$$
\begin{aligned}
&\textbf{class}\ \mathit{Generic}_D\ (\alpha :: \star)\ \textbf{where}\\
&\quad \mathit{Rep}_D\quad\ \ \alpha :: \mathit{Un}_D\\
&\quad \mathit{ThePar}_D\ \alpha :: \star\\
&\quad \mathit{ThePar}_D = \mathit{NoPar}\\
&\quad \mathit{from}_D\ :: \alpha \to [\![\, \mathit{Rep}_D\ \alpha \,]\!]_D\ (\mathit{ThePar}_D\ \alpha)\\
&\quad \mathit{to}_D\quad\ :: [\![\, \mathit{Rep}_D\ \alpha \,]\!]_D\ (\mathit{ThePar}_D\ \alpha) \to \alpha
\end{aligned}
$$

$$
\begin{aligned}
&\textbf{class}\ \mathit{Generic}_{1D}\ (\phi :: \star \to \star)\ \textbf{where}\\
&\quad \mathit{Rep}_{1D}\ \phi :: \mathit{Un}_D\\
&\quad \mathit{from}_{1D} :: \phi\ \rho \to [\![\, \mathit{Rep}_{1D}\ \phi \,]\!]_D\ \rho\\
&\quad \mathit{to}_{1D}\quad :: [\![\, \mathit{Rep}_{1D}\ \phi \,]\!]_D\ \rho \to \phi\ \rho
\end{aligned}
$$

Class $\mathit{Generic}_D$ is used for all supported datatypes, and encodes a simple view on the constructor arguments. For datatypes that abstract over (at least) one type parameter, an instance for $\mathit{Generic}_{1D}$ is also required. The type representation in this instance encodes the more general view of constructor arguments (i.e. using Par_D, Rec_D, and $:\!\circ\!:_D$). Note that $\mathit{Generic}_D$ doesn't currently have ThePar_D in GHC, but we think this is a (minor) improvement. Furthermore, the presence of a type family default makes it backwards-compatible.

Since these two classes represent essentially two different universes in `generic-deriving`, we need to define two distinct conversions from `structured` to `generic-deriving`.

4.2 To *Generic*$_D$

The universe of `structured` has a detailed encoding of constructor arguments. However, many generic functions do not need such detailed information, and are simpler to write by giving a single case for constructor arguments (imagine, for example, a function that counts the number of arguments). For this purpose, `generic-deriving` states that representations from $\mathit{Generic}_D$ contain only the K_D type at the arguments (so no Par_D, Rec_D, and $:\!\circ\!:_D$).

To derive $\mathit{Generic}_D$ instances from *Generic*, we use the following instance:

$$
\begin{aligned}
&\textbf{instance}\ (\mathit{Generic}\ \alpha, \mathit{Convert}_{S_{\to}D_0}\ (\mathit{Rep}\ \alpha)) \Rightarrow \mathit{Generic}_D\ \alpha\ \textbf{where}\\
&\quad \mathit{Rep}_D\quad\ \ \alpha = S_{\to}G_0\ (\mathit{Rep}\ \alpha)\ (\mathit{ThePar}\ \alpha)\\
&\quad \mathit{ThePar}_D\ \alpha = \mathit{ThePar}\ \alpha\\
&\quad \mathit{from}_D\ = s_{\to}g_0 \circ \mathit{from}\\
&\quad \mathit{to}_D\quad = \mathit{to} \circ s_{\leftarrow}g_0
\end{aligned}
$$

In the remainder of this section, we explain the definition of $S_{\to}G_0$, a type family that converts a representation of `structured` into one of `generic-deriving`, and the class $\mathit{Convert}_{S_{\to}D_0}$, whose methods $s_{\to}g_0$ and $s_{\leftarrow}g_0$ perform the value-level conversion.

Type Representation Conversion. To convert between the type representations, we use a type family:

$$S_{\rightarrow}G_0\ (\alpha :: \kappa)\ (\rho :: \star) :: Un_D$$

The kind of $S_{\rightarrow}G_0$ is overly polymorphic; its input is not any κ, but only the kinds that make up the `structured` universe. We could encode this by using multiple type families, one at each "level". For simplicity, however, we use a single type family, which we instantiate only for the `structured` representation types.

The encoding of datatype meta-information is left unchanged:

$$S_{\rightarrow}G_0\ (Data\ \iota\ \alpha)\ \rho = M_D\ (D_D\ \iota)\ (S_{\rightarrow}G_0\ \alpha\ \rho)$$

We then proceed with the conversion of the constructors:

$$\begin{array}{ll} S_{\rightarrow}G_0\ Empty & \rho = V_D \\ S_{\rightarrow}G_0\ (Leaf\ (Con\ \iota\ \alpha))\ & \rho = M_D\ (C_D\ \iota)\ (S_{\rightarrow}G_0\ \alpha\ \rho) \\ S_{\rightarrow}G_0\ (Bin\ \alpha\ \beta) & \rho = (S_{\rightarrow}G_0\ \alpha\ \rho)\ :+:_D\ (S_{\rightarrow}G_0\ \beta\ \rho) \end{array}$$

Again, the structure of the constructors and their meta-information is left unchanged. We proceed similarly for constructor fields:

$$\begin{array}{ll} S_{\rightarrow}G_0\ Empty & \rho = U_D \\ S_{\rightarrow}G_0\ (Leaf\ (Field\ \iota\ \alpha))\ & \rho = M_D\ (F_D\ \iota)\ (S_{\rightarrow}G_0\ \alpha\ \rho) \\ S_{\rightarrow}G_0\ (Bin\ \alpha\ \beta) & \rho = (S_{\rightarrow}G_0\ \alpha\ \rho)\ :\times:_D\ (S_{\rightarrow}G_0\ \beta\ \rho) \end{array}$$

Finally, we arrive at individual fields, where the interesting part of the conversion takes place:

$$\begin{array}{llll} S_{\rightarrow}G_0\ (K\ \iota\ \alpha) & \rho = K_D\ \iota & & \alpha \\ S_{\rightarrow}G_0\ (Rec\ \iota\ \phi) & \rho = K_D\ (R\ \iota) & & (\phi\ \rho) \\ S_{\rightarrow}G_0\ Par & \rho = K_D\ P & & \rho \end{array}$$

Basically, all the information kept about the field is condensed into the first argument of K_D. Composition requires special care, but gets similarly collapsed into a K_D:

$$S_{\rightarrow}G_0\ (\phi :\circ: \alpha)\ \rho = K_D\ U\ (\phi\ (S_{\rightarrow}G_{0_{comp}}\ \alpha\ \rho))$$

$$\begin{array}{ll} S_{\rightarrow}G_{0_{comp}}\ (\alpha :: Arg)\ (\rho :: \star) :: \star & \\ S_{\rightarrow}G_{0_{comp}}\ Par & \rho = \rho \\ S_{\rightarrow}G_{0_{comp}}\ (K\ \alpha) & \rho = \alpha \\ S_{\rightarrow}G_{0_{comp}}\ (Rec\ \iota\ \phi) & \rho = \phi\ \rho \\ S_{\rightarrow}G_{0_{comp}}\ (\phi :\circ: \alpha) & \rho = \phi\ (S_{\rightarrow}G_{0_{comp}}\ \alpha\ \rho) \end{array}$$

The auxiliary type family $S_{\rightarrow}G_{0_{comp}}$ takes care of unwrapping the composition, and re-applying the type to its arguments.

Value Conversion. Having performed the type-level conversion, we have to convert the values in an equally type-directed fashion. We begin with datatypes:

$$
\begin{aligned}
&\textbf{class } \mathit{Convert}_{S_{\rightarrow}D_0}\ (\alpha :: \kappa)\ \textbf{where}\\
&\quad s_{\rightarrow}g_0 :: [\![\alpha]\!]\ \rho \rightarrow [\![S_{\rightarrow}G_0\ \alpha\ \rho]\!]\ \rho\\
&\quad s_{\leftarrow}g_0 :: [\![S_{\rightarrow}G_0\ \alpha\ \rho]\!]\ \rho \rightarrow [\![\alpha]\!]\ \rho\\
&\textbf{instance } (\mathit{Convert}_{S_{\rightarrow}D_0}\ \alpha) \Rightarrow \mathit{Convert}_{S_{\rightarrow}D_0}\ (\mathit{Data}\ \iota\ \alpha)\ \textbf{where}\\
&\quad s_{\rightarrow}g_0\ (D_1\ x) = M_{1D}\ (s_{\rightarrow}g_0\ x)\\
&\quad s_{\leftarrow}g_0\ (D_1\ x) = M_{1D}\ (s_{\leftarrow}g_0\ x)
\end{aligned}
$$

As in the type conversion, we simply traverse the representation, and convert the constructors with another function. From here on, we omit the $s_{\leftarrow}g_0$ direction, as it is entirely symmetrical.

Constructors and selectors simply traverse the meta-information:

$$
\begin{aligned}
&\textbf{instance } (\mathit{Convert}_{S_{\rightarrow}D_0}\ \alpha) \Rightarrow \mathit{Convert}_{S_{\rightarrow}D_0}\ (\mathit{Leaf}\ (\mathit{Con}\ \iota\ \alpha))\ \textbf{where}\\
&\quad s_{\rightarrow}g_0\ (C_1\ x) = M_{1D}\ (s_{\rightarrow}g_0\ x)\\
&\textbf{instance } (\mathit{Convert}_{S_{\rightarrow}D_0}\ \alpha, \mathit{Convert}_{S_{\rightarrow}D_0}\ \beta) \Rightarrow \mathit{Convert}_{S_{\rightarrow}D_0}\ (\mathit{Bin}\ \alpha\ \beta)\ \textbf{where}\\
&\quad s_{\rightarrow}g_0\ (L_1\ x) = L_{1D}\ (s_{\rightarrow}g_0\ x)\\
&\quad s_{\rightarrow}g_0\ (R_1\ x) = R_{1D}\ (s_{\rightarrow}g_0\ x)
\end{aligned}
$$

$$
\begin{aligned}
&\textbf{instance } \mathit{Convert}_{S_{\rightarrow}D_0}\ \mathit{Empty}\ \textbf{where}\\
&\quad s_{\rightarrow}g_0\ U_1 = U_{1D}\\
&\textbf{instance } (\mathit{Convert}_{S_{\rightarrow}D_0}\ \alpha) \Rightarrow \mathit{Convert}_{S_{\rightarrow}D_0}\ (\mathit{Leaf}\ (\mathit{Field}\ \iota\ \alpha))\ \textbf{where}\\
&\quad s_{\rightarrow}g_0\ (S_1\ x)\ \ = M_{1D}\ (s_{\rightarrow}g_0\ x)\\
&\textbf{instance } (\mathit{Convert}_{S_{\rightarrow}D_0}\ \alpha, \mathit{Convert}_{S_{\rightarrow}D_0}\ \beta) \Rightarrow \mathit{Convert}_{S_{\rightarrow}D_0}\ (\mathit{Bin}\ \alpha\ \beta)\ \textbf{where}\\
&\quad s_{\rightarrow}g_0\ (x \mathbin{:\times:} y) = s_{\rightarrow}g_0\ x\ \mathbin{:\times:_D}\ s_{\rightarrow}g_0\ y
\end{aligned}
$$

Finally, at the argument level, we collapse everything into K_{1D}:

$$
\begin{aligned}
&\textbf{instance } \mathit{Convert}_{S_{\rightarrow}D_0}\ (K\ \iota\ \alpha)\ \ \ \textbf{where } s_{\rightarrow}g_0\ (K\ x)\ \ \ = K_{1D}\ x\\
&\textbf{instance } \mathit{Convert}_{S_{\rightarrow}D_0}\ (\mathit{Rec}\ \iota\ \phi)\ \textbf{where } s_{\rightarrow}g_0\ (\mathit{Rec}\ x) = K_{1D}\ x\\
&\textbf{instance } \mathit{Convert}_{S_{\rightarrow}D_0}\ \mathit{Par}\ \ \ \ \ \ \ \ \textbf{where } s_{\rightarrow}g_0\ (\mathit{Par}\ x) = K_{1D}\ x\\
&\textbf{instance } (\mathit{Functor}\ \phi, \mathit{Convert}_{comp}\ \alpha) \Rightarrow \mathit{Convert}_{S_{\rightarrow}D_0}\ (\phi \mathbin{:\circ:} \alpha)\ \textbf{where}\\
&\quad s_{\rightarrow}g_0\ (\mathit{Comp}\ x) = K_{1D}\ (g_{\rightarrow}g_{0_{comp}}\ x)
\end{aligned}
$$

Again, for composition we need to unwrap the representation, removing all representation types within:

$$
\begin{aligned}
&\textbf{class } \mathit{Convert}_{comp}\ (\alpha :: \mathit{Arg})\ \textbf{where}\\
&\quad g_{\rightarrow}g_{0_{comp}} :: \mathit{Functor}\ \phi \Rightarrow \phi\ ([\![\alpha]\!]\ \rho) \rightarrow \phi\ (S_{\rightarrow}G_{0_{comp}}\ \alpha\ \rho)\\
&\textbf{instance } \mathit{Convert}_{comp}\ \mathit{Par}\ \ \ \ \ \ \ \ \textbf{where } g_{\rightarrow}g_{0_{comp}} = \mathit{fmap}\ \mathit{unPar}\\
&\textbf{instance } \mathit{Convert}_{comp}\ (K\ \iota\ \alpha)\ \ \ \textbf{where } g_{\rightarrow}g_{0_{comp}} = \mathit{fmap}\ \mathit{unK}_1\\
&\textbf{instance } \mathit{Convert}_{comp}\ (\mathit{Rec}\ \iota\ \phi)\ \textbf{where } g_{\rightarrow}g_{0_{comp}} = \mathit{fmap}\ \mathit{unRec}\\
&\textbf{instance } (\mathit{Functor}\ \phi, \mathit{Convert}_{comp}\ \alpha) \Rightarrow \mathit{Convert}_{comp}\ (\phi \mathbin{:\circ:} \alpha)\ \textbf{where}\\
&\quad g_{\rightarrow}g_{0_{comp}} = \mathit{fmap}\ (g_{\rightarrow}g_{0_{comp}} \circ \mathit{unComp})
\end{aligned}
$$

With all these instances in place, the $Generic\ \alpha \Rightarrow Generic_D\ \alpha$ shown at the beginning of this section takes care of converting to the simpler representation of **generic-deriving** without syntactic overhead. In particular, all generic functions defined over the $Generic_D$ class, such as *gshow* and *genum* from the **generic-deriving** package, are now available to all types in **structured**, such as *Choice* and $[\alpha]$.

Example: Length. To test the conversion, we define a generic function in **generic-deriving** that computes the number of elements in a structure:

$$
\begin{array}{l}
\mathbf{class}\ GLength_r\ (\alpha :: Un_D)\ \mathbf{where} \\
\quad gLength_r :: [\![\, \alpha \,]\!]_D\ \rho \rightarrow Int
\end{array}
$$

We omit the instances of $GLength_r$ as they are unsurprising: we traverse the representation until we reach the arguments, which are recursively counted and added.

While the $GLength_r$ class works on the generic representation, a user-facing class *GLength* takes care of handling user-defined datatypes. We define a generic default which implements *gLength* generically:

$$
\begin{array}{l}
\mathbf{class}\ GLength\ (\alpha :: \star)\ \mathbf{where} \\
\quad gLength :: \alpha \rightarrow Int \\
\quad \mathbf{default}\ gLength :: (Generic_D\ \alpha, GLength_r\ (Rep_D\ \alpha)) \Rightarrow \alpha \rightarrow Int \\
\quad gLength = gLength_r \circ from_D
\end{array}
$$

Because of the generic default, instantiating *GLength* to datatypes with a $Generic_D$ instance is very simple:

$$
\begin{array}{l}
\mathbf{instance}\ GLength\ [\,\alpha\,] \\
\mathbf{instance}\ GLength\ Choice
\end{array}
$$

Recall, however, that in Sect. 2.4 we have given only *Generic* instances for *Choice* and $[\alpha]$, not $Generic_D$. However, due to the $(Generic\ \alpha, Convert_{S \rightarrow D_0}\ (Rep\ \alpha)) \Rightarrow Generic_D\ \alpha$ instance of the beginning of this section, *Choice* and $[\,\alpha\,]$ automatically get a $Generic_D$ instance, which is being used here.

We can test that this function behaves as expected: $gLength\ [0, 1, 2, 3]$ returns 4, and *gLength D* returns 0. And further: using our previous work (Magalhães and Löh 2014), we also gain all the functionality from other libraries, such as `syb` traversals or a zipper, for example.

4.3 To $Generic_{1D}$

Converting to $Generic_{1D}$ is very similar, only that we preserve more structure. The conversion is similarly performed by two components.

Type Representation Conversion. We define a type family to perform the conversion of the type representation:

$$S_{\rightarrow}G_1\ (\alpha :: \kappa) :: Un_D$$

The type instances for the datatype, constructors, and fields behave exactly like in $S_{\rightarrow}G_0$, so we skip straight to the constructor arguments, which are simple to handle because they are in one-to-one correspondence:

$$\begin{aligned}
S_{\rightarrow}G_1\ (K\ \iota\ \alpha) &= K_D\ \iota\ \alpha \\
S_{\rightarrow}G_1\ (Rec\ \iota\ \alpha) &= Rec_D\ \iota\ \alpha \\
S_{\rightarrow}G_1\ Par &= Par_D \\
S_{\rightarrow}G_1\ (\phi :\!\circ\!: \alpha) &= \phi :\!\circ\!:_D\ S_{\rightarrow}G_1\ \alpha
\end{aligned}$$

Value Conversion. The value-level conversion is as trivial as the type-level conversion, so we omit it from the paper. It is witnessed by a poly-kinded type class:

$$\begin{aligned}
&\textbf{class}\ Converts_{S_{\rightarrow}D_1}\ (\alpha :: \kappa)\ \textbf{where} \\
&\quad s_{\rightarrow}g_1 :: [\![\alpha]\!]\ \rho \rightarrow [\![S_{\rightarrow}G_1\ \alpha]\!]_D\ \rho
\end{aligned}$$

Again, we only give instances of $Converts_{S_{\rightarrow}D_1}$ for the representation types of `structured`.

Using this class we can give instances for each user datatype that we want to convert. For example, the list datatype (instantiated in `structured` in Sect. 2.4) can be transported to `generic-deriving` with the following instance:

$$\begin{aligned}
&\textbf{instance}\ Generic_{1D}\ [\,]\ \textbf{where} \\
&\quad Rep_{1D}\ [\,] = S_{\rightarrow}G_1\ (Rep\ [NoPar]) \\
&\quad from_{1D}\ x = s_{\rightarrow}g_1\ (from\ x)
\end{aligned}$$

We use $Rep\ [NoPar]$ because we need to instantiate the list with some parameter. Any parameter will do, because we know that $\forall \phi\ \alpha\ \beta. Rep\ (\phi\ \alpha) \sim Rep\ (\phi\ \beta)$. However, this means that, unlike in Sect. 4.2, we cannot give a single instance of the form $Generic\ (\phi\ \rho) \Rightarrow Generic_{1D}\ \phi$. The reason for this is the disparity between the kinds of the two classes involved; $Generic_{1D}$ only mentions the parameter ρ in the signature of its methods, where it's impossible to state that said ρ is the same as in the instance head ($Generic\ (\phi\ \rho)$).

This is not a major issue, however, because $Generic_{1D}$ instances are currently derived by the compiler. If these instances were to be replaced by conversions from *Generic*, the behaviour of **deriving** $Generic_{1D}$ would change to mean "derive *Generic*, and define a trivial $Generic_{1D}$ instance".

With the instance above, functionality defined in the `generic-deriving` package over the $Generic_{1D}$ class, such as *gmap*, is now available to $[\alpha]$.

5 Conclusion

Following the lines of generic generic programming, we've shown how to add another level of hierarchy to the current landscape of GP libraries in Haskell. Introducing `structured` allows us to unify the two generic views in `generic-deriving`, and brings the possibility of using different nestings in the constructor and constructor arguments encoding. These developments can help in simplifying the implementation of GP in the compiler, as less code has to be part of the compiler itself (only that for generating `structured` instances), and more code can be moved into the user domain. GP library writers also see their life simplified, by gaining access to multiple generic views without needing to duplicate code.

Should `structured` turn out to be not informative enough to cover a particular approach, then it can always be refined or extended. Since we do not advocate to use `structured` directly, this means that only the direct conversions from `structured` have to be extended, and everything else will just keep working. Our hierarchical approach facilitates a future where GP libraries themselves are as modular and duplication-free as the code they enable end users to write.

Acknowledgements. The first author is funded by EPSRC grant number EP/J0 10995/1. We thank the anonymous reviewers for the helpful feedback.

References

Holdermans, S., Jeuring, J., Löh, A., Rodriguez Yakushev, A.: Generic views on data types. In: Uustalu, T. (ed.) MPC 2006. LNCS, vol. 4014, pp. 209–234. Springer, Heidelberg (2006)

Lämmel, R., Peyton Jones, S.: Scrap your boilerplate: a practical design pattern for generic programming. In: Proceedings of the 2003 ACM SIGPLAN International Workshop on Types in Languages Design and Implementation, pp. 26–37. ACM (2003). doi:10.1145/604174.604179

Lämmel, R., Peyton Jones, S.: Scrap more boilerplate: reflection, zips, and generalised casts. In: Proceedings of the 9th ACM SIGPLAN International Conference on Functional Programming, pp. 244–255. ACM (2004). doi:10.1145/1016850.1016883

Magalhães, J.P.: Less Is More: Generic programming theory and practice. Ph.D. thesis, Universiteit Utrecht (2012)

Magalhães, J.P.: Generic programming with multiple parameters. In: Codish, M., Sumii, E. (eds.) FLOPS 2014. LNCS, vol. 8475, pp. 136–151. Springer, Heidelberg (2014)

Magalhães, J.P., Löh, A.: A formal comparison of approaches to datatype-generic programming. In: Chapman, J., Levy, P.B. (eds.) Proceedings Fourth Workshop on Mathematically Structured Functional Programming. Electronic Proceedings in Theoretical Computer Science, vol. 76, pp. 50–67. Open Publishing Association (2012). doi:10.4204/EPTCS.76.6

Magalhães, J.P., Löh, A.: Generic generic programming. In: Flatt, M., Guo, H.-F. (eds.) PADL 2014. LNCS, vol. 8324, pp. 216–231. Springer, Heidelberg (2014)

Magalhães, J.P., Dijkstra, A., Jeuring, J., Löh, A.: A generic deriving mechanism for Haskell. In: Proceedings of the 3rd ACM Haskell Symposium, pp. 37–48. ACM (2010). doi:10.1145/1863523.1863529

Morris, P: Constructing universes for generic programming. Ph.D. thesis, The University of Nottingham, November 2007

van Noort, T., Rodriguez Yakushev, A., Holdermans, S., Jeuring, J., Heeren, B.: A lightweight approach to datatype-generic rewriting. In: Proceedings of the ACM SIGPLAN Workshop on Generic Programming, pp. 13–24. ACM (2008). doi:10.1145/1411318.1411321

Rodriguez Yakushev, A., Holdermans, S., Löh, A., Jeuring, J.: Generic programming with fixed points for mutually recursive datatypes. In: Proceedings of the 14th ACM SIGPLAN International Conference on Functional Programming, pp. 233–244. ACM (2009). doi:10.1145/1596550.1596585

Schrijvers, T., Peyton Jones, P., Chakravarty, M., Sulzmann, M.: Type checking with open type functions. In: Proceedings of the 13th ACM SIGPLAN International Conference on Functional Programming, pp. 51–62. ACM (2008). doi:10.1145/1411204.1411215

Schrijvers, T., Peyton Jones, P., Sulzmann, M., Vytiniotis, D.: Complete and decidable type inference for GADTs. In: Proceedings of the 14th ACM SIGPLAN International Conference on Functional Programming, pp. 341–352. ACM (2009). doi:10.1145/1596550.1596599

Sheard, T., Peyton Jones, S.: Template meta-programming for Haskell. In: Proceedings of the 2002 ACM SIGPLAN Workshop on Haskell, vol. 37, pp. 1–16. ACM, December 2002. doi:10.1145/581690.581691

Yorgey, B.A., Weirich, S., Cretin, J., Peyton Jones, P., Vytiniotis, D., Magalhães, J.P.: Giving Haskell a promotion. In: Proceedings of the 8th ACM SIGPLAN Workshop on Types in Language Design and Implementation, pp. 53–66. ACM (2012). doi:10.1145/2103786.2103795

Polynomial Functors Constrained by Regular Expressions

Dan Piponi[1] and Brent A. Yorgey[2](✉)

[1] A Neighborhood of Infinity, Oakland, CA, USA
dpiponi@gmail.com
[2] Williams College, Williamstown, MA, USA
byorgey@gmail.com

Abstract. We show that every regular language, via some DFA which accepts it, gives rise to a homomorphism from the semiring of polynomial functors to the semiring of $n \times n$ matrices over polynomial functors. Given some polynomial functor and a regular language, this homomorphism can be used to automatically derive a functor whose values have the same shape as those of the original functor, but whose sequences of leaf types correspond to strings in the language.

The primary interest of this result lies in the fact that certain regular languages correspond to previously studied derivative-like operations on polynomial functors, which have proven useful in program construction. For example, the regular language a^*ha^* yields the *derivative* of a polynomial functor, and b^*ha^* its *dissection*. Using our framework, we are able to unify and lend new perspective on this previous work. For example, it turns out that dissection of polynomial functors corresponds to taking *divided differences* of real or complex functions, and, guided by this parallel, we show how to generalize binary dissection to n-ary dissection.

Keywords: Polynomial · Functors · Regular expressions · Differentiation · Dissection

1 Introduction

Consider the standard polymorphic singly-linked list type, which can be defined in Haskell [8] as:

$$\textbf{data}\ List\ a = Nil$$
$$\qquad\qquad\mid\ Cons\ a\ (List\ a)$$

This type is *homogeneous*, meaning that each element in the list has the same type as every other element.

Suppose, however, that we wanted lists with a different constraint on the types of its elements. For example, we might want lists whose elements alternate between two types a and b, beginning with a and ending with b (Fig. 1).

R. Hinze and J. Voigtländer (Eds.): MPC 2015, LNCS 9129, pp. 113–136, 2015.
DOI:10.1007/978-3-319-19797-5_6

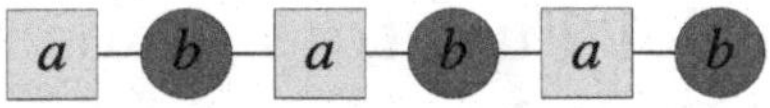

Fig. 1. A list with alternating types

One way to encode such an alternating list is with a pair of mutually recursive types, as follows:

$$
\begin{aligned}
\textbf{data } \mathit{List}_1\ a\ b &= \mathit{Nil}_1 \\
&\mid \mathit{Cons}_1\ a\ (\mathit{List}_2\ a\ b) \\
\textbf{data } \mathit{List}_2\ a\ b &= \mathit{Cons}_2\ b\ (\mathit{List}_1\ a\ b)
\end{aligned}
$$

The required type is $\mathit{List}_1\ a\ b$: a value of type $\mathit{List}_1\ a\ b$ must be either empty (Nil_1) or contain a value of type a, followed by a value of type b, followed recursively by another $\mathit{List}_1\ a\ b$.

In fact, we can think of $\mathit{List}_1\ a\ b$ as containing values whose *shape* corresponds to the original *List* type (that is, there is a natural embedding from $\mathit{List}_1\ a\ a$ into $\mathit{List}\ a$, i.e. an injective polymorphic function $\forall a.\mathit{List}_1\ a\ a \rightarrow \mathit{List}\ a$), but whose sequence of element types corresponds to the *regular expression* $(ab)^*$, that is, any number of repetitions of the sequence ab.

We can easily generalize this idea to regular expressions other than $(ab)^*$ (though constructing the corresponding types may be complicated). We can also generalize to algebraic data types other than *List*, by considering the sequence of element types encountered by a canonical inorder traversal of each data structure [10]. That is, in general, given some algebraic data type and a regular expression, we consider the problem of constructing a corresponding algebraic data type "of the same shape" but with sequences of element types matching the regular expression.

For example, consider the following type *Tree* of nonempty binary trees with data stored in the leaves:

$$
\begin{aligned}
\textbf{data } \mathit{Tree}\ a &= \mathit{Leaf}\ a \\
&\mid \mathit{Fork}\ (\mathit{Tree}\ a)\ (\mathit{Tree}\ a)
\end{aligned}
$$

Consider again the problem of writing down a type whose values have the same shape as values of type $\mathit{Tree}\ a$, but where the data elements alternate between two types a and b, beginning with a leftmost element of type a and ending with a rightmost element of type b. An example can be seen in Fig. 2.

Suppose $\mathit{Tree}_{12}\ a\ b$ is such a type. Values of type $\mathit{Tree}_{12}\ a\ b$ cannot consist solely of a leaf node: there must be at least two elements, one of type a and one of type b. Hence a value of type $\mathit{Tree}_{12}\ a\ b$ must be a fork consisting of two subtrees. There are two ways this could happen. The left subtree could start with a and end with b, in which case the right subtree must also start with a and end with b. Or the left subtree could start with a and end with a, in which case the right subtree must start with b and end with b. So we are led to define

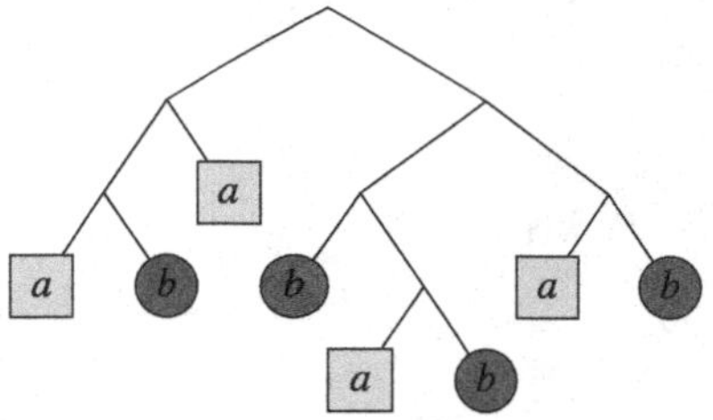

Fig. 2. A tree with alternating leaf types

$$
\begin{aligned}
\textbf{data}\ \mathit{Tree}_{12}\ a\ b \quad = & \quad \mathit{Fork}_{12}\ (\mathit{Tree}_{12}\ a\ b)\ (\mathit{Tree}_{12}\ a\ b) \\
& \mid \mathit{Fork}'_{12}\ (\mathit{Tree}_{11}\ a\ b)\ (\mathit{Tree}_{22}\ a\ b)
\end{aligned}
$$

where $\mathit{Tree}_{11}\ a\ b$ represents alternating trees with left and rightmost elements both of type a, and similarly for Tree_{22}.

Of course, we are now left with the task of defining Tree_{11} and Tree_{22}, but we can carry out similar reasoning: for example a Tree_{11} value can either be a single leaf of type a, or a branch with a Tree_{12} and Tree_{11}, or a Tree_{11} and Tree_{21}. All told, we obtain

$$
\begin{aligned}
\textbf{data}\ \mathit{Tree}_{11}\ a\ b = & \ \mathit{Leaf}_{11}\ a \\
& \mid \mathit{Fork}_{11}\ (\mathit{Tree}_{12}\ a\ b)\ (\mathit{Tree}_{11}\ a\ b) \\
& \mid \mathit{Fork}'_{11}\ (\mathit{Tree}_{11}\ a\ b)\ (\mathit{Tree}_{21}\ a\ b)
\end{aligned}
$$

$$
\begin{aligned}
\textbf{data}\ \mathit{Tree}_{22}\ a\ b = & \ \mathit{Leaf}_{22}\ b \\
& \mid \mathit{Fork}_{22}\ (\mathit{Tree}_{22}\ a\ b)\ (\mathit{Tree}_{12}\ a\ b) \\
& \mid \mathit{Fork}'_{22}\ (\mathit{Tree}_{21}\ a\ b)\ (\mathit{Tree}_{22}\ a\ b)
\end{aligned}
$$

$$
\begin{aligned}
\textbf{data}\ \mathit{Tree}_{21}\ a\ b = & \ \mathit{Fork}_{21}\ (\mathit{Tree}_{22}\ a\ b)\ (\mathit{Tree}_{11}\ a\ b) \\
& \mid \mathit{Fork}'_{21}\ (\mathit{Tree}_{21}\ a\ b)\ (\mathit{Tree}_{21}\ a\ b)
\end{aligned}
$$

Any tree of type $\mathit{Tree}_{12}\ a\ b$ is now constrained to have alternating leaf node types. For example, here are two values of type $\mathit{Tree}_{12}\ \mathit{Int}\ \mathit{Char}$:

$$
\begin{aligned}
& \mathit{ex}_1, \mathit{ex}_2 :: \mathit{Tree}_{12}\ \mathit{Int}\ \mathit{Char} \\
& \mathit{ex}_1 = \mathit{Fork}'_{12}\ (\mathit{Leaf}_{11}\ 1)\ (\mathit{Leaf}_{22}\ \texttt{'a'}) \\
& \mathit{ex}_2 = \mathit{Fork}'_{12}\ (\mathit{Fork}_{11}\ \mathit{ex}_1\ (\mathit{Leaf}_{11}\ 2))\ (\mathit{Leaf}_{22}\ \texttt{'b'})
\end{aligned}
$$

ex_2 can also be seen in pictorial form in Fig. 3.

While this works, the procedure was somewhat *ad hoc*. We reasoned about the properties of the pieces that result when a string matching $(ab)^*$ is split into two substrings, and used this to find corresponding types for the subtrees. One might wonder why we ended up with four mutually recursive types—is there any simpler solution? And how well does this sort of reasoning extend to more complicated structures or regular expressions? Our goal will be to derive a more principled way to do this analysis for any regular language and any suitable (*polynomial*) data type.

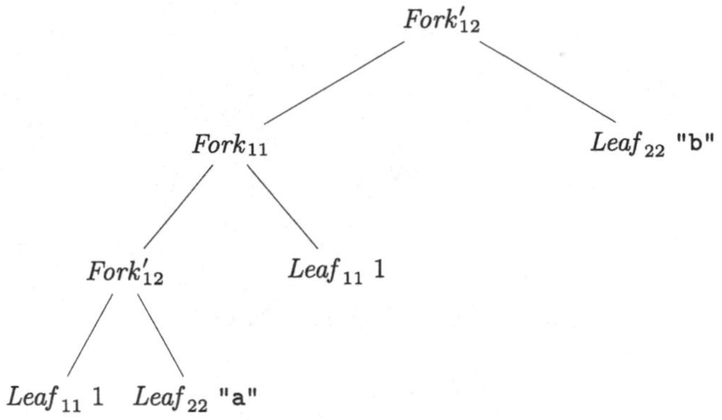

Fig. 3. A tree with alternating leaf types

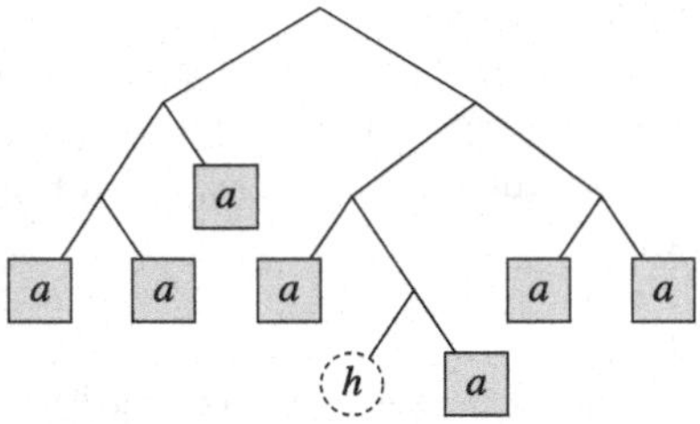

Fig. 4. A tree corresponding to the regular language a^*ha^*

For certain languages, this problem has already been explored in the literature, though without being phrased in terms of regular languages. For example, consider the regular language a^*ha^*. It matches sequences of as with precisely one occurrence of h somewhere in the middle. Data structures whose inorder sequence of element types matches a^*ha^* have all elements of type a, except for one which has type h. This corresponds to a zipper type [5] with elements of type h at the 'focus'; if we substitute the unit type for h, we get the *derivative* of the original type [1] (Fig. 4). Likewise, the regular language b^*ha^* corresponds to *dissection types* [9].

Zippers, derivatives and dissections are usually described using Leibniz rules and their generalizations. We'll show how these rules can be placed in a more general framework applying to any regular language.

In the remainder of the paper, we first review some standard results about regular languages and DFAs (Sect. 2). We describe our framework informally (Sect. 3) and give some examples of its application (Sect. 4) and describe an alternative encoding which can be more convenient in practice (Sect. 5). We conclude with a discussion of derivatives (Sect. 6) and divided differences (Sect. 7).

2 Regular Expressions and DFAs

We begin with a review of the basic theory of regular languages and deterministic finite automata in Sects. 2.1–2.3. Readers already familiar with this theory may safely skip these sections. In Sect. 2.4 we introduce some preliminary material on star semirings which, though not novel, may not be as familiar to readers.

2.1 Regular Expressions

A *regular expression* [6] over an alphabet Σ is a term of the following grammar:

$$R ::= \bullet \mid \varepsilon \mid a \in \Sigma \mid R + R \mid RR \mid R^* \tag{1}$$

When writing regular expressions, we allow parentheses for disambiguation, and adopt the common convention that Kleene star (R^*) has higher precedence than concatenation (RR), which has higher precedence than alternation ($R+R$).

Semantically, we can interpret each regular expression R as a set of strings $[\![R]\!] \subseteq \Sigma^*$, where Σ^* denotes the set of all finite sequences built from elements of Σ. In particular,

- $[\![\bullet]\!] = \varnothing$ denotes the empty set.
- $[\![\varepsilon]\!] = \{\varepsilon\}$ denotes the singleton set containing the empty string.
- $[\![a]\!] = \{a\}$ denotes the singleton set containing the length-1 sequence a.
- $[\![R_1 + R_2]\!] = [\![R_1]\!] \cup [\![R_2]\!]$.
- $[\![R_1 R_2]\!] = [\![R_1]\!]\,[\![R_2]\!]$, where $L_1 L_2$ denotes pairwise concatenation of sets,

$$L_1 L_2 = \{s_1 s_2 \mid s_1 \in L_1, s_2 \in L_2\}.$$

- $[\![R^*]\!] = [\![R]\!]^*$, where L^* denotes the least fixed point solution of

$$L^* = \{\varepsilon\} \cup LL^*.$$

Note that such a least fixed point must exist by the Knaster-Tarski theorem [18], since the mapping $\varphi(S) = \{\varepsilon\} \cup LS$ is monotone, that is, if $S \subseteq T$ then $\varphi(S) \subseteq \varphi(T)$.

Finally, a *regular language* over the alphabet Σ is a set $L \subseteq \Sigma^*$ which is the interpretation $L = [\![R]\!]$ of some regular expression R.

2.2 DFAs

A *deterministic finite automaton* (DFA) is a quintuple $(Q, \Sigma, \delta, q_0, \mathcal{A})$ consisting of

- a nonempty set of states Q,
- a set of input symbols Σ,
- a *transition function* $\delta : Q \times \Sigma \to Q$,
- a distinguished *start state* $q_0 \in Q$, and
- a set $\mathcal{A} \subseteq Q$ of *accept states*. (One often sees F used to represent the set of accept or "final" states, but this would conflict with our use of F to represent functors later.)

We can "run" a DFA on an input string by feeding it symbols from the string one by one. When encountering the symbol s in state q, the DFA changes to state $\delta(q, s)$. If a DFA beginning in its start state q_0 ends in state q' after being fed a string in this way, we say the DFA *accepts* the string if $q' \in \mathcal{A}$, and *rejects* the string otherwise. Thus, a DFA D can be seen as defining a subset $L_D \subseteq \Sigma^*$ of the set of all possible strings, namely, those strings which it accepts.

We can draw a DFA as a directed multigraph where each graph edge is labeled by a symbol from Σ. Each state is a vertex, and an edge is drawn from q_1 to q_2 and labeled with symbol s whenever $\delta(q_1, s) = q_2$. In addition, we indicate accept states with a double circle, and always label the start state as 1. We can think of the state of the DFA as "walking" through the graph each time it receives an input. Figure 5 shows an example.

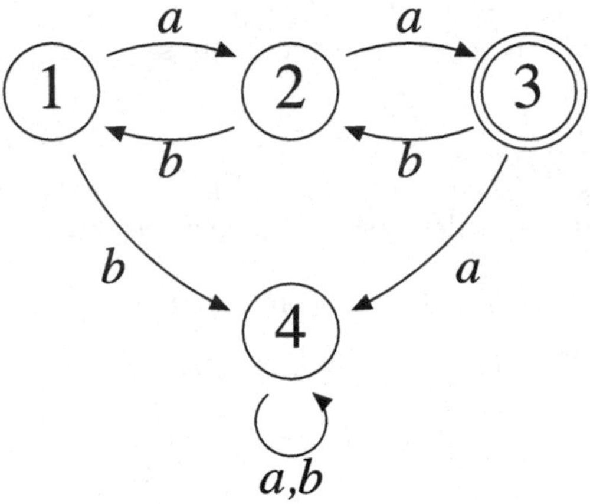

Fig. 5. An example DFA

It is convenient to allow the transition function δ to be partial. Operationally, encountering a state q and input s for which $\delta(q, s)$ is undefined corresponds to the DFA *rejecting* its input. This often simplifies matters, since we may omit "sink states" from which there is no path to any accepting state, making δ undefined whenever it would have otherwise yielded such a sink state. For example, the DFA from Fig. 5 may be simplified to the one shown in Fig. 6, by dropping state 4.

As is standard, we may define $\delta^* : Q \times \Sigma^* \rightharpoonup Q$ as an iterated version of δ:

$$\delta^*(q, \varepsilon) = q \tag{2}$$

$$\delta^*(q, s\omega) = \delta^*(\delta(q, s), \omega) \tag{3}$$

If $\delta^*(q_0, \omega) = q_1$, then we say that the string ω "takes" or "drives" the DFA from state q_0 to state q_1. More generally, given a string ω, we can partially apply δ^* to obtain a "driving function" $\chi : Q \rightharpoonup Q$ which encodes how the string ω drives the DFA: if the DFA starts in state q then after processing ω it will either halt with an error or end in state $\chi(q)$.

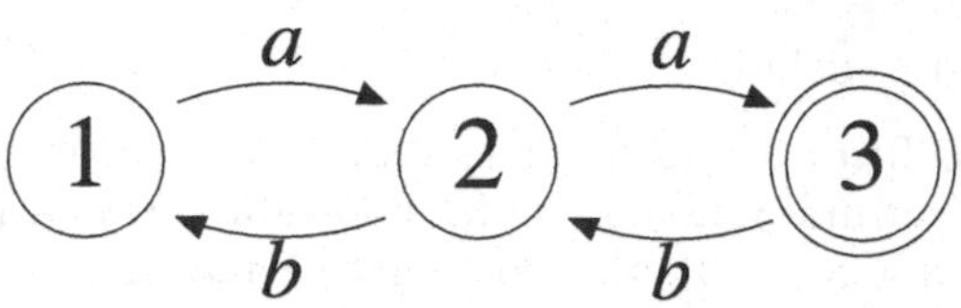

Fig. 6. Example DFA, simplified

2.3 Kleene's Theorem

Connecting the previous two sections is *Kleene's Theorem*, which says that the theory of regular expressions and the theory of DFAs are really "about the same thing". In particular, the set of strings accepted by a DFA is always a regular language, and conversely, for every regular language there exists a DFA which accepts it. Moreover, the proof of the theorem is constructive: given a regular expression, we may algorithmically construct a corresponding DFA, and vice versa. For example, the regular expression b^*ha^* corresponds to the DFA shown in Fig. 7. It is not hard to verify that strings taking the DFA from state 1 to state 2 (the accept state) are precisely those matching the regular expression b^*ha^*.

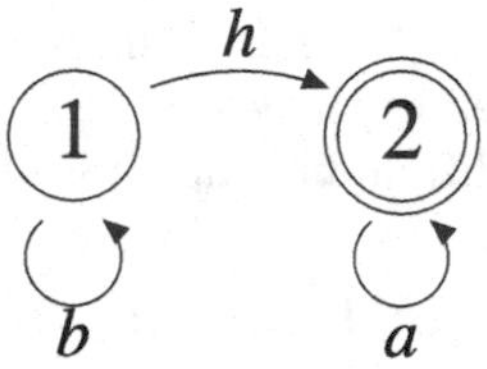

Fig. 7. A DFA for b^*ha^*

The precise details of these constructions are not important for the purposes of this paper; interested readers should consult a reference such as Sipser [15]. We note in passing that one can also associate *nondeterministic* finite automata (NFAs) to regular expressions, and the remainder of the story of this paper could probably be retold using NFAs. However, it is not clear whether we would gain any benefit from making this generalization, so we will stick with the simpler notion of DFAs.

2.4 Semirings

A *semiring* is a set R equipped with two binary operations, $+$ and $\cdot$, and two distinguished elements, $0, 1 \in R$, such that

- $(+, 0)$ is a commutative monoid (that is, 0 is an identity for $+$, and $+$ is commutative and associative),
- $(\cdot, 1)$ is a monoid,
- $\cdot$ distributes over $+$ from both the left and the right, that is, $a \cdot (b+c) = a \cdot b + a \cdot c$ and $(b+c) \cdot a = b \cdot a + c \cdot a$, and
- 0 is an annihilator for $\cdot$, that is $r \cdot 0 = 0 \cdot r = 0$ for all $r \in R$.

Examples of semirings include:

- $(\mathit{Bool}, \vee, \mathit{False}, \wedge, \mathit{True})$, boolean values under disjunction and conjunction;
- $(\mathbb{N}, +, 0, \cdot, 1)$, the natural numbers under addition and multiplication;
- $(\mathbb{R}^+ \cup \{-\infty\}, \max, -\infty, +, 0)$, the nonnegative real numbers (adjoined with $-\infty$) under maximum and addition;
- the set of regular languages forms a semiring under the operations of union and pairwise concatenation, with $0 = \varnothing$ and $1 = \{\varepsilon\}$.

A *star semiring* or *closed semiring* [7] has an additional operation, $(-)^*$, satisfying the axiom

$$r^* = 1 + r \cdot r^* = 1 + r^* \cdot r, \tag{4}$$

for all $r \in R$. Intuitively, $r^* = 1 + r + r^2 + r^3 + \ldots$ (although such infinite sums do not necessarily make sense in all semirings). The semiring of regular languages is closed, via Kleene star.[1]

If R is a semiring, then the set of $n \times n$ matrices with elements in R is also a semiring, where matrix addition and multiplication are defined in the usual manner in terms of addition and multiplication in R. If R is a star semiring, then a star operator can also be defined for matrices; for details see Lehmann [7] and Dolan [2].

Finally, a *semiring homomorphism* is a mapping from the elements of one semiring to another that preserves the semiring structure, that is, that sends 0 to 0, 1 to 1, and preserves addition and multiplication.

3 DFAs and Matrices of Functors

Viewing regular expressions through the lens of DFAs gives us exactly the tools we need to generalize our *ad hoc* analysis from the introduction.

3.1 A More Principled Derivation

Consider again the task of encoding a type with the same shape as

```
data Tree a = Leaf a
            | Fork (Tree a) (Tree a)
```

whose sequence of element types matches the regular expression $(ab)^*$, as in the introduction. This time, however, we will think about it from the point of view of the corresponding DFA, shown in Fig. 8.

The key is to consider not just the data type we are ultimately interested in—in this case, those trees which take the DFA from state 1 to itself—but an entire family of related types. In particular, let $T_{ij}\ a\ b$ denote the type of binary trees whose element type sequences take the DFA from state i to state j. Since the DFA has two states, there are four such types:

[1] In fact, regular languages (and several of the other examples above) form *Kleene algebras*, which essentially add to a star semiring the requirement that $+$ is idempotent ($a + a = a$). However, for our purposes we do not need the extra restriction.

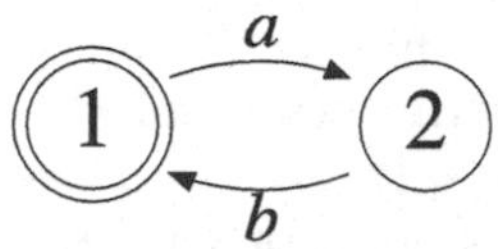

Fig. 8. A DFA for $(ab)^*$

- $T_{11}\ a\ b$ — this is the type of trees we are primarily interested in constructing, whose leaf sequences match $(ab)^*$.
- $T_{12}\ a\ b$ — trees of this type have leaf sequences which take the DFA from state 1 to state 2; that is, they match the regular expression $a(ba)^*$ (or, equivalently, $(ab)^*a$).
- $T_{21}\ a\ b$ — trees matching $b(ab)^*$.
- $T_{22}\ a\ b$ — trees matching $(ba)^*$.

What does a tree of type T_{11} look like? It cannot be a leaf, because a single leaf takes the DFA from state 1 to 2 or vice versa. It must be a pair of trees, which together take the DFA from state 1 to state 1. There are two ways for that to happen: both trees could themselves begin and end in state 1; or the first tree could take the DFA from state 1 to state 2, and the second from state 2 to state 1. We can carry out a similar analysis for the other three types. In fact, we have already carried out this exact analysis in the introduction, but it is now a bit less ad hoc. In particular, we can now see that we end up with four mutually recursive types precisely because the DFA for $(ab)^*$ has two states, and we need one type for each ordered pair of states.

In general, given a DFA with states Q and alphabet $\Sigma = \{a_1, \ldots, a_n\}$, we get a mutually recursive family of types

$$T_{ij}\ a_1\ \ldots\ a_n$$

indexed by a pair of states from Q and by one type argument for each alphabet symbol. We are ultimately interested in types of the form $T_{q_0 k}$ where $k \in \mathcal{A}$, that is, types which are indexed by the start state and some accept state of the DFA.

Though shifting our point of view to DFAs has given us a better framework for determining which types we must define, we still had to reason on a case-by-case basis to determine the definitions of these types. The next two sections show how we can concisely and elegantly formalize this process in terms of *matrices*.

3.2 Polynomial Functors

We now abstract away from the particular details of Haskell data types and work in terms of a simple language of *polynomial functors*. We inductively define the universe **Fun** of polynomial functors as follows, simultaneously giving both syntax and semantics.

- $K_A \in$ **Fun** denotes the constant functor $K_A\ a = A$, which ignores its argument and yields A.

- $X \in \mathbf{Fun}$ denotes the identity functor $X\ a = a$.
- Given $F, G \in \mathbf{Fun}$, we can form their sum, $F + G \in \mathbf{Fun}$, with $(F + G)\ a = F\ a + G\ a$.
- We can also form products of functors, $(F \times G)\ a = F\ a \times G\ a$. We often abbreviate $F \times G$ as FG.
- Finally, we allow functors to be defined by mutually recursive systems of equations

$$\begin{cases} F_1 = \Phi_1(F_1, \dots, F_n) \\ \vdots \\ F_n = \Phi_n(F_1, \dots, F_n), \end{cases}$$

where each Φ_k is a polynomial functor expression with free variables in $\{F_1, \dots, F_n\}$, and interpret them using a standard least fixed point semantics. For example, the single recursive equation $L = 1 + X \times L$ denotes the standard type of (finite) polymorphic lists. As another example, the pair of mutually recursive equations

$$E = K_{Unit} + X \times O$$
$$O = X \times E$$

defines the types of even- and odd-length polymorphic lists. Here, *Unit* denotes the unit type with a single inhabitant.

It is worth pointing out that functors form a semiring (up to isomorphism) under $+$ and $\times$, where $1 = K_{Unit}$ and $0 = K_{Void}$ (*Void* denotes the type with no inhabitants). We therefore will simply write 0 and 1 in place of K_{Unit} and K_{Void}. In fact, functors also form a star semiring, with the polymorphic list type playing the role of the star operator, that is, $F^* = 1 + F \times F^*$.

The above language also generalizes naturally from unary to n-ary functors. We write $\mathbf{Fun}_n$ for the universe of n-ary polynomial functors, so $\mathbf{Fun} = \mathbf{Fun}_1$.

- $K_A\ a_1\ \dots\ a_n = A$.
- The identity functor X generalizes to the family of projections X_m, where

$$X_m\ a_1\ \dots\ a_n = a_m.$$

That is, X_m is the functor which yields its mth argument, and may be regarded as an n-ary functor for any $n \geqslant m$. More generally, the arguments to a functor can be labeled by the elements of some alphabet Σ, instead of being numbered positionally, and we write $\mathbf{Fun}_\Sigma$ for the universe of such functors. In that case, for $a \in \Sigma$ we write X_a for the projection which picks out the argument labeled by a.
- $(F + G)\ a_1\ \dots\ a_n = (F\ a_1\ \dots\ a_n) + (G\ a_1\ \dots\ a_n)$.
- $(F \times G)\ a_1\ \dots\ a_n = (F\ a_1\ \dots\ a_n) \times (G\ a_1\ \dots\ a_n)$.
- Recursion also generalizes straightforwardly.

Of course, n-ary functors also form a semiring for any n.

As an example, the Haskell type

$$\mathbf{data}\ S\ a\ b = \mathit{Apple}\ a \mid \mathit{Banana}\ b \mid \mathit{Fork}\ (S\ a\ b)\ (S\ a\ b)$$

corresponds to the bifunctor (that is, 2-ary functor) $S = X_a + X_b + S \times S$; we may also abbreviate $S \times S$ as S^2.

By induction over functor descriptions, we may define $\mathcal{S} : \mathbf{Fun}_\Sigma \to \mathcal{P}(\Sigma^*)$ which gives the sequences of leaf types that can occur in the values of a given functor. Thinking of values of a given functor as trees, $\mathcal{S}(-)$ corresponds to an inorder traversal. That is:

$$\begin{aligned}
\mathcal{S}(0) &= \varnothing \\
\mathcal{S}(K_A) &= \{\varepsilon\} \quad (A \neq \mathit{Void}) \\
\mathcal{S}(X_a) &= \{a\} \\
\mathcal{S}(F + G) &= \mathcal{S}(F) \cup \mathcal{S}(G) \\
\mathcal{S}(F \times G) &= \mathcal{S}(F)\mathcal{S}(G)
\end{aligned}$$

Finally, given a system $F_m = \Phi_m(F_1, \dots, F_n)$ we simply set

$$\mathcal{S}(F_m) = \mathcal{S}(\Phi_m(F_1, \dots, F_n))$$

for each m, and take the least fixed point (ordering sets by inclusion). For example, given the list functor $L = 1 + XL$, we obtain

$$\mathcal{S}(L) = \{\varepsilon\} \cup \{1\sigma \mid \sigma \in \mathcal{S}(L)\}$$

whose least fixed point is the infinite set $\{\varepsilon, 1, 11, 111, \dots\}$ as expected.

3.3 Matrices of Functors

Now suppose we have a unary functor F and some DFA $D = (Q, \Sigma, \delta, q_0, \mathcal{A})$. Let $F_{ij} \in \mathbf{Fun}_\Sigma$ denote the type with the same shape as F but whose sequences of leaf types take D from state i to state j. We are ultimately interested in constructing

$$\sum_{k \in \mathcal{A}} T_{q_0 k},$$

the sum of all types T_{ij} whose leaf sequences start in state q_0 and tahe the DFA to some accept state. Note that F_{ij} has arity Σ, that is, there is a leaf type corresponding to each alphabet symbol of D. We can deduce F_{ij} compositionally, by recursion on the syntax of functor expressions.

- The constant functor K_A creates structures containing no elements, *i.e.* which do not cause the DFA to transition at all. So the only way a K_A-structure can take the DFA from state i to state j is if $i = j$:

$$(K_A)_{ij} = \begin{cases} K_A & i = j \\ 0 & i \neq j \end{cases} \tag{5}$$

As a special case, the functor $1 = K_{Unit}$ yields

$$1_{ij} = \begin{cases} 1 & i = j \\ 0 & i \neq j \end{cases}. \tag{6}$$

- A value with shape $F+G$ is either a value with shape F or a value with shape G; so the set of $F+G$ shapes taking the DFA from state i to state j is the disjoint sum of the corresponding F and G shapes:

$$(F+G)_{ij} = F_{ij} + G_{ij}. \tag{7}$$

- Products are more interesting. An FG-structure consists of an F-structure paired with a G-structure, whose leaf types drive the DFA in sequence. Hence, in order to take the DFA from state i to state j overall, the F-structure must take the DFA from state i to some state k, and then the G-structure must take it from k to j. This works for any state k, so $(FG)_{ij}$ is the sum over all such possibilities. Thus,

$$(FG)_{ij} = \sum_{k \in Q} F_{ik} G_{kj}. \tag{8}$$

- Finally, for a recursive system of functors

$$\overline{F_m} = \overline{\Phi_m(F_1, \ldots, F_n)},$$

  we may mutually define

$$(F_m)_{ij} = (\Phi_m(F_1, \ldots, F_n))_{ij},$$

  interpreted via the same least fixed point semantics.

The above rules for 1, sums, and products might look familiar: in fact, they are just the definitions of the identity matrix, matrix addition, and matrix product. That is, given some functor F and DFA D, we can arrange all the F_{ij} in a matrix, $[F]_D$, whose (i,j)th entry is F_{ij}. (We also write simply $[F]$ when D can be inferred.) Then we can rephrase (6)–(8) above as

- $[1]_D = I_{|\Sigma|}$, that is, the $|\Sigma| \times |\Sigma|$ identity matrix, with ones along the main diagonal and zeros everywhere else;
- $[F+G]_D = [F]_D + [G]_D$; and
- $[FG]_D = [F]_D [G]_D$.

So far, given a DFA D, we have the makings of a homomorphism from the semiring of arity-1 functors to the semiring of $|Q| \times |Q|$ matrices of arity-Σ functors. However, there is still some unfinished business, namely, the interpretation of $[X]_D$. This gets at the heart of the matter, and to understand it, we must take a slight detour.

3.4 Transition Matrices

Given a simple directed graph G with n nodes, its *adjacency matrix* is an $n \times n$ matrix M_G with a 1 in the i, j position if there is an edge from node i to node j, and a 0 otherwise. It is a standard observation that the mth power of M_G encodes information about length-m paths in G; specifically, the i, j entry of M_G^m is the number of distinct paths of length m from i to j. This is because a path from i to j of length m is the concatenation of a length-$(m-1)$ path from i to some k followed by an edge from k to j, so the total number of length-m paths is the sum of such paths over all possible k; this is exactly what is computed by the matrix multiplication $M_G^{m-1}M = M_G^m$.

However, as observed independently by O'Connor [11] and Dolan [2], and as is standard weighted automata theory [3], this can be generalized by parameterizing the construction over an arbitrary semiring. In particular, we may suppose that the edges of G are labeled by elements of some semiring R, and form the adjacency matrix M_G as before, but using the labels on edges, and $0 \in R$ for missing edges. The mth power of M_G still encodes information about length-m paths, but the interpretation depends on the specific choice of R and on the edge labeling. Choosing the semiring $(\mathbb{N}, +, \cdot)$ with all edges labeled by 1 gives us a count of distinct paths, as before. If we choose $(\mathit{Bool}, \vee, \wedge)$ and label each edge with *True*, the i, j entry of M_G^m tells us whether there exists any path of length m from i to j. Choosing $(\mathbb{R}, \min, +)$ and labeling edges with costs yields the minimum cost of length-m paths; choosing $(\mathcal{P}(\Sigma^*), \cup, \cdot)$ (that is, languages over some alphabet Σ under union and pairwise concatenation) and labeling edges with elements from Σ yields sets of words corresponding to length-m paths.

Moreover, if R is a star semiring, then M_G^* encodes information about paths of *any* length (recall that, intuitively, $M_G^* = I + M_G + M_G^2 + M_G^3 + \dots$). Choosing $R = (\mathbb{R}, \min, +)$ and computing M_G^* thus solves the all-pairs shortest paths problem; $(\mathit{Bool}, \vee, \wedge)$ tells us whether any paths exist between each pair of nodes; and so on. Note that $(\mathbb{N}, +, \cdot)$ is not closed, but we can make it so by adjoining $+\infty$; this corresponds to the observation that the number of distinct paths between a pair of nodes in a graph may be infinite if the graph contains any cycles.

Of course, DFAs can also be thought of as graphs. Suppose we have a DFA D, a semiring R, and a function $\Sigma \to R$ assigning an element of R to each alphabet symbol. In this context, we call the adjacency matrix for D a *transition matrix*.[2] The graph of a DFA may not be simple, that is, there may be multiple edges in a DFA between a given pair of nodes, each corresponding to a different alphabet symbol. We can handle this by summing in R. That is, the transition matrix M_D is the $|Q| \times |Q|$ matrix over R whose component at i, j is the sum, over all edges from i to j, of the R-values corresponding to their labels.

For example, consider the DFA in Fig. 6, and the semiring $(\mathbb{N}, +, \cdot)$. If we send each edge label (i.e. alphabet symbol) to 1, we obtain the transition matrix

[2] Textbooks on automata often define the *transition matrix* for a DFA as the $|Q| \times |\Sigma|$ matrix with its q, s entry equal to $\delta(q, s)$. This is just a particular representation of the function δ, and quite uninteresting, so we co-opt the term *transition matrix* to refer to something more worthwhile.

$$\begin{bmatrix} 0 & 1 & 0 \\ 1 & 0 & 1 \\ 0 & 1 & 0 \end{bmatrix}.$$

The mth power of this matrix tells us how many strings of length m take the DFA from one given state to another. If we instead send each edge label to the singleton language containing only that symbol as a length-1 string, as a member of the semiring of regular languages, we obtain the transition matrix

$$\begin{bmatrix} \varnothing & \{a\} & \varnothing \\ \{b\} & \varnothing & \{a\} \\ \varnothing & \{b\} & \varnothing \end{bmatrix}.$$

The star of this matrix yields the complete set of strings that drives the DFA between each pair of states.

We can now see how to interpret $[X]_D$: it is the transition matrix for D, taken over the semiring of arity-Σ functors, where each transition a is replaced by the functor X_a. That is, in general, each entry of $[X]_D$ will consist of a (possibly empty) sum of functors

$$\sum_{\substack{a \in \Sigma \\ \delta(i,a)=j}} X_a.$$

By definition, these will drive the DFA in the proper way; moreover, sums of X_a are the only functors with the same shape as X.

4 Examples

To make things more concrete, we can revisit some familiar examples using our new framework. As a first example, consider the regular expression $(aa)^*$, corresponding to the DFA shown in Fig. 9, along with the standard polymorphic list type, $L = 1 + XL$. The matrix for L can be written

$$[L] = \begin{bmatrix} L_{11} & L_{12} \\ L_{21} & L_{22} \end{bmatrix}.$$

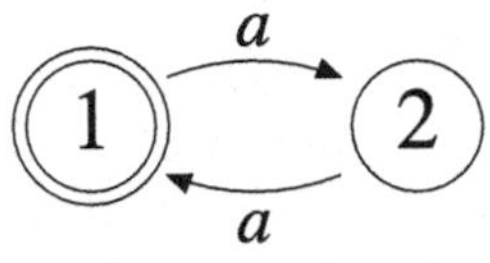

Fig. 9. A DFA for $(aa)^*$

The punchline is that we can take the recursive equation for L and simply apply the homomorphism to both sides, resulting in the matrix equation

$$[L] = [1 + XL] = [1] + [X]\,[L],$$

where $[X]$ is the transition matrix for D, namely

$$[X] = \begin{bmatrix} 0 & X_a \\ X_a & 0 \end{bmatrix}.$$

Expanding out this matrix equation and performing the indicated matrix operations yields

$$\begin{aligned}\begin{bmatrix} L_{11} & L_{12} \\ L_{21} & L_{22} \end{bmatrix} &= \begin{bmatrix} 1 & 0 \\ 0 & 1 \end{bmatrix} + \begin{bmatrix} 0 & X_a \\ X_a & 0 \end{bmatrix} \begin{bmatrix} L_{11} & L_{12} \\ L_{21} & L_{22} \end{bmatrix} \\ &= \begin{bmatrix} 1 + X_a L_{21} & X_a L_{22} \\ X_a L_{11} & 1 + X_a L_{12} \end{bmatrix}.\end{aligned}$$

We can see that L_{11} and L_{22} are isomorphic, as are L_{12} and L_{21}; this is because the DFA D has a nontrivial automorphism (ignoring start and accept states). Thinking about the meaning of paths through the DFA, we see that L_{11} is the type of lists with even length, and L_{21}, lists with odd length. More familiarly:

data *EvenList a* = *EvenNil* | *EvenCons a* (*OddList a*)
data *OddList a* = *OddCons a* (*EvenList a*)

As another example, consider again the recursive tree type given by $T = X + T^2$, along with the two-state DFA for $(ab)^*$ shown in Fig. 8. Applying the homomorphism, we obtain

$$[T] = [X + T^2] = [X] + [T]^2,$$

where

$$[X] = \begin{bmatrix} 0 & X_a \\ X_b & 0 \end{bmatrix}.$$

This yields

$$\begin{aligned}\begin{bmatrix} T_{11} & T_{12} \\ T_{21} & T_{22} \end{bmatrix} &= \begin{bmatrix} 0 & X_a \\ X_b & 0 \end{bmatrix} + \begin{bmatrix} T_{11} & T_{12} \\ T_{21} & T_{22} \end{bmatrix}^2 \\ &= \begin{bmatrix} T_{11}^2 + T_{12} T_{21} & X_a + T_{11} T_{12} + T_{12} T_{22} \\ X_b + T_{21} T_{11} + T_{22} T_{21} & T_{21} T_{12} + T_{22}^2 \end{bmatrix}.\end{aligned}$$

Equating the left- and right-hand sides elementwise yields precisely the definitions for T_{ij} we derived in Sect. 1.

As a final example, consider the type $T = X + T^2$ again, but this time constrained by the regular expression $b^* h a^*$, with transition matrix $\begin{bmatrix} X_b & X_h \\ 0 & X_a \end{bmatrix}$. Applying the homomorphism yields

$$\begin{aligned}\begin{bmatrix} T_{11} & T_{12} \\ T_{21} & T_{22} \end{bmatrix} &= \begin{bmatrix} X_b & X_h \\ 0 & X_a \end{bmatrix} + \begin{bmatrix} T_{11} & T_{12} \\ T_{21} & T_{22} \end{bmatrix}^2 \\ &= \begin{bmatrix} X_b + T_{11}^2 + T_{12} T_{21} & X_h + T_{11} T_{12} + T_{12} T_{22} \\ T_{21} T_{11} + T_{22} T_{21} & X_a + T_{21} T_{12} + T_{22}^2 \end{bmatrix}.\end{aligned}$$

Here something strange happens: looking at the DFA, it is plain that there are no paths from state 2 to state 1, and we therefore expect the corresponding type T_{21} to be empty. However, it does not look empty at first sight: we have $T_{21} = T_{21}\,T_{11} + T_{22}\,T_{21}$. In fact, it *is* empty, since we are interpreting recursively defined functors via a least fixed point semantics, and it is not hard to see that 0 is in fact a fixed point of the above equation for T_{21}. In practice, we can perform a reachability analysis for a DFA beforehand (e.g. by taking the star of its transition matrix under $(\mathit{Bool}, \vee, \wedge)$) to see which states are reachable from which other states; if there is no path from i to j then we know $T_{ij} = 0$, which can simplify calculations. For example, substituting $T_{21} = 0$ into the above equation and simplifying yields

$$\begin{bmatrix} T_{11} & T_{12} \\ T_{21} & T_{22} \end{bmatrix} = \begin{bmatrix} X_b + T_{11}^2 & X_h + T_{11}\,T_{12} + T_{12}\,T_{22} \\ 0 & X_a + T_{22}^2 \end{bmatrix}.$$

5 An Alternative Representation

One way to look at the examples shown so far is that we have essentially had to *duplicate* the initial functor F, resulting in several slightly different copies, each with a slightly different set of "constructors", in order to keep track of which constructors are allowed at which points. Such encodings would be extremely trying to work with in practice, requiring much tedious case analysis. In a language with a sufficiently expressive type system, however, we do not need to duplicate anything, but can instead make use of types to dictate which constructors are allowed in which situations.

What information, exactly, do we need to keep around at the level of types? It is not enough to just index by a pair of DFA states; the problem is that each constructor may correspond to *multiple* possible pairs of states. In fact, what we need is to index by an entire *driving function*. Given some functor T, the idea is to produce just a *single* n-ary functor T_χ indexed by a (total!) driving function $\chi : Q \to Q$. A value of type T_χ is a structure with a shape allowed by T, whose sequence of leaf types, taken together, drives the DFA in the way encoded by χ. The desired type can then be selected as the sum of all types indexed by driving functions taking the start state to some accepting state.

For details of this encoding, see Yorgey [19]. Encoding driving functions and their composition requires only natural numbers and lists, so they can be encoded in any language which allows encoding these at the level of types.

The above approach requires indexing by *total* driving functions. As pointed out by an anonymous reviewer, one can also index by *relations* which can encode partial driving functions. For example, considering again the DFA for b^*ha^* shown in Fig. 7, and the tree type $T = X + T^2$, we have the following Haskell code. *States* encodes the states of the DFA, and *Trans* encodes a relation on states, with each constructor corresponding to an edge in the DFA. The original *Tree a* type is transformed into *Tree′*, where the *Leaf* constructor is parameterized by a transition, and the *Fork* constructor encodes a sum via existential quantification of k. *Tree′* could also be parameterized over an arbitrary relation

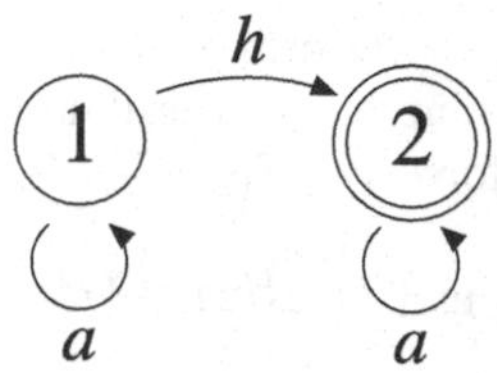

Fig. 10. A DFA for derivatives

of the appropriate kind, which allows constructing *Tree* variants constrained by any DFA over an alphabet of size 3.

```
{-# LANGUAGE DataKinds, GADTs, KindSignatures, PolyKinds #-}
data States = S₁ | S₂
data Trans b h a :: State → State → * where
  B :: b → Trans b h a S₁ S₁
  H :: h → Trans b h a S₁ S₂
  A :: a → Trans b h a S₂ S₂
data Tree' :: * → * → * → State → State → * where
  Leaf :: Trans b h a i j → Tree' r b h a i j
  Fork :: Tree' r b h a i k → Tree' r b h a k j → Tree' r b h a i j
```

6 Derivatives, Again

Now that we have seen the general framework, let's return to the specific application of computing *derivatives* of data types. In order to compute a derivative, we need the DFA for the regular expression a^*ha^*, shown in Fig. 10. The corresponding transition matrix is

$$[X] = \begin{bmatrix} X_a & X_h \\ 0 & X_a \end{bmatrix}.$$

Suppose we start with a functor defined as a product:

$$F = GH$$

Expanding via the homomorphism to matrices of bifunctors, we obtain

$$\begin{bmatrix} F_{11} & F_{12} \\ 0 & F_{22} \end{bmatrix} = \begin{bmatrix} G_{11} & G_{12} \\ 0 & G_{22} \end{bmatrix} \begin{bmatrix} H_{11} & H_{12} \\ 0 & H_{22} \end{bmatrix}$$

(the occurrences of 0 correspond to the observation that there are no paths in the DFA from state 2 to state 1). Let's consider each of the nonzero F_{ij} in turn. First, we have

$$F_{11} = G_{11} \times H_{11}$$

F_{11} is simply the type of structures whose leaves take the DFA from state 1 to itself and so whose leaves match the regular expression a^*; hence we have $F_{11}\ a\ h \cong F\ a$ (and similarly for G_{11}, H_{11}, F_{22}, G_{22}, and H_{22}). We also have

$$F_{12} = G_{11}H_{12} + G_{12}H_{22} \cong GH_{12} + G_{12}H.$$

This looks suspiciously like the usual Leibniz law for the derivative of a product (i.e. the "product rule" for differentiation). We also know that

$$1_{12} = 0$$

and

$$X_{12} = X_h,$$

and if $F = G+H$ then $F_{12} = G_{12}+H_{12}$. If we substitute the unit type for h, these are precisely the rules for differentiating polynomials. So F_{12} is the derivative of F.

There is another way to look at this. Write

$$[X] = \begin{bmatrix} X_a & X_h \\ 0 & X_a \end{bmatrix} = X_a I + d$$

where

$$d = \begin{bmatrix} 0 & X_h \\ 0 & 0 \end{bmatrix}$$

Note that $d^2 = 0$. Note also that

$$(X_a I)d = \begin{bmatrix} 0 & X_a X_h \\ 0 & 0 \end{bmatrix}$$

and

$$d(X_a I) = \begin{bmatrix} 0 & X_h X_a \\ 0 & 0 \end{bmatrix}.$$

Treating the product of functors as commutative is problematic in our setting, since we care about the precise sequence of leaf types. However, in this particular instance, we can specify that X_h commutes with everything, which corresponds to letting the "hole" of type h "float" to the outside—typically, when constructing a zipper structure, one does this anyway, storing the focused element separately from the rest of the structure. Under this interpretation, then, $(X_a I)$ and d commute even though matrix multiplication is not commutative in general. We then note that

$$(X_a I + d)^n = (X_a I)^n + n(X_a I)^{n-1} d,$$

making use of this special commutativity and the fact that $d^2 = 0$, annihilating all the subsequent terms. We can linearly extend this to an entire polynomial f, that is,

$$\begin{aligned} f([X]) &= f(X_a I + d) \\ &= f(X_a I) + f'(X_a I)d \\ &= \begin{bmatrix} f(X_a) & 0 \\ 0 & f(X_a) \end{bmatrix} + \begin{bmatrix} 0 & f'(X_a)X_h \\ 0 & 0 \end{bmatrix} \end{aligned}$$

The matrix d is thus playing a role similar to an "infinitesimal" in calculus, where the expression dx is manipulated informally as if $(dx)^2 = 0$. (Compare with the dual numbers described by [16].)

7 Dissection and Divided Differences

Consider again the regular expression b^*ha^*. Data structures with leaf sequences matching this pattern have a "hole" of type h, with values of type b to the left of the hole and values of type a to the right (Fig. 11).[3]

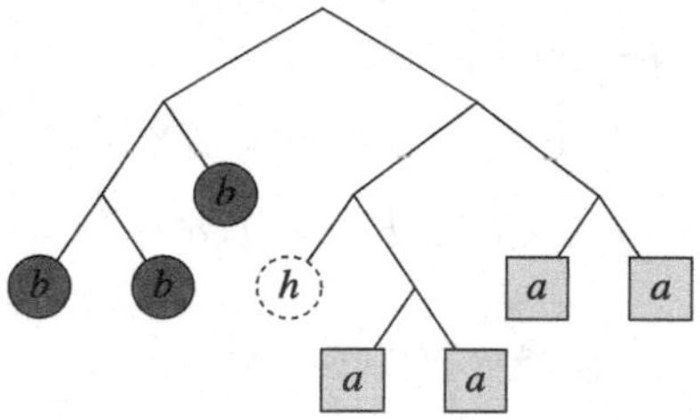

Fig. 11. A tree with leaf sequence matching b^*ha^*

7.1 Dissection

Such structures have been considered by McBride [9], who refers to them as *dissections* and shows how they can be used, for example, to generically derive tail-recursive maps and folds.

Given a functor F, McBride uses $\Delta\!\!\!\!\Delta F$ to denote the bifunctor which is the dissection of F (where the unit type has been substituted for h). We have

$$\Delta\!\!\!\!\Delta X\ b\ a = 1,$$

since a dissected X consists merely of a hole,

$$\Delta\!\!\!\!\Delta 1\ b\ a = 0,$$

and

$$\Delta\!\!\!\!\Delta (F + G)\ b\ a = \Delta\!\!\!\!\Delta F\ b\ a + \Delta\!\!\!\!\Delta G\ b\ a.$$

[3] Typically we substitute the unit type for h, but it makes the theory work more smoothly if we represent it initially with a unique type variable.

The central construction is the Leibniz rule for dissection of a product,

$$\triangle(F \times G) = \angle F \times \triangle G + \triangle F \times \setminus G,$$

where $(\angle F)\ b\ a = F\ b$ and $(\setminus F)\ b\ a = F\ a$. That is, a dissection of an $(F \times G)$-structure consists either of an F-structure containing only elements of the first type paired with a G-dissection, or an F-dissection paired with a G-structure containing only elements of the second type.

As a simple example, consider the polynomial functor $L = 1 + XL$ of finite lists. Intuitively, the dissection of a list should consist of a list of b's, followed by a hole, and then a list of a's, that is,

$$\triangle L\ b\ a \cong L\ b \times L\ a.$$

Applying the rules above, we can derive

$$\begin{aligned}
\triangle L\ b\ a &= \triangle(1 + XL)\ b\ a \\
&= (0 + \angle X \times \triangle L + \triangle X \times \setminus L)\ b\ a \\
&= b \times (\triangle L\ b\ a) + L\ a
\end{aligned}$$

and thus $\triangle L\ b\ a \cong L\ b \times L\ a$ as expected.

7.2 Dissection via Matrices

The DFA recognizing $b^* h a^*$ is illustrated in Fig. 7, and has transition matrix

$$\begin{bmatrix} X_b & X_h \\ 0 & X_a \end{bmatrix}.$$

There are clearly no leaf sequences taking this DFA from state 2 to state 1; leaf sequences matching b^* or a^* keep the DFA in state 1 or state 2, respectively; and leaf sequences matching $b^* h a^*$ take the DFA from state 1 to state 2. That is, under the homomorphism induced by this DFA, the functor F maps to the matrix of bifunctors

$$\begin{bmatrix} \angle F & \triangle F \\ 0 & \setminus F \end{bmatrix}.$$

Taking the product of two such matrices indeed yields

$$\begin{bmatrix} \angle F & \triangle F \\ 0 & \setminus F \end{bmatrix} \begin{bmatrix} \angle G & \triangle G \\ 0 & \setminus G \end{bmatrix} = \begin{bmatrix} \angle F \times \angle G & \angle F \times \triangle G + \triangle F \times \setminus G \\ 0 & \setminus F \times \setminus G \end{bmatrix},$$

as expected.

7.3 Divided Differences

Just as differentiation of types has an analytic analogue, dissection has an analogue as well, known as *divided difference*. Let $f : \mathbb{R} \to \mathbb{R}$ be a real-valued function, and let $b, a \in \mathbb{R}$. Then the *divided difference* of f at b and a, notated[4] $f_{b,a}$, is defined by

$$f_{b,a} = \frac{f_b - f_a}{b - a},$$

where for consistency of notation we write f_b for $f(b)$, and likewise for f_a. In the limit, as $a \to b$, this yields the usual derivative of f.

We now consider the type-theoretic analogue of $f_{b,a}$. We cannot directly interpret subtraction and division of functors. However, if we multiply both sides by $(b - a)$ and rearrange a bit, we can derive an equivalent relationship expressed in terms of only addition and multiplication, namely,

$$f_a + f_{b,a} \times b = a \times f_{b,a} + f_b.$$

This equation corresponds exactly to the isomorphism witnessed by McBride's function *right*,

$$right : F\ a + (\triangle F\ b\ a,\ b) \to (a, \triangle F\ b\ a) + F\ b$$

We can now explain why the letters b and a are "backwards". Intuitively, we can think of a dissection as a "snapshot" of a data structure in the midst of a traversal; values of type a are "unprocessed" and values of type b are "processed". The "current position" moves from left to right through the structure, turning a values into b values. This is exactly what is accomplished by *right*: given a structure full of unprocessed a values, or a dissected F with a focused b value, it moves the focus right by one step, either focusing on the first unprocessed a, or yielding a structure full of bs in the case that all the values have been processed.

7.4 Higher-Order Divided Differences

Higher-order divided differences, corresponding to higher derivatives, are defined by the recurrence

$$f_{x_n \dots x_0} = \frac{f_{x_n \dots x_1} - f_{x_{n-1} \dots x_0}}{x_n - x_0}. \tag{9}$$

Alternatively, the higher-order divided differences of a function f can be arranged in a matrix, as, for example,

$$\triangle_{cba} f = \begin{bmatrix} f_c & f_{c,b} & f_{c,b,a} \\ 0 & f_b & f_{b,a} \\ 0 & 0 & f_a \end{bmatrix} \tag{10}$$

[4] Our notation is actually "backwards" with respect to the usual notation—what we write as $f_{b,a}$ is often written $f[a, b]$ or $[a, b]f$—in order to better align with the combinatorial intuition discussed later.

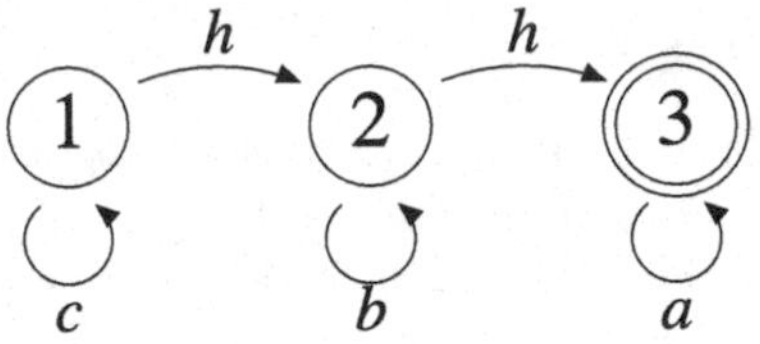

Fig. 12. A DFA for higher-order divided difference

in such a way as to be a semiring homomorphism, that is, $\triangle_{cba}(f+g) = \triangle_{cba}f + \triangle_{cba}g$ and $\triangle_{cba}(fg) = \triangle_{cba}f\triangle_{cba}g$, and so on. Proving that this yields a definition equivalent to the recurrence (9) boils down to showing that if $f = gh$ then

$$f_{x_n \ldots x_0} = \sum_{j=0}^{n} g_{x_n \ldots x_j} h_{x_j \ldots x_0}. \tag{11}$$

Proving (11) is not entirely straightforward; in fact, we conjecture that the computational content of the proof, in the $n = 2$ case, essentially consists of (the interesting part of) the implementation of the isomorphism *right*.

If we now consider the DFA D in Fig. 12, we can see that (10) corresponds to the matrix $[F]_D$. More generally, a DFA consisting of a sequence of n states with self-loops chained together by h transitions will have a transition matrix corresponding to an order-n matrix of divided differences. In general, F_{ij} will consist of $j - i$ holes interspersed among sequences of consecutive alphabet elements.

By analogy with the binary dissection case, we would expect (9) to yield an isomorphism with type

$$\triangle_{x_{n-1} \ldots x_0} F + (\triangle_{x_n \ldots x_0}, x_n) \to (x_0, \triangle_{x_n \ldots x_0} F) + \triangle_{x_n \ldots x_1} F.$$

We have not yet been able to fully make sense of this, but hope to understand it better in the future. In particular, our intuition is that this will yield a tail-recursive implementation of a structure being processed by multiple coroutines.

8 Discussion and Future Work

This paper arose out of several blog posts by both authors [12–14, 19], although the content of this paper is neither a strict subset nor superset of the content of the blog posts. There is much remaining to be explored, in particular understanding the isomorphisms induced by higher-order divided differences, and generalizing this framework to n-ary functors and partial differentiation. It seems likely that q-derivatives can also fruitfully be seen in a similar light [17].

There are several more practical aspects to this work that remain to be explored. At a fundamental level, there would be some interesting engineering work involved in turning this into a practical library. One might also wonder to what extent it is possible to take *operations* on some polynomial functor T and automatically lift them into operations on a constrained version of T. At the very least this would require checking that the operation actually preserves the given constraints.

Some of the ideas in this paper are implicitly present in earlier work; we note in particular Duchon *et al.* [4, p.590], who mention generating Boltzmann samplers for strings corresponding to regular expressions, also via their DFAs. It would be interesting to explore the relationship in more detail.

Acknowledgements. This work was partially supported by the National Science Foundation, under NSF 1218002, CCF-SHF Small: *Beyond Algebraic Data Types: Combinatorial Species and Mathematically-Structured Programming.*

Our sincere thanks to the anonymous reviewers, who had many helpful suggestions. Thanks also to Lukas Mai for pointing out some errors in a draft.

References

1. Abbott, M., Altenkirch, T., McBride, C., Ghani, N.: ∂ for data: differentiating data structures. Fundam. Inform. **65**(1–2), 1–28 (2005)
2. Stephen, D.: Fun with semirings: a functional pearl on the abuse of linear algebra. In: ACM SIGPLAN Notices, vol. 48, pp. 101–110. ACM (2013)
3. Droste, M., Kuich, W., Vogler, H. (eds.): Handbook of Weighted Automata. Springer Science & Business Media, Heidelberg (2009)
4. Duchon, P., Flajolet, P., Louchard, G., Schaeffer, G.: Boltzmann samplers for the random generation of combinatorial structures. Comb. Probab. Comput. **13**(4–5), 577–625 (2004)
5. Huet, G.: Functional pearl: the zipper. J. Funct. Program. **7**, 549–554 (1997)
6. Kleene, S.C.: Representation of events in nerve nets and finite automata. Technical report, DTIC Document (1951)
7. Lehmann, D.J.: Algebraic structures for transitive closure. Theor. Comput. Sci. **4**(1), 59–76 (1977)
8. Marlow, S.: Haskell 2010 language report (2010). https://www.haskell.org/onlinereport/haskell2010/
9. McBride, C.: Clowns to the left of me, jokers to the right (pearl): dissecting data structures. In: POPL, pp. 287–295 (2008)
10. McBride, C., Paterson, R.: Applicative programming with effects. J. Funct. Program. **18**(01), 1–13 (2008)
11. O'Connor, R.: A very general method for computing shortest paths (2011). http://r6.ca/blog/20110808T035622Z.html
12. Piponi, D.: Finite differences of types (2009). http://blog.sigfpe.com/2009/09/finite-differences-of-types.html
13. Piponi, D.: Constraining types with regular expressions (2010). http://blog.sigfpe.com/2010/08/constraining-types-with-regular.html
14. Piponi, D.: Divided differences and the tomography of types (2010). http://blog.sigfpe.com/2010/08/divided-differences-and-tomography-of.html
15. Sipser, M.: Introduction to the Theory of Computation. Cengage Learning, Boston (2012)
16. Siskind, J.M., Pearlmutter, B.A.: Nesting forward-mode ad in a functional framework. High.-Order Symb. Comput. **21**(4), 361–376 (2008)

17. Stay, M., Jokers, Q., Clowns (2014). https://reperiendi.wordpress.com/2014/08/05/q-jokers-and-clowns/
18. Tarski, A.: A lattice-theoretical fixpoint theorem and its applications. Pac. J. Math. **5**(2), 285–309 (1955)
19. Yorgey, B.A.: On a problem of SIGFPE (2010). https://byorgey.wordpress.com/2010/08/12/on-a-problem-of-sigfpe/

A Program Construction and Verification Tool for Separation Logic

Brijesh Dongol[1(✉)], Victor B.F. Gomes[2], and Georg Struth[2]

[1] Department of Computer Science, Brunel University, Uxbridge, UK
Brijesh.Dongol@brunel.ac.uk
[2] Department of Computer Science, University of Sheffield, Sheffield, UK
{v.gomes,g.struth}@sheffield.ac.uk

Abstract. An algebraic approach to the design of program construction and verification tools is applied to separation logic. The control-flow level is modelled by power series with convolution as separating conjunction. A generic construction lifts resource monoids to assertion and predicate transformer quantales. The data domain is captured by concrete store-heap models. These are linked to the separation algebra by soundness proofs. Verification conditions and transformation or refinement laws are derived by equational reasoning within the predicate transformer quantale. This separation of concerns makes an implementation in the Isabelle/HOL proof assistant simple and highly automatic. The resulting tool is itself correct by construction; it is explained on three simple examples.

1 Introduction

Separation logic is an approach to program verification that has received considerable attention over the last decade. It is designed for local reasoning about a system's states or resources, by isolating the part of a system that is affected by an action from the remainder. This capability is provided by its separating conjunction operator together with the frame inference rule, which makes local reasoning modular. A key application is the verification of programs with pointers [32,35]; but the method has also been used for modular concurrency verification [6,30,38].

Separation logic is currently supported by a large number of tools, some of which are discussed in Sect. 9. Implementations in higher-order interactive proof assistants [8,23,37,40] are particularly relevant to this article. In comparison to automated tools or tools for decidable fragments, they can express more program properties, but are less effective for proof search. Ultimately, an integration of these different proof methods is desirable.

This article adds to this tool chain (and presents another implementation in the Isabelle/HOL theorem-proving environment [29]). However, our approach is different in several respects. It focusses almost entirely on making the control-flow layer as simple as possible and on separating it cleanly from the data layer. This supports the integration of various data models and modular reasoning

R. Hinze and J. Voigtländer (Eds.): MPC 2015, LNCS 9129, pp. 137–158, 2015.
DOI:10.1007/978-3-319-19797-5_7

about these two layers, with assignment laws providing an interface. To achieve this separation of concerns, we develop an algebraic approach to separation logic which aims to combine the simplicity of the original logical approaches [32] with the abstractness and elegance of O'Hearn and Pym's categorical logic of bunched implications [31] in a way suitable for formalisation in Isabelle. Our approach is based on *power series* [15], which have found applications in formal language and automata theory [5,16]. Their use in the context of separation logic is a contribution in itself.

In a nutshell, a *power series* is a function $f : M \to Q$ from a partial monoid M into a quantale Q. Defining addition of power series by lifting addition pointwise from the quantale, and multiplication as *convolution*

$$(f \otimes g)\ x = \sum_{x = y \circ z} f\ y \odot g\ z,$$

where $\cdot$ acts on M, $\odot$ on Q and $\otimes$ on Q^M, it turns out that the function space Q^M of power series itself forms a quantale [15]. If M is commutative (a *resource monoid* [7]) and Q formed by the booleans $\mathbb{B}$ (with $\odot$ as meet), one can interpret power series as assertions or predicates over M. Separating conjunction then arises as a special case of convolution, and, in fact, as a language product over resources. The function space $\mathbb{B}^M$ is the assertion quantale of separation logic. The approach generalises to power series over program states modelled by store-heap pairs, which is needed for programming applications.

Using lifting results for power series again, we construct the quantale-like algebraic semantics of predicate transformers over assertion quantales in the style of boolean algebras with operators, following a previous approach by O'Hearn and Yang [41]. We characterise the monotone predicate transformers and derive the inference rules of Hoare logic for partial correctness (without assignment) within this subalgebra. We also derive the frame rule of separation logic on the subalgebra of local monotone predicate transformers. We use these rules for automated verification condition generation. Formalising Morgan's specification statement [27] on the transformer quantale yields tools for program construction and refinement with a frame refinement law with minimal effort. The predicate transformer approach, instead of the more common state transformer one [7], fits well into the power series approach and simplifies the development.

The formalisation of the algebraic hierarchy from resource monoids to predicate transformer algebras benefits from the functional programming approach imposed by Isabelle and its integration of automated theorem provers and SMT-solvers via the Sledgehammer tool. These are optimised for equational reasoning, which makes the entire development highly automatic. In addition, Isabelle's reconstruction of proof outputs provided by the external tools makes our tools correct by construction.

At the data-domain level, we currently use Isabelle's extant infrastructure for the store, the heap and pointer-based data structures. An interface to the control-flow algebra is provided by the standard assignment laws of separation logic and their refinement counterparts. Isabelle's concrete data domain models are linked formally with our abstract separation algebra by soundness proofs. Algebraic facts are then picked up automatically by Isabelle for reasoning in the concrete model. Our verification examples show that, at the concrete layer, proofs may require some user interaction, but an integration of domain-specific provers and solvers for the data domain is an avenue of future work. The entire technical development has been formalised in Isabelle; all proofs have been formally verified. We therefore show only some example proofs which demonstrate the simplicity of algebraic reasoning. The complete executable Isabelle theories can be found online[1].

2 Partial Monoids, Quantales and Power Series

This section presents the algebraic structures that underlie our approach to separation logic. Further details on power series and lifting constructions can be found in [15].

A *partial semigroup* is defined as a set S with a composition $\cdot : D \rightarrow S$ for some $D \subseteq S \times S$ that satisfies the usual associativity law in the sense that if either side is defined then so is the other side and both are equal [4]. A *partial monoid* M is an obvious extension by an unit 1 such that $x \cdot 1 = x$ for all $(x, 1) \in D$ and $1 \cdot x = x$ for all $(1, x) \in D$. A partial monoid M is *commutative* if $x \cdot y = y \cdot x$ for all $(x, y) \in D$. Henceforth $\cdot$ is used for a general and $*$ for a commutative multiplication.

A *quantale* (or *standard Kleene algebra*) is a structure $(Q, \leq, \cdot, 1)$ such that $(Q, \leq)$ is a complete lattice, $(Q, \cdot, 1)$ is a monoid and the distributivity axioms

$$(\sum_{i\in I} x_i) \cdot y = \sum_{i\in I} x_i \cdot y, \qquad x \cdot (\sum_{i\in I} y_i) = \sum_{i\in I}(x \cdot y_i)$$

hold, where $\sum X$ denotes the supremum of a set $X \subseteq Q$. Similarly, we write $\bigsqcap X$ for the infimum of X, and 0 for the least and U for the greatest element of the lattice. The monotonicity laws

$$x \leq y \Rightarrow z \cdot x \leq z \cdot y, \qquad x \leq y \Rightarrow x \cdot z \leq y \cdot z$$

follow from distributivity. The two annihilation laws $x \cdot 0 = 0 = 0 \cdot x$ follow from $\sum_{i\in\emptyset} x_i = \sum \emptyset = 0$. A quantale is *commutative* and *partial* if the underlying monoid is. It is *distributive* if

$$x \sqcap (\sum_{i\in I} y_i) = \sum_{i\in I}(x \sqcap y_i), \qquad x + (\bigsqcap_{i\in I} y_i) = \bigsqcap_{i\in I}(x + y_i).$$

A *boolean quantale* is a complemented distributive quantale. The boolean quantale $\mathbb{B}$ of the booleans, where multiplication coincides with join, is an important example.

[1] https://github.com/vborgesfer/sep-logic.

We call a *power series* a function $f : M \to Q$, from a partial monoid M into a quantale Q. For $f, g : M \to Q$ and a family of functions $f_i : M \to Q$, $i \in I$ we define

$$(f \cdot g)\ x = \sum_{x = y \cdot z} f\ y \cdot g\ z, \qquad (\sum_{i \in I} f_i)\ x = \sum_{i \in I} f_i\ x,$$

where $y, z \in M$. The composition $f \cdot g$ is called *convolution*; the multiplication symbol is often overloaded to be used on M, Q and the function space Q^M.

The idea behind convolution is simple: element x is split into y and z, the functions f and g are applied in parallel to y and z to calculate the values $f\ y$ and $g\ z$, and their results are composed to form a value for the summation with respect to all possible splits of x.

Because x ranges over M, the constant $\bot \notin M$ is excluded as a value; undefined splittings of x do not contribute to convolutions. In addition, $(f + g)\ x = f\ x + g\ x$ arises as a special case of the supremum. Finally, we define the power series $\mathbb{0} : M \to Q$ and $\mathbb{1} : M \to Q$ as

$$\mathbb{0} = \lambda x.\ 0, \qquad \mathbb{1} = \lambda x.\ \begin{cases} 1, & \text{if } x = 1, \\ 0, & \text{otherwise.} \end{cases}$$

The quantale structure lifts from Q to the function space Q^M of power series.

Theorem 2.1 ([15]). Let M be a partial monoid. If Q is a boolean quantale, then so is $(Q^M, \leq, \cdot, \mathbb{1})$. If M and Q are commutative, then so is Q^M.

The power series approach generalises from one to n dimensions [15]. For separation logic, the two-dimensional case with power series $f : S \times M \to Q$ from set S and partial commutative monoid M into the commutative quantale Q is needed. Now

$$(f * g)\ (x, y) = \sum_{y = y_1 * y_2} f\ (x, y_1) * g\ (x, y_2), \qquad (\sum_{i \in I} f_i)\ (x, y) = \sum_{i \in I} f_i\ (x, y).$$

The convolution $f * g$ acts solely on the second coordinate. Finally, we define two-dimensional units as

$$\mathbb{0} = \lambda x, y.\ 0, \qquad \mathbb{1} = \lambda x, y.\ \begin{cases} 1, & \text{if } y = 1, \\ 0, & \text{otherwise.} \end{cases}$$

Theorem 2.2 ([15]). Let S be a set and M a partial commutative monoid. If Q is a commutative boolean quantale, then so is $Q^{S \times M}$.

We have implemented partial monoids and quantales by using Isabelle's type class and locale infrastructure, building on existing libraries for monoids, quantales and complete lattices. The implementation of power series uses Isabelle's well developed libraries for functions. This makes proofs in this setting simple and highly automatic.

3 Assertion Quantales

In language theory, power series have been introduced for modelling formal languages. Here, M is the free monoid X^* and Q can be taken as a semiring $(Q, +, \cdot, 0, 1)$, because there are only finitely many ways of splitting words into prefix/suffix pairs in convolutions. Infinite suprema in the definition of convolution are therefore not needed. In the particular case of the boolean semiring $\mathbb{B}$, where composition $\cdot$ is meet $\sqcap$, power series $f : X^* \to \mathbb{B}$ are interpreted as characteristic functions (i.e., predicates) that indicate whether a word is in a set. In this case, sets are languages, and hence, convolution specialises to

$$(f \cdot g)\, x = \sum_{x=yz} f\, x \sqcap g\, y,$$

identifying predicates with their extensions to the language product

$$p \cdot q = \{yz \mid y \in p \wedge z \in q\}.$$

More generally, we consider power series $S \to \mathbb{B}$ from a partial monoid S into the boolean quantale $\mathbb{B}$ and set up a connection with separation logic. There, one is interested in modelling assertions or predicates over the memory heap. The heap can be represented abstractly by a so-called *resource monoid* [7], which is a partial commutative monoid. By analogy to the language case, an assertion p of separation logic is a boolean-valued function from a resource monoid M, hence a power series $p : M \to \mathbb{B}$. Thus Theorem 2.1 applies.

Corollary 3.1. *The assertions $\mathbb{B}^M$ over resource monoid M form a commutative boolean quantale with convolution as separating conjunction.*

The logical structure of the assertion quantale $\mathbb{B}^M$ is as follows. The predicate $\mathbb{0}$ is a contradiction whereas $\mathbb{1}$ holds of the empty resource 1 and is false otherwise. The operations $\sum$ and $\sqcap$ correspond to existential and universal quantification; their finite cases yield conjunctions and disjunctions. The order $\leq$ is implication. Convolution becomes

$$(p * q)\, x = \sum_{x=y*z} p\, y \sqcap q\, z,$$

and it gives a simple algebraic account of separating conjunction. By $x = y * z$, resource x is separated into resources y and z. By $p\, y \sqcap q\, z$, the value of predicate p on y is conjoined with that of q on z. Finally, the supremum is true if one of the conjunctions holds for some splitting of x.

As for languages, one can again identify predicates with their extensions. Then

$$p * q = \{y * z \in M \mid y \in p \wedge z \in q\},$$

and separating conjunction becomes a language product over resources (cf. [21]). The analogy to language theory is even more striking when considering the paradigmatic kind of resource: multisets or Parikh vectors over a finite set X. These form the free commutative monoids over X.

Applications of separation logic, however, require program states which are store-heap pairs. Now Theorem 2.2 applies.

Corollary 3.2. *The assertions $\mathbb{B}^{S \times M}$ over set S (the store) and resource monoid M form a commutative boolean quantale with convolution as separating conjunction. For all $p, q : S \times M \to \mathbb{B}$, $s \in S$ and $h \in M$,*

$$(p * q)\ (s, h) = \sum_{h = h_1 * h_2} p\ (s, h_1) \sqcap\ q\ (s, h_2).$$

Written in language-product style, therefore,

$$p * q = \{(s, h_1 * h_2) \in S \times M \mid (s, h_1) \in p \wedge (s, h_2) \in q\}.$$

The definition of convolution and the associated lifting is obviously flexible enough to encompass situations where pairs are extended to tuples or where the store as well as the heap are split by convolution. Isabelle also supports uncurried representations of such tuples and translations between them. The constructive functional approach of power series is very convenient for the functional programming style of Isabelle.

Quantales carry a rich algebraic structure. Their distributivity laws give rise to continuity or co-continuity properties. Therefore, many functions constructed from the quantale operations have adjoints as well as fixpoints, which can be iterated to the first limit ordinal. This is well known in denotational semantics and important for our approach to program verification. In particular, separating conjunction $*$ distributes over arbitrary suprema in $\mathbb{B}^M$ and $\mathbb{B}^{S \times M}$ and therefore has an upper adjoint: the *magic wand* operation $-\!\!*$, which is widely used in separation logic. In the quantale setting, the adjunction gives us theorems for the magic wand for free. This and other residuals that arise on the assertion quantale of separation logic have been studied, for instance, in [9].

One can think of the power series approach to separation logic as a simpler account of the category-theoretical approach in O'Hearn and Pym's logic of bunched implication [31] in which convolution generalises to coends and the quantale lifting is embodied by Day's construction [12]. For the design of verification tools and our implementation in Isabelle, the simplicity of the power series approach is certainly an advantage.

4 Predicate Transformer Quantales

Our algebraic approach to separation logic is based on predicate transformers (cf. [3]). This is in contrast to most previous state-transformer-based approaches and implementations [7,23,37], with [20,41] being exceptions. First of all, predicate transformers are more amenable to algebraic reasoning [3]—simply because their source and target types are both at powerset level. Second, the approach is coherent and easily implementable within our framework. Predicate transformers can be seen once more as power series and instances of Theorem 2.1 describe their algebras.

A *state transformer* $f_R : A \to 2^B$ is often associated with a relation $R \subseteq A \times B$ from set A to set B by defining

$$f_R\ a = \{b \mid (a, b) \in R\}.$$

It can be lifted to a function $\langle R\rangle : 2^A \to 2^B$ defined by

$$\langle R\rangle X = \bigcup_{a\in X} f_R\ a$$

for all $X \subseteq A$. More importantly, state transformers are lifted to *predicate transformers* $[R] : 2^B \to 2^A$ by defining

$$[R]\ Y = \{x \mid f_R\ x \subseteq Y\}$$

for all $Y \subseteq B$. The modal box and diamond notation is justified by the correspondence between diamond operators and Hoare triples as well as box operators and weakest liberal precondition operators in the context of modal semirings and modal Kleene algebras [26]. In fact we obtain the adjunction

$$\langle R\rangle X \subseteq Y \Leftrightarrow X \subseteq [R]Y$$

from the above definitions.

Predicate transformers in $(2^A)^{2^B}$ form complete distributive lattices [3]. In the power series setting, this follows from Theorem 2.1 in two steps, ignoring the monoidal structure. Since $\mathbb{B}$ forms a complete distributive lattice, so do $2^B \cong \mathbb{B}^B$ and $2^A \cong \mathbb{B}^A$ in the first step, and so does $(2^A)^{2^B}$ in the second one.

In addition, predicate transformers in $(2^A)^{2^A}$ form a monoid under function composition with the identity function as the unit. Such predicate transformers form a distributive *near-quantale*, which is a quantale such that the left distributivity law, $x \cdot (\sum_{i\in I} y_i) = \sum_{i\in I}(x \cdot y_i)$, need not hold. The monotone predicate transformers in $(2^A)^{2^A}$, which satisfy $p \leq q \Rightarrow f\ p \leq f\ q$, form a distributive *pre-quantale* [3], which is a near-quantale in which the left monotonicity law, $x \leq y \Rightarrow z \cdot x \leq z \cdot y$, holds. In these cases, the monoidal parts of the lifting are not obtained with the power series technique. The monoidal operation on predicate transformers is function composition and not convolution. Of course it is associative and the identity function is a unit of composition.

Adapting these results to separation logic requires the consideration of assertion quantales $\mathbb{B}^M$ or $\mathbb{B}^{S\times M}$ with store S and resource monoid M instead of the powerset algebra over a set A. Instead of lifting these quantales, we only lift their boolean algebra reduct, disregarding separation conjuction by not lifting it to a convolution on predicate transformers, which is not needed for separation logic. The quantale structure of predicate transformers is again obtained by considering function composition as the monoidal operation. This yields the following result.

Theorem 4.1. *Let S be a set, M a resource monoid and $\mathbb{B}^{S\times M}$ an assertion quantale. The monotone predicate transformers over $\mathbb{B}^{S\times M}$ form a distributive pre-quantale.*

The proof consists of showing that the predicate transformers over $\mathbb{B}^{S\times M}$ form a near quantale and checking that the monotone predicate transformers form a subalgebra of this near-quantale. In fact, the unit predicate transformer has to be monotone—which is the case—and the quantale operations of suprema, infima and composition have to preserve monotonicity. This is implied by properties such as

$$[R \cup S] = [R] \sqcap [S], \qquad [R; S] = [R] \cdot [S] = \lambda x.\ [R]\ ([S]\ x).$$

Monotone predicate transformers can be used to derive the standard inference rules of Hoare logic as verification conditions (Sect. 5) and the usual rules of Morgan's refinement calculus (Sect. 6). Derivation of the frame rule of separation logic, however, requires a smaller class of predicate transformers defined as follows.

A state transformer f is *local* [7] if

$$f\ (x * y) \leq (f\ x) * \{y\}$$

whenever $x * y$ is defined. Intuitively, this means that the effect of such a state transformer is restricted to a part of the heap; see [7] for further discussion. Analogously, and similarly to [20], we call a predicate transformer F *local* if

$$(F\ p) * q \leq F\ (p * q).$$

It is easy to show that the two definitions are compatible.

Lemma 4.2. *State transformer f_R is local iff predicate transformer $[R]$ is local.*

The final theorem in this section establishes the local monotone predicate transformers as a suitable algebraic framework for separation logic.

Theorem 4.3. *Let S be a set and M a resource monoid. The local monotone predicate transformers over the assertion quantale $\mathbb{B}^{S \times M}$ form a distributive pre-quantale.*

The fact that the local monotone predicate transformers form a complete lattice has been observed previously [41]. Once again it must be checked that the zero predicate transformer is local—which is the case—and that the quantale operations preserve locality and monotonicity.

We have implemented the whole approach in Isabelle; all theorems have been formally verified, mainly using Theorem 2.1 for the lifting to predicate transformers. Apart from the local case, ours is not the first Isabelle formalisation of predicate transformers; it is based on previous work by Preoteasa [34].

5 Verification Conditions

The pre-quantale of local monotone predicate transformers supports the derivation of verification conditions by equational reasoning. A standard set of such conditions are the inference rules of Hoare logic. For sequential programs, Hoare logic provides one inference rule per program construct, and these can be applied by and large in non-deterministic fashion to simple while-programs. This suffices to eliminate the control structure of a program and generate verification conditions at the data level.

The quantale setting also guarantees that the finite iteration F^* of a predicate transformer is well defined. Writing **skip** for the quantale unit (the identity function) we obtain the greatest fixpoint laws

$$\mathbf{skip} \sqcap F \cdot F^* = F^*, \qquad G \leq H \sqcap F \cdot G \Rightarrow G \leq F^* \cdot H,$$
$$\mathbf{skip} \sqcap F^* \cdot F = F^*, \qquad G \leq H \sqcap G \cdot F \Rightarrow G \leq H \cdot F^*,$$

from the explicit definition

$$F^* = \bigsqcap_{i \in \mathbb{N}} F^i$$

by iteration to the first infinite ordinal where $F^0 = \mathbf{skip}$ and $F^{n+1} = F \cdot F^n$, as usual. This supports a shallow algebraic embedding of a simple while language with the standard intermediate language for the verification of while-programs.

First we lift predicates to predicate transformers [3]:

$$[p] = \lambda q.\ \overline{p} + q,$$

where $\overline{p}$ denotes the boolean complement of p. With predicates modelled as relational subidentities, this definition is justified by the lifting from the previous section: $(s, s) \in [p]\ q$ iff $(s, s) \in p \Rightarrow (s, s) \in q$.

Second, we change notation to use descriptive while program syntax for predicate transformers. We now write ; for function composition, and we encode the algebraic semantics of conditionals and while loops as

$$\mathbf{if}\ p\ \mathbf{then}\ F\ \mathbf{else}\ G\ \mathbf{fi} = [p] \cdot F \sqcap [\overline{p}] \cdot G,$$
$$\mathbf{while}\ p\ \mathbf{do}\ F\ \mathbf{od} = ([p] \cdot F)^* \cdot [\overline{p}].$$

Third, we provide the usual assertions notation for programs via Hoare triple syntax:

$$\{p\}\ F\ \{q\} \Leftrightarrow p \leq F\ q.$$

Box notation shows that

$$\{p\}\ [R]\ \{q\} \Leftrightarrow p \leq [R]q$$

for relational program R. Thus $[R]q = wlp(R, q)$ is the well-known weakest liberal precondition of program R and postcondition q. This encoding is standard in the context of Kleene algebras with domain [26]. It also explains our slight abuse of relational or imperative notation for predicate transformers: e.g. we write $[R]; [S]$ instead of $[R] \cdot [S]$ because the latter expression is equal to $[R; S]$, as indicated in the previous section.

Proposition 5.1. *Let $p, q, r, p', q' \in \mathbb{B}^{S \times M}$ be predicates. Let F, G, H be monotone predicate transformers over $\mathbb{B}^{S \times M}$, with H being local. Then the rules* *of propositional Hoare logic (no assignment rule) and the frame rule of separation logic are derivable.*

$$
\begin{aligned}
p \le q &\Rightarrow \{p\}\ \mathbf{skip}\ \{q\},\\
p \le p' \wedge q' \le q\ \wedge \{p'\}\ F\ \{q'\} &\Rightarrow \{p\}\ F\ \{q\},\\
\{p\}\ F\ \{r\}\ \wedge \{r\}\ G\ \{q\} &\Rightarrow \{p\}\ F;G\ \{q\},\\
\{p \sqcap r\}\ F\ \{G\}\ \wedge \{p \sqcap \overline{r}\}\ G\ \{q\} &\Rightarrow \{p\}\ \mathbf{if}\ r\ \mathbf{then}\ F\ \mathbf{else}\ G\ \mathbf{fi}\ \{q\},\\
\{p \sqcap q\}\ F\ \{p\} &\Rightarrow \{p\}\ \mathbf{while}\ q\ \mathbf{do}\ F\ \mathbf{od}\ \{\overline{b} \sqcap p\},\\
\{p\}\ H\ \{q\} &\Rightarrow \{p * r\}\ H\ \{q * r\}.
\end{aligned}
$$

Proof. We derive the frame rule as an example. Suppose $p \le H\ q$. Then, by isotonicity of $*$ and locality, $p * r \le (H\ q) * r \le H(q * r)$. □

The remaining derivations are just as simple and fully automatic in Isabelle. In fact, these laws can be derived within the predicate transformer quantale [20,26], but pragmatically, in the context of verification condition generation, this abstraction leads to less applicable Isabelle tactics.

Beyond the simple fixpoints studied in this section, the quantale setting guarantees the existence of least fixpoints of arbitrary monotone functions. A verification condition for parameterless recursive procedures can therefore be derived as well, supporting the verification of more general recursive programs. A Hoare logic for recursive imperative programs has already been implemented in the quantale setting in Isabelle[2], however, it remains to be combined with the assertion quantale of separation logic.

6 Refinement Laws

To demonstrate the power of the predicate transformer approach to separation logic we now outline its applicability to local reasoning in program construction and transformation. We show that the standard laws of Morgan's refinement calculus [27] plus an additional framing law for resources can be derived and programmed in Isabelle with little effort. It only requires defining one single additional concept—Morgan's specification statement—which is definable in every predicate transformer quantale.

Formally, for predicates $p, q \in \mathbb{B}^{S \times M}$, we define the *specification statement* as

$$[\![p, q]\!] = \sum \{F \mid p \le F\ q\}.$$

It models the most general predicate transformer or program that links postcondition q with precondition p. It is easy to see that

$$\{p\}\ F\ \{q\} \Leftrightarrow F \le [\![p, q]\!],$$

[2] http://www.dcs.shef.ac.uk/~victor/verification.

which entails the characteristic properties

$$\{p\}\ [\![p,q]\!]\ \{q\}, \qquad \{p\}\ F\ \{q\} \Rightarrow F \leq [\![p,q]\!]$$

of the specification statement: program $[\![p,q]\!]$ relates precondition p with post-condition q whenever it terminates; and it is the largest program with that property. It is easy to check that specification statements over the pre-quantale of local monotone predicate transformers are themselves local and monotone.

Like Hoare logic, Morgan's basic refinement calculus provides one refinement law per program construct. Once more we ignore assignments at this stage. We also switch to standard refinement notation with refinement order $\sqsubseteq$ being the converse of $\leq$.

Proposition 6.1. *For $p, q, r, p', q' \in \mathbb{B}^{S \times M}$, and predicate transformer F the following refinement laws are derivable in the algebra of local monotone predicate transformers.*

$$\begin{aligned}
p \leq q \Rightarrow [\![p,q]\!] &\sqsubseteq \mathbf{skip},\\
p' \leq p \wedge q \leq q' \Rightarrow [\![p,q]\!] &\sqsubseteq [\![p',q']\!],\\
[\![p,q]\!] &\sqsubseteq [\![p,r]\!]; [\![r,q]\!],\\
[\![p,q]\!] &\sqsubseteq \mathbf{if}\ b\ \mathbf{then}\ [\![b \sqcap p, q]\!]\ \mathbf{else}\ [\![\bar{b} \sqcap p, q]\!]\ \mathbf{fi},\\
[\![p, \bar{b} \sqcap p]\!] &\sqsubseteq \mathbf{while}\ b\ \mathbf{do}\ [\![b \sqcap p, p]\!]\ \mathbf{od},\\
[\![p * r, q * r]\!] &\sqsubseteq [\![p,q]\!],\\
[\![0,1]\!] &\sqsubseteq F,\\
F &\sqsubseteq [\![1,0]\!].
\end{aligned}$$

Proof. Using the frame rule, we derive the framing law, the sixth law in Proposition 6.1, as an example:

$$\{p\}\ [\![p,q]\!]\ \{q\} \Rightarrow \{p * r\}\ [\![p,q]\!]\ \{q * r\} \Leftrightarrow [\![p * r, q * r]\!] \sqsubseteq [\![p,q]\!].$$

The first step uses the frame rule from Proposition 5.1, the second one the Galois connection for the specification statement. The proofs of the other refinement laws are equally simple, using the corresponding Hoare rules in their proofs. They are fully automatic in Isabelle. A refinement law for recursive programs can be derived as well. □

The entire theory hierarchy discussed so far, from partial monoids to predicate transformer quantales, has been formalised in modular fashion as algebraic components in Isabelle/HOL, much of which was highly automatic and required only a moderate effort. It benefits, to a large extent, from Isabelle's integrated first-order theorem proving, SMT-solving and counterexample generation technology. These tools are highly optimised for equational reasoning, interacting efficiently with the algebraic layer.

7 Data Domain Integration

This section describes the integration of the data domain layer into our Isabelle tools for program construction and verification. It uses an important Isabelle feature, namely that the mathematical structures formalised in Isabelle are all polymorphic. We can therefore instantiate the abstract algebras for the control flow with various concrete models by soundness proofs, that is, quantales with predicate transformers, predicate transformers with binary relations and functions which update program states. In particular, abstract resource monoids are linked with various concrete models for resources, including store-heap pairs.

Our data domain integration can build on excellent Isabelle libraries and decades-long experience in reasoning with functions and relations, all sorts of data structures and data types. In particular, for program construction and verification with separation logic, Isabelle already provides support for reasoning with pointers and the heap [25,40]. This is predominantly based on set theory.

As previously mentioned, program states in separation logic are store-heap pairs (s, h). Program stores are implemented in Isabelle as records of program variables, each of which has a *retrieve* and an *update* function. On the one hand, this approach is polymorphic and supports variables of any Isabelle type. For instance, Isabelle's built-in list data type and list libraries can be used to reason about list-based programs. On the other hand, Isabelle records are static, which makes it difficult to accommodate dynamic features such as variable scoping, as considered in the framing laws of Morgan's refinement calculus.

Heaps have been modelled in Isabelle as partial functions on $\mathbb{N}$ [25,40]; they therefore have type **nat** → **nat option**.

We implement assignments first as functions from states to states,

$$(`x := e) = \lambda(s,h).\ (x\_update\ s\ e, h),$$

where $`x$ is a program variable, $x\_update$ the update function for $`x$, (s, h) a state and e an evaluated expression of the same type as $`x$.

Next, we implement the typical commands for heap manipulation of separation logic. For heap allocation, we use Hilbert's ε operator, where $\varepsilon\ x.\ P\ x$ denotes some x such that $P\ x$ provided it exists. Heap allocation is then programmed as

$$\begin{aligned}(`x := \mathbf{cons}\ e) = \lambda(s,h).\ &\mathbf{let}\ n = \varepsilon\ y.\ (\forall x \in dom\ h.\ x < y\ ) \\ &\mathbf{in}\ (x\_update\ s\ n, h[n \mapsto e]),\end{aligned}$$

where $dom\ h$ is the domain of the heap h, expression e is of natural number type, and $h[n \mapsto e]$ maps n to e and is the same as h for all other parameters, namely, $h[n \mapsto e] = \lambda m.\ \mathbf{if}\ m = n\ \mathbf{then}\ e\ \mathbf{else}\ h(m)$. In a similar fashion, we implement deallocation and mutation as

$$(\mathbf{dispose}\ e) = \lambda(s,h).\ (s, h[e := None]),$$
$$(@e := e') = \lambda(s,h).\ (s, h[e \mapsto e']),$$

where the expressions e and e' evaluate to natural numbers and $h[e := \mathit{None}]$ removes e from the domain of h.

We lift these atomic commands to predicate transformers as

$$[f] = \lambda q.\ q \cdot f,$$

where $\cdot$ denotes function composition, as usual. This definition is consistent with the definition of lifting in Sect. 4. As previously, we generally do not write the lifting brackets explicitly, identifying program pseudocode with predicate transformers to simplify verification notation. It then remains to show that these atomic commands are local and monotonic.

With this infrastructure in place we can prove Hoare's assignment rule and Reynolds' local axioms for allocation, deallocation and mutation of separation logic [35] in the concrete heap model. We write $q[e/`x]$ for the substitution of variable $`x$ by expression e in q, write $e \mapsto e'$ for the singleton heap mapping e to e', write $e \mapsto -$ for the singleton heap mapping e to any value, and write **emp** for the empty heap.

Proposition 7.1. *Suppose $p \le q[e/`x]$ holds. Then the following local rules are derivable in the store-heap model:*

$$\begin{gathered}\{p\}\ `x := e\ \{q\},\\ \{\mathbf{emp}\}\ `x := \mathbf{cons}\ e\ \{`x \mapsto e\},\\ \{e \mapsto -\}\ \mathbf{dispose}\ e\ \{\mathbf{emp}\},\\ \{e \mapsto -\}\ @e := e'\ \{e \mapsto e'\}.\end{gathered}$$

Variants of these rules can easily be derived. For example, global rules, obtained by a simple application of the frame rule, emphasise that anything in the heap different from the mutated location is left unchanged:

$$\begin{gathered}\{r\}\ `x := \mathbf{cons}\ e\ \ \{(`x \mapsto e) * r\},\\ \{(e \mapsto -) * r\}\ \ \mathbf{dispose}\ e\ \{r\},\\ \{(e \mapsto -) * r\}\ @e := e'\ \{(e \mapsto e') * r\}.\end{gathered}$$

Applying the weakening rule and the identity $p*(p \mathrel{-\!\!*} q) \le q$ yields another global mutation rule for backward reasoning, which is more suitable for automation

$$\{(e \mapsto -) * ((e \mapsto e') \mathrel{-\!\!*} q)\}\ \ @e := e'\ \ \{q\}.$$

Note that the magic wand operation $\mathrel{-\!\!*}$ has been discussed briefly at the end of Sect. 3. The resulting set of control-flow and data-domain inference rules for separation logic allows us to program the Isabelle proof tactic *hoare*, which generates verification conditions automatically and eliminates the entire control structure when the invariants of while loops are annotated.

One can also use the assignment rules to derive their refinement counterparts:

$$\begin{aligned}p \le q[e/`x] &\Rightarrow [\![p, q]\!] \sqsubseteq (`x := e),\\ q' \le q[e/`x] &\Rightarrow [\![p, q]\!] \sqsubseteq [\![p, q']\!];\ (`x := e),\\ p' \le p[e/`x] &\Rightarrow [\![p, q]\!] \sqsubseteq (`x := e);\ [\![p', q]\!].\end{aligned}$$

The second and third laws are called the *following* and *leading* refinement law for assignments [27]. They are useful for program construction. We have derived analogous laws for heap allocation, deallocation and mutation. We have also programmed the tactic *refinement*, which automatically tries to apply all the rules of this refinement calculus in construction steps of pointer programs.

8 Examples

To show our approach at work, we present three examples, among them the obligatory correctness proof of the classical in situ linked-list reversal algorithm. The post-hoc verification of this algorithm in Isabelle has been considered before [25,40]. However, we follow Reynolds [35], who gave an informal annotated proof, and reconstruct his proof step-by-step in refinement style. As usual for verification with interactive theorem provers, functional specifications are related to imperative data structures. The former are defined recursively in functional programming style and hence amenable to proof by induction. Such detailed functional specifications of data structures used in separation logic are usually not amenable to pure first-order reasoning and therefore beyond the scope of first-order tools such as SMT solvers.

First, we define two inductive predicates on the heap. The first one creates a contiguous heap from a position e using Isabelle's functional lists as its representation, i.e., by induction on the structure of the list,

$$\begin{aligned} e \; [\mapsto] \; [\,] &= \; \mathbf{emp}, \\ e \; [\mapsto] \; (t\#ts) &= \; (e \;\mapsto\; t) * (e + 1 \; [\mapsto] \; ts), \end{aligned}$$

where $[\,]$ denotes the empty list, $t\#ts$ denotes concatenation of element t with list ts, $e \;\mapsto\; t$ is again a singleton heap predicate and **emp** states that the heap is empty.

The second predicate indicates whether a heap, starting from position i,' contains the linked list represented as a functional list:

$$\begin{aligned} \mathit{list}\; i\; [\,] &= \; (i = 0) \wedge \mathbf{emp}, \\ \mathit{list}\; i\; (j\#js) &= \; (i \neq 0) \wedge (\exists k.\; i \; [\mapsto] \; [j, k] \; * \; \mathit{list}\; k\; js). \end{aligned}$$

This is Reynolds' definition; it uses separating conjunction instead of plain conjunction.

Example 1: Constructing a linked list reversal algorithm. We reconstruct Reynolds' classical proof relative to the standard recursive function *rev* for functional list reversal. The initial specification statement is

$$[\![\mathit{list}\; `i\; A_0, \mathit{list}\; `j\; (\mathit{rev}\; A_0)]\!],$$

where A_0 is the input list and $`i$ and $`j$ are pointers to the head of the list on the heap.

The main idea behind Reynolds' proof is to split the heap into two lists, initially A_0 and an empty list, and then iteratively swing the pointer of the first element of the first list to the second list. The full proof is shown in Fig. 1; we now explain its details.

In (1), we strengthen the precondition, splitting the heap into two lists A and B, and inserting a variable $`j$ initially assigned to 0 (or *null*). The equation

$$(rev\ A_0) = (rev\ A)\ @\ B$$

then holds of these lists, where @ denotes the append operation on linked lists. Justifying this step in Isabelle requires calling the *refinement* tactic from Sect. 6, which applies the leading law for assignment. This obliges us to prove that the lists A and B *de facto* exist, which is discharged automatically by calling Isabelle's *force* tactic. In fact, 8 out of the 10 proof steps in our construction are essentially automatic: they only require calling *refinement* followed by Isabelle's *force* or *auto* provers.

The new precondition generated then becomes the loop invariant of the algorithm. It allows us to refine our specification statement to a while loop in step (2), where we iterate $`i$ until it becomes 0. Calling the *refinement* tactic applies the while law for refinement. From step (3) to (10), we refine the body of the while loop and do not display the outer part of the program.

Because now $`i \neq 0$, the list $A$ has at least an element $a$. We can thus expand the definition of *list* in step (3). Next, we assign the value pointed to by $`i + 1$ to $`k$—our first list now starts at $`k$ and $`i$ points to $[a, `k]$. Isabelle then struggles to discharge the generated proof goal automatically. In this predicate, the heap is divided in three parts. One needs to prove first that $`i + 1$ really points to the same value when considering just the first part of the heap or the entire heap. After that, the proof is automatic.

Step (5) performs a mutation on the heap, changing the cell $`i + 1$ to $`j$, consequently $`i$ now points to $[a, `j]$. Because $*$ is commutative, we can strengthen the precondition accordingly in step (6). We now work backwards, folding the definition of *list* in step (7) and removing the existential of a in step (8). This step again requires interaction: we need to indicate to Isabelle how to properly split the heap. This amounts to finding the right summand in the convolution expressing separating conjunction, or alternatively finding the existential split in the corresponding logical formulation. Last, to establish the invariant, we only need to swap the pointers $`j$ to $`i$ and $`i$ to $`k$ in steps (9) and (10). The resulting algorithm is highlighted in Fig. 1 and shown in Fig. 2. □

Example 2: Verifying a list deallocation algorithm. We verify an algorithm for list deallocation in a post-hoc fashion. It has been annotated in the standard way by a precondition, a postcondition and a loop invariant. The latter simply states that there exists a list xs starting from the pointer $`x$ while $`x$ is not equal to 0. The Isabelle code is shown in Fig. 3. Calling the *hoare* tactic generates several verification conditions for the data domain , which are easily discharged by *auto* and other Isabelle proof tools. □

$$
\begin{array}{ll}
\llbracket\, \mathit{list}\ \text{‘}i\ A_0,\ \mathit{list}\ \text{‘}j\ (\mathit{rev}\ A_0)\, \rrbracket & \\
\quad \sqsubseteq & (1) \\
\text{‘}j := 0; & \\
\llbracket\, \exists A\ B.\ (\mathit{list}\ \text{‘}i\ A * \mathit{list}\ \text{‘}j\ B\ ) \wedge (\mathit{rev}\ A_0) = (\mathit{rev}\ A)\ @\ B,\ \mathit{list}\ \text{‘}j\ (\mathit{rev}\ A_0)\, \rrbracket & \\
\quad \sqsubseteq & (2) \\
\text{‘}j := 0; & \\
\mathbf{while}\ \text{‘}i \neq 0\ \mathbf{do} & \\
\quad \llbracket (\exists A\ B.\ (\mathit{list}\ \text{‘}i\ A * \mathit{list}\ \text{‘}j\ B) \wedge (\mathit{rev}\ A_0) = (\mathit{rev}\ A)\ @\ B) \wedge \text{‘}i \neq 0, & \\
\quad\ \exists A\ B.\ (\mathit{list}\ \text{‘}i\ A * \mathit{list}\ \text{‘}j\ B) \wedge (\mathit{rev}\ A_0) = (\mathit{rev}\ A)\ @\ B\, \rrbracket & \\
\mathbf{od} &
\end{array}
$$

$$
\begin{array}{ll}
\llbracket (\exists A\ B.\ (\mathit{list}\ \text{‘}i\ A * \mathit{list}\ \text{‘}j\ B) \wedge (\mathit{rev}\ A_0) = (\mathit{rev}\ A)\ @\ B) \wedge \text{‘}i \neq 0, & \\
\ \exists A\ B.\ (\mathit{list}\ \text{‘}i\ A * \mathit{list}\ \text{‘}j\ B) \wedge (\mathit{rev}\ A_0) = (\mathit{rev}\ A)\ @\ B\, \rrbracket & \\
\quad \sqsubseteq & (3) \\
\llbracket (\exists a\ A\ B\ k.\ (\text{‘}i\ [\mapsto]\ [a,k] * \mathit{list}\ k\ A * \mathit{list}\ \text{‘}j\ B) \wedge (\mathit{rev}\ A_0) = (\mathit{rev}\ (a\#A))\ @\ B) \wedge \text{‘}i \neq 0, & \\
\ \exists A\ B.\ (\mathit{list}\ \text{‘}i\ A * \mathit{list}\ \text{‘}j\ B) \wedge (\mathit{rev}\ A_0) = (\mathit{rev}\ A)\ @\ B\, \rrbracket & \\
\quad \sqsubseteq & (4) \\
\text{‘}k := @(\text{‘}i + 1); & \\
\llbracket (\exists a\ A\ B.\ (\text{‘}i\ [\mapsto]\ [a,\text{‘}k] * \mathit{list}\ \text{‘}k\ A * \mathit{list}\ \text{‘}j\ B) \wedge (\mathit{rev}\ A_0) = (\mathit{rev}\ (a\#A))\ @\ B) \wedge \text{‘}i \neq 0, & \\
\ \exists A\ B.\ (\mathit{list}\ \text{‘}i\ A * \mathit{list}\ \text{‘}j\ B) \wedge (\mathit{rev}\ A_0) = (\mathit{rev}\ A)\ @\ B\, \rrbracket & \\
\quad \sqsubseteq & (5) \\
\text{‘}k := @(\text{‘}i + 1); & \\
@(\text{‘}i + 1) := \text{‘}j; & \\
\llbracket (\exists a\ A\ B.\ (\text{‘}i\ [\mapsto]\ [a,\text{‘}j] * \mathit{list}\ \text{‘}k\ A * \mathit{list}\ \text{‘}j\ B) \wedge (\mathit{rev}\ A_0) = (\mathit{rev}\ (a\#A))\ @\ B) \wedge \text{‘}i \neq 0, & \\
\ \exists A\ B.\ (\mathit{list}\ \text{‘}i\ A * \mathit{list}\ \text{‘}j\ B) \wedge (\mathit{rev}\ A_0) = (\mathit{rev}\ A)\ @\ B\, \rrbracket & \\
\quad \sqsubseteq & (6) \\
\text{‘}k := @(\text{‘}i + 1); & \\
@(\text{‘}i + 1) := \text{‘}j; & \\
\llbracket (\exists a\ A\ B.\ (\mathit{list}\ \text{‘}k\ A * \text{‘}i\ [\mapsto]\ [a,\text{‘}j] * \mathit{list}\ \text{‘}j\ B) \wedge (\mathit{rev}\ A_0) = (\mathit{rev}\ (a\#A))\ @\ B) \wedge \text{‘}i \neq 0, & \\
\ \exists A\ B.\ (\mathit{list}\ \text{‘}i\ A * \mathit{list}\ \text{‘}j\ B) \wedge (\mathit{rev}\ A_0) = (\mathit{rev}\ A)\ @\ B\, \rrbracket & \\
\quad \sqsubseteq & (7) \\
\text{‘}k := @(\text{‘}i + 1); & \\
@(\text{‘}i + 1) := \text{‘}j; & \\
\llbracket (\exists a\ A\ B.\ (\mathit{list}\ \text{‘}k\ A * \mathit{list}\ \text{‘}i\ (a\#B)\ ) \wedge (\mathit{rev}\ A_0) = (\mathit{rev}\ A)\ @\ (a\#B)\ ) \wedge \text{‘}i \neq 0, & \\
\ \exists A\ B.\ (\mathit{list}\ \text{‘}i\ A * \mathit{list}\ \text{‘}j\ B) \wedge (\mathit{rev}\ A_0) = (\mathit{rev}\ A)\ @\ B\, \rrbracket & \\
\quad \sqsubseteq & (8) \\
\text{‘}k := @(\text{‘}i + 1); & \\
@(\text{‘}i + 1) := \text{‘}j; & \\
\llbracket (\exists A\ B.\ (\mathit{list}\ \text{‘}k\ A * \mathit{list}\ \text{‘}i\ B) \wedge (\mathit{rev}\ A_0) = (\mathit{rev}\ A)\ @\ B) \wedge \text{‘}i \neq 0, & \\
\ \exists A\ B.\ (\mathit{list}\ \text{‘}i\ A * \mathit{list}\ \text{‘}j\ B) \wedge (\mathit{rev}\ A_0) = (\mathit{rev}\ A)\ @\ B\, \rrbracket & \\
\quad \sqsubseteq & (9) \\
\text{‘}k := @(\text{‘}i + 1); & \\
@(\text{‘}i + 1) := \text{‘}j; & \\
\text{‘}j := \text{‘}i; & \\
\llbracket (\exists A\ B.\ (\mathit{list}\ \text{‘}k\ A * \mathit{list}\ \text{‘}j\ B) \wedge (\mathit{rev}\ A_0) = (\mathit{rev}\ A)\ @\ B) \wedge \text{‘}i \neq 0, & \\
\ \exists A\ B.\ (\mathit{list}\ \text{‘}i\ A * \mathit{list}\ \text{‘}j\ B) \wedge (\mathit{rev}\ A_0) = (\mathit{rev}\ A)\ @\ B\, \rrbracket & \\
\quad \sqsubseteq & (10) \\
\text{‘}k := @(\text{‘}i + 1); & \\
@(\text{‘}i + 1) := \text{‘}j; & \\
\text{‘}j := \text{‘}i; & \\
\text{‘}i := \text{‘}k &
\end{array}
$$

Fig. 1. *In situ* list reversal by refinement. The first block shows the refinement up to the introduction of the while-loop. The second block shows the refinement of the body of that loop.

```
⟦ list ‘i A₀, list ‘j (rev A₀) ⟧
  ⊑
‘j := 0;
while ‘i ≠ 0 do
  ‘k := @(‘i + 1);
  @(‘i + 1) := ‘j;
  ‘j := ‘i;
  ‘i := ‘k
od
```

Fig. 2. *In situ* reversal list algorithm

```
lemma list-dealloc: ⊢ {| list ‘x xs |}
  while ‘x ≠ 0
  inv {| ∃xs. list ‘x xs |}
  do
    ‘y := ‘x;
    ‘x := @(‘x + 1);
    dispose ‘y;
    dispose (‘y + 1)
  od
  {| emp |}
apply hoare
apply (auto simp: mono-def dispose-comm-def list-i-null, frule list-i-not-null-var)
by (auto simp: is-singleton-def, auto intro!: list-cong-ex heap-div-the heap-ortho-div2)
```

Fig. 3. List deallocation algorithm

Example 3: Explicit application of the frame rule. The frame rule is hidden as part of the global mutation rule in Example 1 and is not required in Example 2. Therefore we present a third (somewhat artificial) example designed to demonstrate its use. The following Isabelle code fragment shows the Hoare triple used for verification.

$$\{\ x\ [\mapsto]\ [-\ ,\ j]\ \ *\ \ list\ j\ as\ \}\ @x := a\ \{x\ [\mapsto]\ [a,\ j]\ \ *\ \ list\ j\ as\ \}$$

Calling the *hoare* tactic for verification condition generation was sufficient for proving the correctness of this simple example automatically. Internally, the frame and the global mutation rule have been applied. □

In addition, we have performed a post-hoc verification of the list reversal algorithm (Fig. 1) using two different approaches. The first, previously taken by Weber [40], uses Reynolds' list predicate, as we have used it in the above refinement proof. The second follows Nipkow in using separating conjunction in the pre- and postcondition, but not in the definition of the list predicate. Since our approach is modular with respect to the underlying data model, it was straightforward to replay Nipkow's proof in our setting. The degree of proof automation with our tool is comparable to Nipkow's proof.

In summary, our approach supports the program construction and verification of pointer-based programs with separation logic, but larger case studies need to be performed to assess the performance of our tool. The algebraic approach has been used, apart from soundness proofs and the certification of the tool itself, predominantly in the derivation and implementation of tactics for program construction and verification. In the future, a Sledgehammer-style integration of optimised provers and solvers for the data level seems desirable for increasing the general degree of automation. This can be obtained, first of all, by integrating decision procedures for data types and data structures in Isabelle. Alternatively, it is sometimes possible to represent such data structures algebraically and use automated reasoning for their analysis. Dang and Möller [10], for instance, have shown how pointer structures with a generalised notion of separating conjunction can be modelled within modal Kleene algebras. Implementing this approach in Isabelle seems promising.

9 Related Work

This section discusses two main lines of related work: tool support for separation logic and similar algebraic approaches.

Numerous tools supporting separation logic have been created for various purposes. Some tools (e.g., *Predator* [17], *JStar* [14] and *VeriFast* [36]) are able to reason automatically about shape properties of real-word programming languages like C and Java. These are often highly optimised by using decision procedures and SMT solver at the data domain level. However, soundness of these tools is not guaranteed as they have not been verified relative to a small core, as provided by an LCF-style proof assistant such as Isabelle/HOL. Additionally, these tools have different degrees of interactivity and generally do not allow the user to prove any remaining verification conditions by interactively; they also cannot deal with higher-order aspects of data types and the store.

Other verification tools have been built on top of proof assistants such as Coq or Isabelle/HOL. These are generally less automatic, but allow more precise properties of the store and heap to be proved. *Smallfoot* [37], for example, has been implemented within HOL4; it supports concurrent separation logic in an approach based on [7]. Several formalisations of separation logic have been obtained in Coq, of which *YNot* [8] seems to provide the highest level of proof automation. Formalisations in Isabelle/HOL include that of Kolanski and Klein [23], which is targeted towards a subset of C, and Weber [40], which uses a shallow embedding of a simple imperative language similar to ours, but without allocation and deallocation laws. None of these tools has a lightweight middle layer formed by an algebraic semantics, which provides more modularity and flexibility when changing the programming language, logic or even the semantics. In particular, none of these tools supports program construction and refinement.

An early algebraic approach to separation logic is O'Hearn and Pym's logic of bunched implication, which describes the assertion quantale of separation logic in a category-theoretic setting [31]. Another source of inspiration is the abstract separation logic of Calgagno, O'Hearn and Yang [7], where the role of the resource monoid, the power set lifting to an assertion algebra, and the role of locality for deriving the frame rule have been elaborated within a state transformer approach. Our predicate transformer approach based on convolution seems conceptually simpler; it is certainly more suitable for implementation. Aspects of predicate transformers and the role of locality have also been investigated in [20], but a coherent approach has not been developed.

The assertion quantale structure of separation logic has also been observed by Dang, Höfner and Möller [9], and several subclasses of assertions are studied by these authors. In addition, a relational characterisation of separating conjunction is given, and a so-called frame property, which seems similar to locality, is used for deriving the frame rule in an approach based on Kleene algebra with tests [24] and a relational semantics. The precise relationship to the approach of [7] remains to be explored. It seems to be conceptually more involved and less straightforward to implement in Isabelle than ours. Using ideas from concurrent Kleene algebra [20,21], Dang and Möller have extended their approach to concurrent separation logic [11] with a relational semantics that seems compatible with our predicate transformer approach. This might support an extension of our formalisation to separation-based concurrency verification.

Another approach to concurrency verification is the Views framework [13,39], which aims to provide a metatheory consisting of several parameters. Specific instantiations of these parameters gives rise to formalisms such as Owicki-Gries [33] and rely/guarantee [22]. One of the parameters of the Views framework is a resource semigroup or monoid, which supports the parallel or concurrent application of predicates to a resource, as in [21]. However, there is no coherent algebraic approach and more interesting algebraic structures, such as quantales, are never explored. Instead, reasoning proceeds by using the operational semantics of their simple language, and hence, the approach, and even the aims, of the Views framework differ significantly from ours.

Finally, we could have adapted modal Kleene algebra [26], which has already been formalised in Isabelle [19], to program our predicate transformer approach. We could have defined this algebra over the assertion quantale instead of the usual boolean algebra, with the definition of locality (Sect. 4) providing the interaction of separating conjunction with the modal box operator. In practice, however, this leads to more complex and less automatic Isabelle proofs due to the increased distance between the abstract algebraic level and the concrete store-heap model. An optimisation is left for future work.

10 Conclusion

A principled approach to the design of program verification and construction tools for separation logic with the Isabelle theorem proving environment has been presented. The general approach has been used previously for implementing tools for the construction and verification of simple while programs [2] and rely-guarantee based concurrent programs [1]. It aims at a clean separation of concerns between the control flow and the data domain of programs and focusses on developing a lightweight algebraic layer from which verification conditions or transformation and refinement laws can be developed by simple equational reasoning. Previously, in the case of while programs, this layer has been provided by Kleene algebras with tests; in the rely-guarantee case, new algebraic foundations based on concurrent Kleene algebras were required.

Our approach to separation logic is a conceptual reconstruction of separation logic beyond a mere implementation as well, which forms a contribution in its own right. To make an Isabelle implementation as small, automatic and modular as possible, we have once more aimed at finding a conceptually minimalist setting from which powerful verification conditions as well as transformation and refinement laws can be derived. Though strongly inspired by abstract separation logic [7] and the logic of bunched implications [31], we use a different combination of simplicity and mathematical abstraction. In contrast to the logic of bunched implication, we use power series instead of higher categories, and in contrast to abstract separation logic we follow [41] in using predicate transformers in the style of boolean algebras with operators instead of state transformers. These design choices allow us to use power series, quantales and generic lifting constructions throughout the approach, which leads indeed to a very small and highly automated Isabelle implementation. A particular feature of this approach is the view on separating conjunction as a notion of convolution over resources.

Our tool prototype has so far allowed us to verify some simple pointer-based programs with a relatively high degree of automation. So far it is certainly useful for educational and research purposes, but extensions and optimisations beyond the mere proof of concept are desirable. This includes the consideration of recursive procedures [20], for variable framing laws in our refinement calculus or of error states [7], the development of more sophisticated proof tactics, and the integration of tools and techniques for automatic data-level reasoning in Sledgehammer style.

Other opportunities for future work lie in the integration of categorial approaches to data type constructions [18,28], the consolidation with Preoteasa's approach to predicate transformers in Isabelle [34], in a further abstraction of the control-flow layer by defining modal Kleene algebras over assertion quantales [26] for which some Isabelle infrastructure already exists [19], in a combination with our rely-guarantee tool into RGSep-style tools for concurrency verification [38], and in the exploration of the language connection of separating conjunction in terms of representability and decidability results.

Acknowledgements. We are grateful for support by EPSRC grant EP/J003727/1 and the CNPq. The third author would like to thank Tony Hoare, Peter O'Hearn and Matthew Parkinson for discussions on concurrent Kleene algebra and separation logic.

References

1. Armstrong, A., Gomes, V.B.F., Struth, G.: Algebraic principles for rely-guarantee style concurrency verification tools. In: Jones, C., Pihlajasaari, P., Sun, J. (eds.) FM 2014. LNCS, vol. 8442, pp. 78–93. Springer, Heidelberg (2014)
2. Armstrong, A., Gomes, V.B.F., Struth, G.: Lightweight program construction and verification tools in Isabelle/HOL. In: Giannakopoulou, D., Salaün, G. (eds.) SEFM 2014. LNCS, vol. 8702, pp. 5–19. Springer, Heidelberg (2014)
3. Back, R.-J., von Wright, J.: Refinement Calculus. Springer, Berlin (1999)
4. Bergelson, V., Blass, A., Hindman, N.: Partition theorems for spaces of variable words. Proc. London Math. Soc. **68**(3), 449–476 (1994)
5. Berstel, J., Reutenauer, C.: Les séries rationnelles et leurs langagues. Masson (1984)
6. Bornat, R., Calcagno, C., O'Hearn, P.W., Parkinson, M.J.: Permission accounting in separation logic. In: Palsberg, J., Abadi, M. (eds.) POPL, pp. 259–270. ACM (2005)
7. Calcagno, C., O'Hearn, P.W., Yang, H.: Local action and abstract separation logic. In: LICS 2007, pp. 366–378. IEEE Computer Society (2007)
8. Chlipala, A., Malecha, J.G., Morrisett, G., Shinnar, A., Wisnesky, R.: Effective interactive proofs for higher-order imperative programs. In: Hutton, G., Tolmach, A.P. (eds.) ICFP 2009, pp. 79–90. ACM (2009)
9. Dang, H.-H., Höfner, P., Möller, B.: Algebraic separation logic. J. Log. Algebr. Program. **80**(6), 221–247 (2011)
10. Dang, H.-H., Möller, B.: Transitive separation logic. In: Kahl, W., Griffin, T.G. (eds.) RAMICS 2012. LNCS, vol. 7560, pp. 1–16. Springer, Heidelberg (2012)
11. Dang, H.-H., Möller, B.: Concurrency and local reasoning under reverse interchange. Sci. Comput. Program. **85**, 204–223 (2014)
12. Day, B.: On closed categories of functors. In: MacLane, S., et al. (eds.) Reports of the Midwest Category Seminar IV. Lecture Notes in Mathematics, vol. 137, pp. 1–38. Springer, Heidelberg (1970)
13. Dinsdale-Young, T., Birkedal, L., Gardner, P., Parkinson, M.J., Yang, H.: Views: compositional reasoning for concurrent programs. In: Giacobazzi, R., Cousot, R. (eds.) POPL, pp. 287–300. ACM (2013)
14. Distefano, D., Parkinson, M.J.: jstar: towards practical verification for Java. In: Harris, G.E. (ed.) OOPSLA 2008, pp. 213–226. ACM (2008)
15. Dongol, B., Hayes, I.J., Struth, G.: Convolution, separation and concurrency. CoRR, abs/1312.1225 (2014)
16. Droste, M., Kuich, W., Vogler, H.: Handbook of Weighted Automata, 1st edn. Springer, Heidelberg (2009)
17. Dudka, K., Peringer, P., Vojnar, T.: Byte-precise verification of low-level list manipulation. In: Logozzo, F., Fähndrich, M. (eds.) Static Analysis. LNCS, vol. 7935, pp. 215–237. Springer, Heidelberg (2013)
18. Gardiner, P.H.B., Martin, C.E., de Moor, O.: An algebraic construction of predicate transformers. Sci. Comput. Program. **22**(1–2), 21–44 (1994)
19. Guttmann, W., Struth, G., Weber, T.: Automating algebraic methods in Isabelle. In: Qin, S., Qiu, Z. (eds.) ICFEM 2011. LNCS, vol. 6991, pp. 617–632. Springer, Heidelberg (2011)
20. Hoare, C.A.R., Hussain, A., Möller, B., O'Hearn, P.W., Petersen, R.L., Struth, G.: On locality and the exchange law for concurrent processes. In: Katoen, J.-P., König, B. (eds.) CONCUR 2011. LNCS, vol. 6901, pp. 250–264. Springer, Heidelberg (2011)

21. Hoare, T., Möller, B., Struth, G., Wehrman, I.: Concurrent kleene algebra and its foundations. J. Log. Algebr. Program. **80**(6), 266–296 (2011)
22. Jones, C.B.: Tentative steps toward a development method for interfering programs. ACM TOPLAS **5**(4), 596–619 (1983)
23. Klein, G., Kolanski, R., Boyton, A.: Mechanised separation algebra. In: Beringer, L., Felty, A. (eds.) ITP 2012. LNCS, vol. 7406, pp. 332–337. Springer, Heidelberg (2012)
24. Kozen, D.: On Hoare logic and Kleene algebra with tests. ACM TOCL **1**(1), 60–76 (2000)
25. Mehta, F., Nipkow, T.: Proving pointer programs in higher-order logic. Inf. Comput. **199**(1–2), 200–227 (2005)
26. Möller, B., Struth, G.: Algebras of modal operators and partial correctness. Theoret. Comput. Sci. **351**(2), 221–239 (2006)
27. Morgan, C.: Programming from Specifications. Prentice-Hall, New York (1998)
28. Naumann, D.A.: Beyond fun: order and membership in polytypic imperative programming. In: Jeuring, J. (ed.) MPC 1998. LNCS, vol. 1422, pp. 286–314. Springer, Heidelberg (1998)
29. Nipkow, T., Paulson, L.C., Wenzel, M. (eds.): Isabelle/HOL – A Proof Assistant for Higher-Order Logic. LNCS, vol. 2283. Springer, Heidelberg (2002)
30. O'Hearn, P.W.: Resources, concurrency, and local reasoning. Theoret. Comput. Sci. **375**(1–3), 271–307 (2007)
31. O'Hearn, P.W., Pym, D.J.: The logic of bunched implications. Bull. Symbolic Logic **5**(2), 215–244 (1999)
32. O'Hearn, P.W., Reynolds, J.C., Yang, H.: Local reasoning about programs that alter data structures. In: Fribourg, L. (ed.) CSL 2001 and EACSL 2001. LNCS, vol. 2142, pp. 1–19. Springer, Heidelberg (2001)
33. Owicki, S.S., Gries, D.: An axiomatic proof technique for parallel programs I. Acta Inf. **6**, 319–340 (1976)
34. Preoteasa, V.: Algebra of monotonic boolean transformers. In: Simao, A., Morgan, C. (eds.) SBMF 2011. LNCS, vol. 7021, pp. 140–155. Springer, Heidelberg (2011)
35. Reynolds, J.C.: Separation logic: A logic for shared mutable data structures. In: LICS, pp. 55–74. IEEE Computer Society (2002)
36. Smans, J., Jacobs, B., Piessens, F.: VeriFast for Java: a tutorial. In: Clarke, D., Noble, J., Wrigstad, T. (eds.) Aliasing in Object-Oriented Programming. LNCS, vol. 7850, pp. 407–442. Springer, Heidelberg (2013)
37. Tuerk, T.: A Separation Logic Framework for HOL. Ph.D. thesis, Computer Laboratory, University of Cambridge (2011)
38. V. Vafeiadis. Modular Fine-Grained Concurrency Verificaiton. PhD thesis, Computer Laboratory, University of Cambridge, 2007
39. van Staden, S.: Constructing the Views Framework. In: Naumann, D. (ed.) UTP 2014. LNCS, vol. 8963, pp. 62–83. Springer, Heidelberg (2015)
40. Weber, T.: Towards Mechanized Program Verification with Separation Logic. In: Marcinkowski, J., Tarlecki, A. (eds.) CSL 2004. LNCS, vol. 3210, pp. 250–264. Springer, Heidelberg (2004)
41. Yang, H., O'Hearn, P.W.: A Semantic Basis for Local Reasoning. In: Nielsen, M., Engberg, U. (eds.) FOSSACS 2002. LNCS, vol. 2303, pp. 402–416. Springer, Heidelberg (2002)

Calculating Certified Compilers for Non-deterministic Languages

Patrick Bahr(✉)

Department of Computer Science, University of Copenhagen, Copenhagen, Denmark
paba@di.ku.dk

Abstract. Reasoning about programming languages with non-deterministic semantics entails many difficulties. For instance, to prove correctness of a compiler for such a language, one typically has to split the correctness property into a soundness and a completeness part, and then prove these two parts separately. In this paper, we present a set of proof rules to prove compiler correctness by a *single* proof in calculational style. The key observation that led to our proof rules is the fact that the soundness and completeness proof follow a similar pattern with only small differences. We condensed these differences into a single side condition for one of our proof rules. This side condition, however, is easily discharged automatically by a very simple form of proof search. We implemented this calculation framework in the Coq proof assistant. Apart from verifying a given compiler, our proof technique can also be used to formally derive – from the semantics of the source language – a compiler that is *correct by construction*. For such a derivation to succeed it is crucial that the underlying correctness argument proceeds as a single calculation, as opposed to separate calculations of the two directions of the correctness property. We demonstrate our technique by deriving a compiler for a simple language with interrupts.

1 Introduction

Formally verifying the correctness of compilers is a difficult and expensive endeavour [9]. However, the need for formally verified compilers is a corollary of the need for formal verification of critical pieces of software; for what good is your formally verified program if it is garbled by a defective compiler.

These challenges notwithstanding, we pursue an even more ambitious goal than post hoc verification: Not only do we wish to formally verify the correctness of a given compiler implementation. Beyond that we aim to derive a compiler implementation that is *correct by construction* [3]. That is, given the semantics of the source language and a high-level specification of the compiler, we wish to systematically derive a compiler implementation that satisfies the specification.

This work was supported by the Danish Council for Independent Research, Grant 12-132365, "Efficient Programming Language Development and Evolution through Modularity".

R. Hinze and J. Voigtländer (Eds.): MPC 2015, LNCS 9129, pp. 159–186, 2015.
DOI:10.1007/978-3-319-19797-5_8

This idea has been explored in the literature for quite some time; e.g. by Wand [14], Meijer [10], Ager et al. [1]. Recently, Bahr and Hutton [4] have shown an approach that derives a compiler *directly* from the statement of the compiler correctness property by performing a calculation proof. The parts of the setup that are not given up front, i.e. the compiler itself and the target language, reveal themselves during the calculation proof. Taken together, the final result of the calculation process is a compiler implementation and a (machine checked) formal proof of its correctness. In addition, also the virtual machine, which defines the semantics of the target language, falls out of the calculation proof.

A crucial ingredient of the calculation technique of Bahr and Hutton [4] is that the correctness argument proceeds "in one go", such that at any point during the proof we have a complete view of the computational context and its invariants. So far it was not known how this technique can be applied to languages with an inherent non-deterministic semantics. The problem that arises for these languages is that the compiler correctness property is typically split into a *soundness* and a *completeness* part, which are proved independently; for example, see Hutton and Wright [7]. Another technical challenge arises from the fact that the equational reasoning style used by Bahr and Hutton [4] is incompatible with non-deterministic semantics, which are typically given in the form of a relation (big-step/small-step operational semantics).

In this paper we improve and generalise the technique of Bahr and Hutton [4] for calculating compilers such that we can calculate compilers for non-deterministic languages. While our approach works also for proofs by hand, it works particularly well in a proof assistant such as Coq. In addition to calculating compilers that are correct by construction, our approach is also applicable to post hoc compiler verification.

The key contributions of this paper are the following:

- We devise a set of general proof rules for compiler correctness proofs for non-deterministic languages. These proof rules have been verified in Coq and are the basis for a calculation framework that we developed in Coq.
- We demonstrate the power of our calculation framework on a number of examples both in this paper and in the accompanying Coq development. In particular, we show its effectiveness for both post hoc verification of compilers as well as derivation of correct-by-construction compilers in the style of Bahr and Hutton [4].
- Apart from the ability to deal with non-determinism, the distinguishing feature of our approach is that we are able to *directly* derive a small-step operational semantics of the virtual machine (instead of a tail-recursive function).

To illustrate the problem we are trying to solve we begin with a simple non-deterministic toy language along with a compiler for it, which we define in Sect. 2. In Sect. 3, we present our calculation framework and apply it to the toy language to prove its compiler correct. In Sect. 4, we illustrate how our framework can also be used to derive a correct-by-construction compiler from the specification of its correctness property. We then use this knowledge in Sect. 5 in order to derive a compiler for a simple language with interrupts (taken from Hutton and Wright [7]). Finally, in Sect. 7 we discuss limitations of our approach and outline further work to address these limitations.

This paper uses Coq as a meta language and as a tool for proof automation. However, familiarity with Coq is not required to follow this paper. Section 6 covers some technical details of the use of proof automation in Coq, which does require some familiarity with Coq. Nonetheless, the core idea and the main contributions of this paper are independent of these technical details.

All calculation proofs in this paper can be found in the accompanying Coq source code, which is available from the author's web site[1]. The Coq development also includes calculations for languages that extend the interrupt language of Hutton and Wright [7] with state. Moreover, the source code contains proofs for all theorems presented in this paper including the correctness of the proof rules of our calculation framework.

2 A Simple Non-deterministic Language

To illustrate the problem we aim to solve, we begin with a very simple toy language. The syntax is given by the following inductive type definition:

```
Inductive Expr : Set := Val (n : ℤ) | Add (e₁ e₂ : Expr) | Rnd (e : Expr).
```

Apart from integer literals (represented using the type $\mathbb{Z}$) and an addition operator Add, the language has a construct Rnd to generate a random number. The intended semantics of an expression Rnd e is that it evaluates e to a number n and then returns a number m with $0 \leq m \leq |n|$, where $|n|$ denotes the absolute value of n. For instance, the expression Add (Rnd (Val 5)) (Val 42) generates a random number between 0 and 5 and adds 42 to it.

We formally define the semantics of this simple language by a big-step operational semantics [8], writing $e \Downarrow n$ to mean that the expression e evaluates to the number n. The binary relation $\Downarrow$ is given by the following inductive type definition:

```
Inductive eval : Expr → ℤ → Prop :=
| evalVal n       : Val n ⇓ n
| evalAdd x y m n : x ⇓ m  → y ⇓ n           → Add x y ⇓ (m + n)
| evalRnd x n m   : x ⇓ n  → 0 ⩽ m ⩽ |n|    → Rnd x ⇓ m
where "x ⇓ y" := (eval x y).
```

A more readable form of this definition is given in Fig. 1. From now on, we shall give inductive semantic definitions in this form, with the tacit understanding that it can be easily turned into an inductive type family definition in Coq.

Our compiler for the Expr language will target a simple machine that can store integer values on a stack and that executes the instruction set given by the following type definition:

[1] Or directly from https://github.com/pa-ba/calc-comp-rel.

$$\frac{}{\mathsf{Val\ n \Downarrow n}}\ \textsc{Val} \qquad \frac{\mathsf{x \Downarrow n} \quad \mathsf{y \Downarrow m}}{\mathsf{Add\ x\ y \Downarrow (n{+}m)}}\ \textsc{Add} \qquad \frac{\mathsf{x \Downarrow n} \quad 0 \leqslant \mathsf{m} \leqslant |\mathsf{n}|}{\mathsf{Rnd\ x \Downarrow m}}\ \textsc{Rnd}$$

Fig. 1. Semantics of the language.

```
Inductive Instr : Set := PUSH (n : ℤ) | ADD | RND.
```

The intuitive semantics of the three instructions above is that

- PUSH n pushes the number n onto the stack,
- ADD replaces the two topmost numbers on the stack with their sum, and
- RND replaces the topmost number n on the stack with a number m such that $0 \leqslant m \leqslant |n|$.

A program for this target machine is simply a sequence of these instructions, which is captured by the type Code defined below:

```
Definition Code := list Instr.
```

We will use the notation [] for the empty list, and x :: xs for the list with a head element x and a tail list xs.

Before we give the formal semantics of the target language, we present the compiler that translates expressions of the source language Expr into the target language Code. To this end, the compilation function takes an additional argument of type Code that represents a continuation (cf. Hutton [6], Chap. 13):

```
Fixpoint comp' (e : Expr) (c : Code) : Code :=
  match e with
    | Val n   ⇒ PUSH n :: c
    | Add x y ⇒ comp' x (comp' y (ADD :: c))
    | Rnd x   ⇒ comp' x (RND :: c)
  end.
```

The final compiler is obtained by supplying the empty list of instructions [] as the initial value of the continuation argument:

```
Definition comp (e : Expr) : Code := comp' e [].
```

For instance, the expression Add (Rnd (Val 5)) (Val 42) compiles to the code [PUSH 5; RND; PUSH 42; ADD].

Finally, we give the semantics of the target language Code in the form of a *virtual machine* [1,2]: a small-step operational semantics, given by the reflexive, transitive closure $\overset{*}{\Longrightarrow}$ of a binary relation $\Longrightarrow$ on the type of machine configurations Conf, which is defined below:

```
Definition Stack : Set := list ℤ.
Inductive Conf : Set  := conf (c : Code) (s : Stack).
Notation "⟨ c , s ⟩" := (conf c s).
```

$$\langle \mathsf{PUSH\ n :: c\ ,\ s} \rangle \Longrightarrow \langle \mathsf{c\ ,\ n :: s} \rangle \qquad \text{(VM-Push)}$$

$$\langle \mathsf{ADD :: c,\ m :: n :: s} \rangle \Longrightarrow \langle \mathsf{c,\ (n + m) :: s} \rangle \qquad \text{(VM-Add)}$$

$$\langle \mathsf{RND :: c,\ n :: s} \rangle \Longrightarrow \langle \mathsf{c,\ m :: s} \rangle \qquad \text{if } 0 \leqslant \mathsf{m} \leqslant |\mathsf{n}| \qquad \text{(VM-Rnd)}$$

Fig. 2. Semantics of the language.

A configuration describes the state of the virtual machine; it is a pair $\langle \mathsf{c, s} \rangle$ consisting of a code c and a stack s. The binary relation $\Longrightarrow$ describes a single computation step in the virtual machine: if $\mathsf{C} \Longrightarrow \mathsf{C'}$, then the virtual machine transitions from configuration C to configuration C' in one step. The inductive definition of the relation $\Longrightarrow$ is presented in Fig. 2.

For example, the code [PUSH 5; RND; PUSH 42; ADD] is executed by the virtual machine starting with the empty stack as follows:

$$\begin{aligned} &\langle [\mathsf{PUSH\ 5; RND; PUSH\ 42; ADD}], [] \rangle \\ &\quad \Longrightarrow \langle [\mathsf{RND; PUSH\ 42; ADD}], [5] \rangle && \text{(by VM-Push)} \\ &\quad \Longrightarrow \langle [\mathsf{PUSH\ 42; ADD}], [3] \rangle && \text{(by VM-Rnd)} \\ &\quad \Longrightarrow \langle [\mathsf{ADD}], [42; 3] \rangle && \text{(by VM-Push)} \\ &\quad \Longrightarrow \langle [], [45] \rangle && \text{(by VM-Add)} \end{aligned}$$

That is, $\langle [\mathsf{PUSH\ 5; RND; PUSH\ 42; ADD}], [] \rangle \overset{*}{\Longrightarrow} \langle [], [45] \rangle$. However, this is not the only possible execution. The rule for RND allows for more than one successor configuration after $\langle [\mathsf{RND; PUSH\ 42; ADD}], [5] \rangle$.

3 Correctness Property

Intuitively, the correctness property of the compiler comp states that for each expression e, running the code comp e produced by the compiler on the virtual machine yields the same result as evaluating e according to $\Downarrow$. Taking account of the non-determinism, the correctness property reads as follows: an expression e *may* evaluate to n iff running comp e on the virtual machine *may* produce the result n. The "only if" direction of this equivalence expresses completeness, whereas the "if" direction expresses soundness.

More formally, for the completeness property we want that whenever $\mathsf{e} \Downarrow \mathsf{n}$, then the virtual machine computes the result n as well, i.e. $\langle \mathsf{comp\ e}, [] \rangle \overset{*}{\Longrightarrow} \langle []\ ,\ [\mathsf{n}] \rangle$. In order to prove this property by induction, we have to generalise it to arbitrary c :: Code and s :: Stack as follows:

$$\forall\, \mathsf{e\ n\ c\ s},\quad \mathsf{e} \Downarrow \mathsf{n} \rightarrow \langle \mathsf{comp'\ e\ c,\ s} \rangle \overset{*}{\Longrightarrow} \langle \mathsf{c\ ,\ n :: s} \rangle \qquad \text{(COMPLETENESS)}$$

Conversely, for the soundness property we want that whenever we have $\langle \mathsf{comp\ e}, [] \rangle \overset{*}{\Longrightarrow} \mathsf{C}$, then we find some $\mathsf{C} \overset{*}{\Longrightarrow} \langle [], [\mathsf{n}] \rangle$ with $\mathsf{e} \Downarrow \mathsf{n}$. In other words, any run of the virtual machine can be extended such that it ends in a result value n such that $\mathsf{e} \Downarrow \mathsf{n}$. We have to generalise this property as well, in order to be able to prove it by induction. However, simply generalising it to arbitrary $\mathsf{c} :: \mathsf{Code}$ and $\mathsf{s} :: \mathsf{Stack}$ as follows is not enough:

$$\forall\, \mathsf{e\ c\ s},\quad \langle \mathsf{comp'}\ \mathsf{e}\ \mathsf{c}, \mathsf{s}\rangle \overset{*}{\Longrightarrow} C \rightarrow \exists\, \mathsf{n}, C \overset{*}{\Longrightarrow} \langle \mathsf{c}, \mathsf{n} :: \mathsf{s}\rangle \wedge \mathsf{e} \Downarrow \mathsf{n}$$

Unfortunately, the above straightforward generalisation is not true. The problem is that the given run $\langle \mathsf{comp'}\ \mathsf{e}\ \mathsf{c}, \mathsf{s}\rangle \overset{*}{\Longrightarrow} C$ may have gone already beyond the configuration $\langle \mathsf{c}, \mathsf{n} :: \mathsf{s}\rangle$. That is, we have that

$$\langle \mathsf{comp'}\ \mathsf{e}\ \mathsf{c}, \mathsf{s}\rangle \overset{*}{\Longrightarrow} \langle \mathsf{c}, \mathsf{n} :: \mathsf{s}\rangle \overset{*}{\Longrightarrow} C$$

Thus, we can in general not expect that $C \overset{*}{\Longrightarrow} \langle \mathsf{c}, \mathsf{n} :: \mathsf{s}\rangle$.

To take this situation into account, we follow an approach similar to Hutton and Wright [7]: we formulate the soundness property in terms of the notion that a machine configuration $C : \mathsf{Conf}$ is *barred* by a set of configurations $S : \mathsf{ConfSet}$, denoted $C \lhd S$, cf. Troelstra and van Dalen [13]. For the moment, we shall remain informal about what the type $\mathsf{ConfSet}$ of sets of configurations is and appeal to the intuitive notion of sets. Intuitively, $C \lhd S$ means that any sequence of $\Longrightarrow$-steps starting from C can be extended such that it *passes through* a configuration that is in S, i.e. for any sequence $C_0 \Longrightarrow C_1 \Longrightarrow \ldots \Longrightarrow C_n$ with $C = C_0$, we find a sequence $C_n \Longrightarrow \ldots \Longrightarrow C_m$ such that $C_i \in S$ for some $0 \leqslant i \leqslant m$. Formally, we define $\lhd$ by the following inductive rules:

$$\frac{C \in S}{C \lhd S}\ \textsc{Here-}\lhd \qquad \frac{\forall\, D, C \Longrightarrow D \rightarrow D \lhd S \qquad \exists\, D, C \Longrightarrow D}{C \lhd S}\ \textsc{Step-}\lhd$$

We can then use the relation $\lhd$ to capture the soundness property of the compiler by the following statement:

$$\forall\, \mathsf{e\ c\ s},\quad \langle \mathsf{comp'}\ \mathsf{e}\ \mathsf{c}, \mathsf{s}\rangle \lhd \{\mathsf{n}, \langle \mathsf{c}, \mathsf{n} :: \mathsf{s}\rangle \mid \mathsf{e} \Downarrow \mathsf{n}\} \qquad (\textsc{soundness})$$

Here we use the set comprehension notation $\{\mathsf{n}, \langle \mathsf{c}, \mathsf{n} :: \mathsf{s}\rangle \mid \mathsf{e} \Downarrow \mathsf{n}\}$, which explicitly mentions the existentially quantified variable n, whose scope ranges over both the expression $\langle \mathsf{c}, \mathsf{n} :: \mathsf{s}\rangle$ and the predicate $\mathsf{e} \Downarrow \mathsf{n}$. That is, $\{\mathsf{n}, \langle \mathsf{c}, \mathsf{n} :: \mathsf{s}\rangle \mid \mathsf{e} \Downarrow \mathsf{n}\}$ denotes the set of configurations of the form $\langle \mathsf{c}, \mathsf{n} :: \mathsf{s}\rangle$ such that $\mathsf{e} \Downarrow \mathsf{n}$ holds. In an informal set comprehension notation, one would typically write $\{\langle \mathsf{c}, \mathsf{n} :: \mathsf{s}\rangle \mid \mathsf{e} \Downarrow \mathsf{n}\}$ instead.

The above property (SOUNDNESS) does indeed imply the desired soundness property, i.e. that $\langle \mathsf{comp}\ \mathsf{e}, []\rangle \overset{*}{\Longrightarrow} C$ implies both $C \overset{*}{\Longrightarrow} \langle [], [\mathsf{n}]\rangle$ and $\mathsf{e} \Downarrow \mathsf{n}$ for some n. This implication is a consequence of the following *general* property of the relation $\lhd$:

Proposition 1. *Let S : ConfSet be such that all $C \in S$ are in normal form, i.e. there is no D with $C \Longrightarrow D$. If $C_1 \lhd S$ and $C_1 \overset{*}{\Longrightarrow} C_2$, then $C_2 \overset{*}{\Longrightarrow} C_3$ and $C_3 \in S$ for some C_3.*

In other words, any sequence of $\Longrightarrow$-steps starting from a configuration that is barred by a set of normal forms S can be extended such that it ends in a configuration in S. This property is general in the sense that it is independent of the definition of Conf and $\Longrightarrow$.

For the above proposition to be true, it is important that we have $\exists \mathsf{D}, \mathsf{C} \Longrightarrow \mathsf{D}$ as a second antecedent in the STEP-$\lhd$ rule. Without it, we would have $\mathsf{C} \lhd \mathsf{S}$ for any normal form C, i.e. any C for which there is no D with $\mathsf{C} \Longrightarrow \mathsf{D}$. In other words, we would be able to prove "soundness" even though some computations in the virtual machine might get stuck.[2] We will review an example that illustrates this in Sect. 7.

Our goal is to prove soundness and completeness for the compiler in a calculational style. Such a proof was given by Hutton and Wright [7] (for a much more interesting language, which we will consider in Sect. 5). They combine the two relations $\overset{*}{\Longrightarrow}$ and $\lhd$ into a single relation $\lll$ by defining $\mathsf{C} \lll \mathsf{S}$ iff $\mathsf{C} \overset{*}{\Longrightarrow} \mathsf{S}$ and $\mathsf{C} \lhd \mathsf{S}$, where $\overset{*}{\Longrightarrow}$ is lifted to sets by defining $\mathsf{C} \overset{*}{\Longrightarrow} \mathsf{S}$ iff $\mathsf{C} \overset{*}{\Longrightarrow} \mathsf{D}$ for all $\mathsf{D} \in \mathsf{S}$. Nonetheless, this calculational proof of Hutton and Wright [7] considers soundness and completeness separately. In order to combine the two proofs into a single calculational proof, we have to overcome two obstacles:

Firstly, the completeness proof proceeds by induction on (the proof of) $\mathsf{e} \Downarrow \mathsf{n}$, whereas the soundness proof proceeds by induction on e. This technical hurdle is overcome by transforming the completeness proof into an induction on e. While this approach may not be possible for every language, it does not restrict the applicability of the proof method as a whole, since the soundness proof is already an induction on e (cf. discussion in Sect. 7).

Secondly, the two proofs utilise proof principles that are not valid for both $\overset{*}{\Longrightarrow}$ and $\lhd$. For example, the completeness proof makes use of the fact that $\mathsf{C} \Longrightarrow \mathsf{S}$ implies $\mathsf{C} \overset{*}{\Longrightarrow} \mathsf{S}$, whereas $\mathsf{C} \Longrightarrow \mathsf{S}$ does not necessarily imply that $\mathsf{C} \lhd \mathsf{S}$. Conversely, the soundness proof makes use of the fact that if $\mathsf{S} \subseteq \mathsf{T}$, then $\mathsf{C} \lhd \mathsf{S}$ implies $\mathsf{C} \lhd \mathsf{T}$, whereas $\mathsf{C} \overset{*}{\Longrightarrow} \mathsf{S}$ does not necessarily imply $\mathsf{C} \overset{*}{\Longrightarrow} \mathsf{T}$.

Overcoming the second hurdle is more difficult. But the underlying idea to solve this problem is quite simple. We take the completeness proof as our basis. The only thing that is missing in order to turn this proof into a soundness proof as well is the fact that we do not have that $\mathsf{C} \Longrightarrow \mathsf{S}$ implies $\mathsf{C} \lhd \mathsf{S}$ in general. However, the additional proof obligation necessary to conclude $\mathsf{C} \lhd \mathsf{S}$ can be discharged automatically by proof search on the $\Longrightarrow$ relation.

The automation of proofs of $\mathsf{C} \lhd \mathsf{S}$ is particularly important for our goal of deriving a compiler by calculation, which we will discuss in Sect. 5. In that setting, both compiler and virtual machine are not fully defined in the beginning of the calculation. As the calculation progresses we flesh out the definition of the compiler and the virtual machine. In particular, we add new rules for $\Longrightarrow$. As a consequence, a statement of the form $\mathsf{C} \lhd \mathsf{S}$, might not hold anymore after we have extended the virtual machine relation $\Longrightarrow$. The proof search will extend

[2] The side condition $\exists\, \mathsf{D}, \mathsf{C} \Longrightarrow \mathsf{D}$ is absent in the definition of $\lhd$ by Hutton and Wright [7]. However, their proofs need little change to account for this stronger version of $\lhd$, which then yields proper soundness and completeness for their compiler.

the proof to account for the modified definition of $\Longrightarrow$. If this fails, we are immediately informed that the added rule for $\Longrightarrow$ breaks the existing proof.

3.1 Proof Principles

In this section we will present the proof principles for constructing soundness and completeness proofs. These proof principles are general lemmas about the two relations $\overset{*}{\Longrightarrow}$ and $\lhd$, i.e. they are independent of the definition of Conf and $\Longrightarrow$. Our goal is to have a combined relation, similar to $\lessdot$ of Hutton and Wright [7], that will allow us to prove soundness and completeness in one go. To this end we will lift both relations to sets of configurations; recall that $\overset{*}{\Longrightarrow}$ is of type $\mathsf{Conf} \rightarrow \mathsf{Conf} \rightarrow \mathsf{Prop}$, whereas $\lhd$ is of type $\mathsf{Conf} \rightarrow \mathsf{ConfSet} \rightarrow \mathsf{Prop}$.

As a first step we have to define, what a set of configurations is. The corresponding type $\mathsf{ConfSet}$, has to provide the *element* relation $\in$ of type $\mathsf{Conf} \rightarrow \mathsf{ConfSet} \rightarrow \mathsf{Prop}$. Moreover, we need to be able to construct the empty set $\emptyset$, form the union $\mathsf{S} \cup \mathsf{T}$ of two sets, and define sets using set comprehension notation like the example $\{\mathsf{n}\ ,\ \langle \mathsf{c}\ ,\ \mathsf{n} :: \mathsf{s}\rangle \mid \mathsf{e} \Downarrow \mathsf{n}\}$ above. In general, set comprehensions have the form $\{\overline{x}, \mathsf{C} \mid \mathsf{P}\}$, where $\overline{x}$ is a list of variables that are existentially quantified in C and P, which in turn are of type Conf and Prop, respectively. We shall return to the technical issue of representing this notation in Coq – and more importantly how to reason over it – in Sect. 6.

The inductive rules below lift $\overset{*}{\Longrightarrow}$ and $\lhd$ to the relations $\Rrightarrow$ and $\lll$, respectively, both of type $\mathsf{ConfSet} \rightarrow \mathsf{ConfSet} \rightarrow \mathsf{Prop}$:

$$\frac{\forall\, \mathsf{D},\ \mathsf{D} \in \mathsf{T} \rightarrow (\exists\, \mathsf{C},\ \mathsf{C} \in \mathsf{S} \wedge \mathsf{C} \overset{*}{\Longrightarrow} \mathsf{D})}{\mathsf{S} \Rrightarrow \mathsf{T}} \qquad \frac{\forall\, \mathsf{C},\ \mathsf{C} \in \mathsf{S} \rightarrow \mathsf{C} \lhd \mathsf{T}}{\mathsf{S} \lll \mathsf{T}}$$

That is, $\mathsf{S} \Rrightarrow \mathsf{T}$ iff all configurations in T are reachable from some configuration in S; and $\mathsf{S} \lll \mathsf{T}$ iff all configurations in S are barred by T.

Using these two relations, we can reformulate the soundness and completeness properties as follows:

$$\forall\, \mathsf{e\ c\ s},\quad \{\langle \mathsf{comp'\ e\ c,\ s}\rangle\} \Rrightarrow \{\mathsf{n},\ \langle \mathsf{c}\ ,\ \mathsf{n} :: \mathsf{s}\rangle \mid \mathsf{e} \Downarrow \mathsf{n}\} \qquad \text{(COMPLETENESS)}$$

$$\forall\, \mathsf{e\ c\ s},\quad \{\langle \mathsf{comp'\ e\ c,\ s}\rangle\} \lll \{\mathsf{n},\ \langle \mathsf{c}\ ,\ \mathsf{n} :: \mathsf{s}\rangle \mid \mathsf{e} \Downarrow \mathsf{n}\} \qquad \text{(SOUNDNESS)}$$

We then combine the two relations $\Rrightarrow$ and $\lll$ into the relation $\Rightarrow\!\!\rhd$ by forming their intersection, which we express by the following inductive rule:

$$\frac{\mathsf{S} \Rrightarrow \mathsf{T} \qquad \mathsf{S} \lll \mathsf{T}}{\mathsf{S} \Rightarrow\!\!\rhd \mathsf{T}}$$

This combined relation allows us to succinctly formulate the correctness property of the compiler:

$$\forall\, \mathsf{e\ c\ s},\quad \{\langle \mathsf{comp'\ e\ c,\ s}\rangle\} \Rightarrow\!\!\rhd \{\mathsf{n},\ \langle \mathsf{c}\ ,\ \mathsf{n} :: \mathsf{s}\rangle \mid \mathsf{e} \Downarrow \mathsf{n}\} \qquad \text{(CORRECTNESS)}$$

$$\frac{}{\mathsf{S} \Rrightarrow \mathsf{S}}\ \textsc{Refl} \qquad \frac{\mathsf{S} \Rrightarrow \mathsf{T} \quad \mathsf{T} \Rrightarrow \mathsf{U}}{\mathsf{S} \Rrightarrow \mathsf{U}}\ \textsc{Trans} \qquad \frac{\mathsf{S} \equiv \mathsf{T}}{\mathsf{S} \Rrightarrow \mathsf{T}}\ \textsc{Iff}$$

$$\frac{\mathsf{S} \Rrightarrow \mathsf{T} \quad \mathsf{S'} \Rrightarrow \mathsf{T'}}{\mathsf{S} \cup \mathsf{S'} \Rrightarrow \mathsf{T} \cup \mathsf{T'}}\ \textsc{Union}$$

$$\frac{\forall \overline{\mathsf{x}}, \mathsf{P} \rightarrow \mathsf{C} \Longrightarrow \mathsf{D} \qquad \forall \overline{\mathsf{x}}, \mathsf{P} \rightarrow \mathsf{C} \lhd \{\overline{\mathsf{x}}, \mathsf{D} \mid \mathsf{P}\} \cup \mathsf{T}}{\{\overline{\mathsf{x}}, \mathsf{C} \mid \mathsf{P}\} \cup \mathsf{T} \Rrightarrow \{\overline{\mathsf{x}}, \mathsf{D} \mid \mathsf{P}\} \cup \mathsf{T}}\ \textsc{Step}$$

Fig. 3. Proof principles.

Figure 3 lists the proof rules that we can derive from the definition of $\Rrightarrow$. Most importantly we have reflexivity, transitivity and closure under union, which will allow us do calculational proofs. In addition, $\Rrightarrow$ is closed under extensional set equality $\equiv$, which is defined as follows:

$$\frac{\forall \mathsf{C}, \mathsf{C} \in \mathsf{S} \leftrightarrow \mathsf{C} \in \mathsf{T}}{\mathsf{S} \equiv \mathsf{T}}$$

Lastly, the Step proof rule in Fig. 3 is the only *non-structural* rule. It allows us to "advance" one step according to the $\Longrightarrow$ relation. In order to avoid syntactic clutter, the rule is stated somewhat informally: it is implicitly assumed that P, C and $\mathsf{C'}$ are *expressions* that may contain free variables from $\overline{\mathsf{x}}$. Formalising and proving this rule in Coq requires some technical effort. But we defer discussion of this aspect until Sect. 6.

The Step rule should feel intuitively true: in order to derive

$$\{\overline{\mathsf{x}}, \mathsf{C} \mid \mathsf{P}\} \cup \mathsf{T} \Rrightarrow \{\overline{\mathsf{x}}, \mathsf{D} \mid \mathsf{P}\} \cup \mathsf{T},$$

we have to show a step $\mathsf{C} \Longrightarrow \mathsf{D}$ and that $\mathsf{C} \lhd \{\overline{\mathsf{x}}, \mathsf{D} \mid \mathsf{P}\} \cup \mathsf{T}$ – both in a context of free variables $\overline{\mathsf{x}}$ and assuming that P is true. In practice, the latter proof obligation can be discharged automatically by a simple proof search: check whether C is in the set $\{\overline{\mathsf{x}}, \mathsf{D} \mid \mathsf{P}\} \cup \mathsf{T}$; if not, then recursively check whether $\mathsf{C'} \lhd \{\overline{\mathsf{x}}, \mathsf{D} \mid \mathsf{P}\} \cup \mathsf{T}$ for every $\mathsf{C'}$ with $\mathsf{C} \Longrightarrow \mathsf{C'}$.

3.2 Correctness Proof

To illustrate the proof rules we shall prove correctness of the compiler from Sect. 2. Recall that the correctness property of the compiler is formulated using $\Rrightarrow$ as follows:

$$\forall\, \mathsf{e\ c\ s},\quad \{\langle \mathsf{comp'\ e\ c, s}\rangle\} \Rrightarrow \{\mathsf{n}, \langle \mathsf{c}, \mathsf{n} :: \mathsf{s}\rangle \mid \mathsf{e} \Downarrow \mathsf{n}\}$$

Unfortunately, this correctness property is not suitable for an induction proof. The induction hypothesis is not general enough, since the left-hand side of the

$\Rightarrow$ is always a singleton set. To avoid this issue we generalise the correctness property as follows:

$$\forall\, \mathsf{e}\ \mathsf{c}\ \mathsf{P},\quad \{\mathsf{s},\ \langle \mathsf{comp'}\ \mathsf{e}\ \mathsf{c},\ \mathsf{s}\rangle \mid \mathsf{P}\ \mathsf{s}\} \Rightarrow \{\mathsf{s}\ \mathsf{n},\ \langle \mathsf{c}\ ,\ \mathsf{n} :: \mathsf{s}\rangle \mid \mathsf{e} \Downarrow \mathsf{n} \wedge \mathsf{P}\ \mathsf{s}\}$$

Instead of quantifying over stacks $\mathsf{s} : \mathsf{Stack}$, we quantify over predicates on stacks $\mathsf{P} : \mathsf{Stack} \to \mathsf{Prop}$. We can then prove the above property by induction on $\mathsf{e} : \mathsf{Expr}$. The calculations are given in Fig. 4. Note that we calculate "backwards", i.e. from the right-hand side to the left-hand side. This approach has the benefit that we can let the semantics $\mathsf{e} \Downarrow \mathsf{n}$ guide the calculation. Later in Sect. 5, performing the calculation in this direction becomes crucial: it allows us to start the proof without having defined the compiler nor the virtual machine. Instead the definitions of the compiler and the virtual machine fall out of the calculation process itself.

Before we go into the details of the calculation, we review how it is built up using the proof rules of Fig. 3. Each step of the calculation corresponds to a relation $\mathsf{X} \Leftarrow \mathsf{Y}$, together with its justification. Instead of $\Leftarrow$, we use the symbol $\equiv$ or $\Longleftarrow$, if the relation is justified by proof rule IFF or STEP, respectively. For instance, in the calculation for $\mathsf{Val}\ \mathsf{n}$, we first observe that we have the set equivalence

$$\{\mathsf{s}\ \mathsf{n'},\ \langle \mathsf{c},\ \mathsf{n'} :: \mathsf{s}\rangle \mid \mathsf{Val}\ \mathsf{n} \Downarrow \mathsf{n'} \wedge \mathsf{P}\ \mathsf{s}\} \equiv \{\mathsf{s},\ \langle \mathsf{c},\ \mathsf{n} :: \mathsf{s}\rangle \mid \mathsf{P}\ \mathsf{s}\}$$

This equivalence is justified by the rule VAL of the semantics (cf. Fig. 1), according to which $\mathsf{Val}\ \mathsf{n} \Downarrow \mathsf{n'}$ is equivalent to the equation $\mathsf{n} = \mathsf{n'}$. Applying rule IFF, then yields that

$$\{\mathsf{s}\ \mathsf{n'},\ \langle \mathsf{c},\ \mathsf{n'} :: \mathsf{s}\rangle \mid \mathsf{Val}\ \mathsf{n} \Downarrow \mathsf{n'} \wedge \mathsf{P}\ \mathsf{s}\} \Leftarrow \{\mathsf{s},\ \langle \mathsf{c},\ \mathsf{n} :: \mathsf{s}\rangle \mid \mathsf{P}\ \mathsf{s}\}$$

Similarly, the second step of the calculation indicates that

$$\langle \mathsf{c},\ \mathsf{n} :: \mathsf{s}\rangle \Longleftarrow \langle \mathsf{PUSH}\ \mathsf{n} :: \mathsf{c},\ \mathsf{s}\rangle$$

due to the rule VM-PUSH of the virtual machine (cf. Fig. 2). Using proof rule STEP we then obtain the desired relation

$$\{\mathsf{s},\ \langle \mathsf{c},\ \mathsf{n} :: \mathsf{s}\rangle \mid \mathsf{P}\ \mathsf{s}\} \Leftarrow \{\mathsf{s},\ \langle \mathsf{PUSH}\ \mathsf{n} :: \mathsf{c},\ \mathsf{s}\rangle \mid \mathsf{P}\ \mathsf{s}\}$$

given that the following side condition is met:

$$\forall\, \mathsf{s},\ \mathsf{P}\ \mathsf{s} \to \langle \mathsf{PUSH}\ \mathsf{n} :: \mathsf{c},\ \mathsf{s}\rangle \triangleleft \{\mathsf{s},\ \langle \mathsf{c},\ \mathsf{n} :: \mathsf{s}\rangle \mid \mathsf{P}\ \mathsf{s}\}$$

As we have mentioned earlier, these side conditions are trivial to check. In this particular case, $\langle \mathsf{c},\ \mathsf{n} :: \mathsf{s}\rangle$ is the only successor configuration of $\langle \mathsf{PUSH}\ \mathsf{n} :: \mathsf{c},\ \mathsf{s}\rangle$. Thus the side condition is met.

Finally, the individual calculation steps are combined via the TRANS proof rule, which yields that

$$\{\mathsf{s},\ \langle \mathsf{comp'}\ (\mathsf{Val}\ \mathsf{n})\ \mathsf{c},\ \mathsf{s}\rangle \mid \mathsf{P}\ \mathsf{s}\} \Rightarrow \{\mathsf{s}\ \mathsf{n'},\ \langle \mathsf{c},\ \mathsf{n'} :: \mathsf{s}\rangle \mid \mathsf{Val}\ \mathsf{n} \Downarrow \mathsf{n'} \wedge \mathsf{P}\ \mathsf{s}\}$$

– Val n:

$$
\begin{array}{ll}
& \{\mathsf{s\ n'},\ \langle \mathsf{c},\ \mathsf{n'} :: \mathsf{s}\rangle \mid \mathsf{Val\ n} \Downarrow \mathsf{n'} \wedge \mathsf{P\ s}\} \\
\equiv & \quad \{ \text{ by } \textsc{Val} \} \\
& \{\mathsf{s},\ \langle \mathsf{c},\ \mathsf{n} :: \mathsf{s}\rangle \mid \mathsf{P\ s}\} \\
\Longleftarrow & \quad \{ \text{ by VM-}\textsc{Push} \} \\
& \{\mathsf{s},\ \langle \mathsf{PUSH\ n} :: \mathsf{c},\ \mathsf{s}\rangle \mid \mathsf{P\ s}\} \\
\equiv & \quad \{ \text{ definition of comp' } \} \\
& \{\mathsf{s},\ \langle \mathsf{comp'\ (Val\ n)\ c},\ \mathsf{s}\rangle \mid \mathsf{P\ s}\}
\end{array}
$$

– Add e_1 e_2:

$$
\begin{array}{ll}
& \{\mathsf{s\ n},\ \langle \mathsf{c},\ \mathsf{n} :: \mathsf{s}\rangle \mid \mathsf{Add}\ e_1\ e_2 \Downarrow \mathsf{n} \wedge \mathsf{P\ s}\} \\
\equiv & \quad \{ \text{ by } \textsc{Add} \} \\
& \{\mathsf{s\ n\ m},\ \langle \mathsf{c},\ (\mathsf{n}+\mathsf{m}) :: \mathsf{s}\rangle \mid e_1 \Downarrow \mathsf{n} \wedge e_2 \Downarrow \mathsf{m} \wedge \mathsf{P\ s}\} \\
\Longleftarrow & \quad \{ \text{ by VM-}\textsc{Add} \} \\
& \{\mathsf{s\ n\ m},\ \langle \mathsf{ADD} :: \mathsf{c},\ \mathsf{m} :: \mathsf{n} :: \mathsf{s}\rangle \mid e_1 \Downarrow \mathsf{n} \wedge e_2 \Downarrow \mathsf{m} \wedge \mathsf{P\ s}\} \\
\equiv & \quad \{ \text{ move existential quantifier } \} \\
& \{\mathsf{s'\ m},\ \langle \mathsf{ADD} :: \mathsf{c},\ \mathsf{m} :: \mathsf{s'}\rangle \mid e_2 \Downarrow \mathsf{m} \wedge (\exists\, \mathsf{s\ n},\ e_1 \Downarrow \mathsf{n} \wedge \mathsf{s'} = \mathsf{n} :: \mathsf{s} \wedge \mathsf{P\ s})\} \\
\Leftarrow & \quad \{ \text{ induction hypothesis for } e_2 \} \\
& \{\mathsf{s},\ \langle \mathsf{comp'}\ e_2\ (\mathsf{ADD} :: \mathsf{c})\ ,\ \mathsf{s}\rangle \mid \exists\, \mathsf{s'\ n},\ e_1 \Downarrow \mathsf{n} \wedge \mathsf{s} = \mathsf{n} :: \mathsf{s'} \wedge \mathsf{P\ s'}\} \\
\equiv & \quad \{ \text{ move existential quantifier } \} \\
& \{\mathsf{s\ n},\ \langle \mathsf{comp'}\ e_2\ (\mathsf{ADD} :: \mathsf{c})\ ,\ \mathsf{n} :: \mathsf{s}\rangle \mid e_1 \Downarrow \mathsf{n} \wedge \mathsf{P\ s}\} \\
\Leftarrow & \quad \{ \text{ induction hypothesis for } e_1 \} \\
& \{\mathsf{s},\ \langle \mathsf{comp'}\ e_1\ (\mathsf{comp'}\ e_2\ (\mathsf{ADD} :: \mathsf{c}))\ ,\ \mathsf{s}\rangle \mid \mathsf{P\ s}\} \\
\equiv & \quad \{ \text{ definition of comp' } \} \\
& \{\mathsf{s},\ \langle \mathsf{comp'\ (Add}\ e_1\ e_2)\ \mathsf{c},\ \mathsf{s}\rangle \mid \mathsf{P\ s}\}
\end{array}
$$

– Rnd e:

$$
\begin{array}{ll}
& \{\mathsf{s\ m},\ \langle \mathsf{c},\ \mathsf{m} :: \mathsf{s}\rangle \mid \mathsf{Rnd\ e} \Downarrow \mathsf{m} \wedge \mathsf{P\ s}\} \\
\equiv & \quad \{ \text{ by } \textsc{Rnd} \} \\
& \{\mathsf{s\ m\ n},\ \langle \mathsf{c},\ \mathsf{m} :: \mathsf{s}\rangle \mid \mathsf{e} \Downarrow \mathsf{n} \wedge 0 \leqslant \mathsf{m} \leqslant |\mathsf{n}| \wedge \mathsf{P\ s}\} \\
\Longleftarrow & \quad \{ \text{ by VM-}\textsc{Rnd} \} \\
& \{\mathsf{s\ m\ n},\ \langle \mathsf{RND} :: \mathsf{c},\ \mathsf{n} :: \mathsf{s}\rangle \mid \mathsf{e} \Downarrow \mathsf{n} \wedge 0 \leqslant \mathsf{m} \leqslant |\mathsf{n}| \wedge \mathsf{P\ s}\} \\
\equiv & \quad \{ \text{ eliminate tautology } \exists\, \mathsf{m},\ 0 \leqslant \mathsf{m} \leqslant |\mathsf{n}| \} \\
& \{\mathsf{s\ n},\ \langle \mathsf{RND} :: \mathsf{c},\ \mathsf{n} :: \mathsf{s}\rangle \mid \mathsf{e} \Downarrow \mathsf{n} \wedge \mathsf{P\ s}\} \\
\Leftarrow & \quad \{ \text{ induction hypothesis for e } \} \\
& \{\mathsf{s},\ \langle \mathsf{comp'\ e\ (RND} :: \mathsf{c})\ ,\ \mathsf{s}\rangle \mid \mathsf{P\ s}\} \\
\equiv & \quad \{ \text{ definition of comp' } \} \\
& \{\mathsf{s},\ \langle \mathsf{comp'\ (Rnd\ e)\ c},\ \mathsf{s}\rangle \mid \mathsf{P\ s}\}
\end{array}
$$

Fig. 4. Correctness proof for compiler comp'

and therefore proves the correctness property for the case Val n.

The calculation for Add e_1 e_2 illustrates the need to generalise the correctness property in order to obtain an induction hypothesis that is strong enough. The induction hypotheses for this case are the following, for each $i \in \{1, 2\}$:

$$\forall\, \mathsf{c'}\ \mathsf{P'},\quad \{\mathsf{s},\ \langle \mathsf{comp'}\ e_i\ \mathsf{c'},\ \mathsf{s}\rangle \mid \mathsf{P'\ s}\} \Rightarrow \{\mathsf{s\ n},\ \langle \mathsf{c'}\ ,\ \mathsf{n} :: \mathsf{s}\rangle \mid e_i \Downarrow \mathsf{n} \wedge \mathsf{P'\ s}\}$$

The calculation step that uses the induction hypothesis for e_2 instantiates c' with ADD :: c, and the predicate P' with

$$\exists\, s\, n,\ e_1 \Downarrow n \ \wedge\ s' \ =\ n :: s \ \wedge\ P\, s.$$

The instantiation of the predicate P' in the induction hypothesis allows us to preserve invariants of the stack. The ability to express such invariants is crucial for reasoning about compiler correctness.

Our Coq library that implements this calculational reasoning provides a syntax that is very close to the idealised syntax we used in Fig. 4. To illustrate this, Fig. 5 shows the full Coq proof of the correctness theorem.

4 Calculating a Compiler

In the previous section, we started out with the definition of the semantics of both the source and the target language of the compiler, together with the definition of the compiler itself. We then set out to to prove that this compiler is correct. The proof by calculation, however, also lends itself to a different setup: given the source language (including its semantics), we want to derive a suitable target language and a compiler that satisfies the correctness property. The idea of deriving a compiler from its specification has been explored in detail by a number of authors [1,4,10,14]. Recently, Bahr and Hutton [4] have shown that such a derivation can be performed by simply stating the correctness property of the compiler and then performing the calculation proof. The parts of the setup that are not defined yet, i.e. the compiler itself and the target language, reveal themselves during the calculation proof.

To illustrate this idea of Bahr and Hutton [4] we reconsider the correctness proof from Sect. 3.2. We need to prove the following property about comp':

$$\forall\, e\, c\, P,\quad \{s,\ \langle \text{comp' } e\ c,\ s\rangle \ \mid\ P\, s\} \ \Rightarrow\!\!\!\triangleright\ \{s\, n,\ \langle c\ ,\ n :: s\rangle \ \mid\ e \Downarrow n \ \wedge\ P\, s\}$$

To do so, the calculation proof "transforms" the right-hand side into the left-hand side using the proof rules for $\Rightarrow\!\!\!\triangleright$. How can we do this, if the compiler comp' is not defined yet? The idea is to transform the right-hand side into the form $\{s,\ \langle c',\ s\rangle \ \mid\ P\, s\}$ for some c' : Code, and then simply take comp' e c = c' as a *defining* equation for comp'. The calculation proof in Fig. 4, can be read this way by simply removing the last step in each case of the proof. These are the only steps that make reference to the definition of comp'. Instead of using the definition of comp', we can interpret the final calculation step as the discovery of how comp' must be defined such that the calculation proof can be completed.

However, we do not only derive the compiler from the calculation but also the target language and its semantics. The idea is quite simple: we introduce new elements into the type Instr of instructions and corresponding rules for $\Longrightarrow$ such that we can use the STEP proof rule to manipulate the state of the configuration such that we can apply the induction hypothesis or arrive at the target pattern $\{s,\ \langle c',\ s\rangle \ \mid\ P\, s\}$.

```
Theorem correctness : forall e P c,
  {s, ⟨comp' e c, s ⟩ | P s} =|> {s n, ⟨c , n :: s ⟩ | e ⇓ n /\ P s}.
Proof.
induction e;intros.

begin
  ({s n', ⟨c, n' :: s ⟩ | Val n ⇓ n' /\ P s}).
=  { by_eval }
  ({s, ⟨c, n :: s ⟩ | P s}) .
<== { apply vm_push }
  ({s, ⟨PUSH n :: c, s ⟩ | P s}) .
 [].

begin
  ({s n, ⟨c, n :: s ⟩ | Add e1 e2 ⇓ n /\ P s }) .
=  { by_eval }
  ({s n m, ⟨c, (n + m) :: s ⟩ | e1 ⇓ n /\ e2 ⇓ m /\ P s}) .
<== { apply vm_add }
  ({s n m, ⟨ADD :: c, m :: n :: s ⟩ | e1 ⇓ n  /\ e2 ⇓ m /\ P s}).
=  { eauto }
  ({s' m, ⟨ADD :: c, m :: s' ⟩ | e2 ⇓ m
                             /\ (exists s n, e1 ⇓ n /\ s' = n :: s /\ P s)}).
<|= { apply IHe2 }
  ({s, ⟨comp' e2 (ADD :: c), s ⟩ | exists s' n, e1 ⇓ n /\ s = n :: s' /\ P s'}).
=  { eauto }
  ({s n, ⟨comp' e2 (ADD :: c),  n :: s ⟩ | e1 ⇓ n /\ P s }).
<|= { apply IHe1 }
  ({s, ⟨comp' e1 (comp' e2 (ADD :: c)),  s ⟩ | P s }).
 [].

begin
  ({s m, ⟨c, m :: s ⟩ | Rnd e ⇓ m /\ P s }) .
=  { by_eval }
  ({s m n, ⟨c, m :: s ⟩ | e ⇓ n /\ 0 <= m <= abs n /\ P s }) .
<== {apply vm_rnd}
  ({s (m:Z) n, ⟨RND :: c, n :: s ⟩ | e ⇓ n  /\ 0 <= m <= abs n /\ P s }).
=  {dist' auto}
  ({s n, ⟨RND :: c, n :: s ⟩ | e ⇓ n /\ P s }) .
<|= { apply IHe }
  ({s, ⟨comp' e (RND :: c), s ⟩ | P s }) .
 [].
Qed.
```

Fig. 5. Correctness proof for compiler comp' in Coq.

For example, consider the calculation for the case e = Val n. We start, as in Fig. 4, by using the semantics of the source language:

$$
\begin{aligned}
&\{\mathsf{s\ n'}, \langle \mathsf{c, n' :: s} \rangle \mid \mathsf{Val\ n} \Downarrow \mathsf{n'} \wedge \mathsf{P\ s}\} \\
\equiv \quad &\{ \text{ by } \textsc{Val} \} \\
&\{\mathsf{s}, \langle \mathsf{c, n :: s} \rangle \mid \mathsf{P\ s}\}
\end{aligned}
$$

Our goal is to transform $\{s, \langle c, n :: s\rangle \mid P\ s\}$ by a sequence of calculation steps into the form $\{s, \langle c', s\rangle \mid P\ s\}$. That means that we have to get rid of the n on top of the stack. That is, we need to find a c' such that $\langle c', s\rangle \Longrightarrow \langle c, n :: s\rangle$. We could achieve that by simply adding a rule $\langle c, s\rangle \Longrightarrow \langle c, n :: s\rangle$ to the definition of $\Longrightarrow$, i.e. n is chosen non-deterministically. However, if we added this rule, we would not be able to derive

$$\{s, \langle c, s\rangle \mid P\ s\} \Rightarrow \{s, \langle c, n :: s\rangle \mid P\ s\}$$

since $\langle c, s\rangle$ is not barred by $\{s, \langle c, n :: s\rangle \mid P\ s\}$ for all s with P s; cf. rule Step in Fig. 3. The reason why this fails is that we have $\langle c, s\rangle \Longrightarrow \langle c, m :: s\rangle$ for *any* m, not only $m = n$.

Hence, we have to restrict the rule such that it only pushes integers m onto the stack that are equal to n. The only way we can achieve this is by "storing" the relevant information – i.e. n itself – in the code part of the configuration. Hence, we add a constructor PUSH : $\mathbb{Z} \rightarrow$ Instr to the type of instructions, which allows us to add the rule $\langle \mathsf{PUSH}\ n :: c, s\rangle \Longrightarrow \langle c, n :: s\rangle$ to the definition of $\Longrightarrow$. Using the Step proof rule, we can then conclude the calculation as follows:

$$
\begin{array}{ll}
& \{s, \langle c, n :: s\rangle \mid P\ s\} \\
\Longleftarrow & \quad \{\text{ define } \langle \mathsf{PUSH} :: c, s\rangle \Longrightarrow \langle c, n :: s\rangle \ \} \\
& \{s, \langle \mathsf{PUSH} :: c, n :: s\rangle \mid P\ s\} \\
\equiv & \quad \{\text{ define comp}'\ (\mathsf{Val}\ n)\ c = \mathsf{PUSH}\ n :: c \ \} \\
& \{s, \langle \mathsf{comp}'\ (\mathsf{Val}\ n)\ c, n :: s\rangle \mid P\ s\}
\end{array}
$$

The same can be done for the other two cases of the Expr language. The only difference is that we need to use the induction hypothesis. In order to be able to apply the induction hypothesis, we need to transform the configuration into the right shape. We do this by adding rules to the definition of $\Longrightarrow$ and then applying the Step proof rule accordingly. In the end, we arrive at the very same calculation proof as in Fig. 4. But instead of having comp' and $\Longrightarrow$ defined beforehand and using it in the proof, we discover the definition of comp' and $\Longrightarrow$ as we do the calculation. In fact, the compiler and the target language presented in Sect. 2 have been derived using this approach.

The setup used by Bahr and Hutton [4] to do the calculation is slightly different: the semantics of the source language is given by a structurally recursive evaluation function eval of type Exp $\rightarrow$ Value and the virtual machine is formulated as a tail recursive function exec of type Code $\rightarrow$ Conf' $\rightarrow$ Conf'. The compiler correctness property is then formulated as an equation that relates eval and exec. Moreover, the Code type is not part of the configuration type Conf'. However, it is easy to translate eval into a big-step operational semantics – in fact, Bahr and Hutton [4] do this in their treatment of higher-order languages – and exec into a big-step operational semantics [5]. The methodology of Bahr and Hutton [4] can be adapted to the use of a small-step virtual machine $\Longrightarrow$ by reasoning over the reflexive transitive closure $\overset{*}{\Longrightarrow}$ instead of equational reasoning. But this approach does not work for non-deterministic languages, since the calculation would only prove completeness but not soundness.[3] The example below illustrates this.

Assume that during the calculation we derived a more general rule for the RND instruction (instead of the rule VM-RND in Fig. 2):

$$\langle \mathsf{RND} :: \mathsf{c}, \mathsf{n} :: \mathsf{s} \rangle \implies \langle \mathsf{c}, \mathsf{m} :: \mathsf{s} \rangle \qquad \text{(VM-RND')}$$

The above rule omits the side condition $0 \leqslant \mathsf{m} \leqslant |\mathsf{n}|$. As a result the compiler comp is not sound anymore. The virtual machine now allows the following execution, even though we do not have that $\mathsf{RND}\ (\mathsf{Val}\ 0) \Downarrow 1$:

$$\langle \mathsf{comp}\ (\mathsf{RND}\ (\mathsf{Val}\ 0)), []\rangle = \langle [\mathsf{PUSH}\ 0; \mathsf{RND}], []\rangle \implies \langle [\mathsf{RND}], [0]\rangle \implies \langle [], [1]\rangle$$

The calculation approach of Bahr and Hutton [4] (naively lifted to relational semantics as describes above) will not detect that the rule VM-RND' breaks the soundness property. This calculation approach roughly corresponds to the use of the proof rules in Fig. 3 but with the "barred" side condition removed from the STEP proof rule. The resulting proof system is admissible for $\Rrightarrow$ but not for $\Rightarrow\!\!\rhd$. Thus the corresponding calculation only proves the completeness property, but not the soundness property.

The problem illustrated above is easy to recognise for the simple toy language that we considered here. However, in the next section we will consider a more complex language, where such problems are much more subtle as we will see. The novelty of our approach lies in the proof rules for the $\Rightarrow\!\!\rhd$ relation that combines the two relations $\overset{*}{\implies}$ and $\lhd$. As a consequence, we are able to formulate soundness and completeness in one compact statement and calculate in a style similar to Bahr and Hutton [4].

5 Calculating a Compiler for a Language with Interrupts

We now turn to a more interesting source language that features asynchronous exceptions, also known as *interrupts*. What distinguishes interrupts from ordinary exceptions is the fact that interrupts can potentially arise at (almost) any point in the execution of a program. As a consequence, the language's semantics is non-deterministic – a program's execution may proceed successfully or be interrupted by an asynchronous exception at any point. In addition, we consider language constructs that allow the programmer to limit the scope of interrupts, i.e. blocking interrupts from interfering with some parts of the program.

We consider the language with interrupts of Hutton and Wright [7]. The syntax of the language is given by the following inductive type:

[3] Bahr and Hutton [4] acknowledge that this problem already occurs if the semantics is not total, e.g. for the untyped lambda calculus. However, if the semantics is at least deterministic, soundness for the defined fragment of the language can be achieved easily by ensuring that the derived virtual machine is deterministic.

$$\textbf{Inductive}\ \mathsf{Expr} : \mathsf{Set} := \mathsf{Val}\ (\mathsf{n} : \mathbb{Z}) \mid \mathsf{Add}\ (\mathsf{e_1}\ \mathsf{e_2} : \mathsf{Expr}) \mid \mathsf{Throw} \mid \mathsf{Catch}\ (\mathsf{e}\ \mathsf{h} : \mathsf{Expr}) \mid \mathsf{Seqn}\ (\mathsf{e_1}\ \mathsf{e_2} : \mathsf{Expr}) \mid \mathsf{Block}\ (\mathsf{e} : \mathsf{Expr}) \mid \mathsf{Unblock}\ (\mathsf{e} : \mathsf{Expr}).$$

As before, we have integer literals (Val) and an addition operation (Add). Furthermore, the language allows us to throw (synchronous) exceptions with Throw and to catch (synchronous or asynchronous) exceptions with Catch. An expression of the form Catch e h behaves like e, in case e does not throw any exceptions; otherwise it behaves like h. We can also sequentially compose expressions using Seqn. Finally, Block and Unblock are used to control asynchronous exceptions: in an expression Block e or Unblock e, we allow respectively disallow interruption in e by asynchronous exceptions.

We give the semantics of the language as a big-step operational semantics as presented by Hutton and Wright [7]. To describe blocking and unblocking of interrupts, the semantics uses the following type to indicate the blocking status, where B indicates that interrupts are blocked and U that interrupts are allowed:

$$\textbf{Inductive}\ \mathsf{Status} : \mathsf{Set} := \mathsf{B} \mid \mathsf{U}.$$

The relation that embodies the big-step operational semantics is denoted by $\Downarrow_s$, where the subscript s indicates the status, i.e. either blocked or unblocked. The judgement $\mathsf{e} \Downarrow_s \mathsf{v}$ means that $\mathsf{e} : \mathsf{Expr}$ evaluates to $\mathsf{v} : \mathsf{option}\ \mathbb{Z}$ given the status $\mathsf{s} : \mathsf{Status}$. Values of type option $\mathbb{Z}$ are either of the form Some n, indicating the result value n, or of the form None, indicating that an exception occurred. The inference rules for the semantics are shown in Fig. 6.

$$\frac{}{\mathsf{Val}\ n \Downarrow_i \mathsf{Some}\ n}\ \textsc{Val} \qquad \frac{}{\mathsf{Throw} \Downarrow_i \mathsf{None}}\ \textsc{Throw} \qquad \frac{}{x \Downarrow_U \mathsf{None}}\ \textsc{Int}$$

$$\frac{x \Downarrow_i \mathsf{Some}\ n \qquad y \Downarrow_i \mathsf{Some}\ m}{\mathsf{Add}\ x\ y \Downarrow_i \mathsf{Some}\ (n{+}m)}\ \textsc{Add1} \qquad \frac{x \Downarrow_i \mathsf{None}}{\mathsf{Add}\ x\ y \Downarrow_i \mathsf{None}}\ \textsc{Add2}$$

$$\frac{x \Downarrow_i \mathsf{Some}\ n \qquad y \Downarrow_i \mathsf{None}}{\mathsf{Add}\ x\ y \Downarrow_i \mathsf{None}}\ \textsc{Add3}$$

$$\frac{x \Downarrow_i \mathsf{Some}\ n \qquad y \Downarrow_i v}{\mathsf{Seqn}\ x\ y \Downarrow_i v}\ \textsc{Seqn1} \qquad \frac{x \Downarrow_i \mathsf{None}}{\mathsf{Seqn}\ x\ y \Downarrow_i \mathsf{None}}\ \textsc{Seqn2}$$

$$\frac{x \Downarrow_i \mathsf{Some}\ n}{\mathsf{Catch}\ x\ y \Downarrow_i \mathsf{Some}\ n}\ \textsc{Catch1} \qquad \frac{x \Downarrow_i \mathsf{None} \qquad y \Downarrow_i v}{\mathsf{Catch}\ x\ y \Downarrow_i v}\ \textsc{Catch2}$$

$$\frac{x \Downarrow_B v}{\mathsf{Block}\ x \Downarrow_i v}\ \textsc{Block} \qquad \frac{x \Downarrow_U v}{\mathsf{Unblock}\ x \Downarrow_i v}\ \textsc{Unblock}$$

Fig. 6. Semantics of the language.

Our goal is to derive a compiler and virtual machine such that the compiler is correct with respect to the source language's semantics. To this end, we follow the calculation approach outlined in Sect. 4.

We start with a partially defined compiler comp:

```
Fixpoint comp' (e : Expr) (c : Code) : Code :=
  match e with
    | _ ⇒ Admit
  end.
Definition comp (e : Expr) : Code := comp' e [].
```

Similarly to the compiler in Sect. 2, the above compiler is defined with an additional argument representing the code that is supposed to be executed after the generated code. As we do the calculation proof, we will discover equations of the form comp' p c = c' for some pattern p. We will then add a corresponding clause p ⇒ c' to the **match** statement in the above definition. For now it uses the term Admit as a placeholder. It serves a similar role as the term `undefined` in Haskell.

Likewise, we start with an empty definition of the target language and the virtual machine:

```
Inductive Instr : Set :=.
Inductive VM : Conf → Conf → Prop :=
  where "x ==> y" := (VM x y).
```

We then have to formulate the correctness property of the compiler. We adopt the same form of correctness property as in Sect. 3:

$$\forall\, e\, c\, i\, P,\quad \{s, \langle \text{comp' } e\, c, s\rangle \mid P\, s\} \Rightarrow \{s\, n, \langle c\,, n :: s\rangle \mid e \Downarrow_i \text{Some } n \wedge P\, s\}$$

We could now start the calculation. However, we would soon realise that the above property is not appropriate. We would encounter three problems:

1. The status indicator i only appears in the semantics of the source language but not the virtual machine. Thus, the above property is too general.
2. We only consider the case that $e \Downarrow_i$ Some n, but not the case that $e \Downarrow_i$ None. Thus, the above property is not general enough.
3. As we do the calculation, we realise that we need to store data other than just integers on the stack.

The calculation technique of Bahr and Hutton [4] anticipates these problems and suggests corresponding generalisations, which we shall adopt here as well. However, we have to translate their approach, which uses a tail-recursive function instead of a small-step relation for defining the virtual machine. We have to make the following amendments:

1. Extend the type of configurations with a component of type Status:

```
Inductive Conf : Set := conf (c : Code) (s : Stack) (i : Status).
Notation "⟨ c , s , i ⟩" := (conf c s i).
```

2. Add a case to the type of configurations that corresponds to the None case:[4]

```
Inductive Conf : Set := conf (c : Code) (s : Stack) (i : Status)
                      | fail (s : Stack) (i : Status).
Notation "⟨ c , s , i ⟩" := (conf c s i).
Notation "⟪ s , i ⟫"     := (fail s i).
```

3. Generalise the type of stack elements and extend the type as necessary:

```
Inductive Elem : Set := VAL (n : ℤ).
Definition Stack : Set := list Elem.
```

The type Stack as defined above is isomorphic to the previous definition. However, we can now extend the type Elem with additional constructors to store other kinds of data on the stack.

With these changes we can formulate the correctness property as follows:

$$\begin{aligned}\forall\, \mathsf{e}\ \mathsf{c}\ \mathsf{P},\quad \{\mathsf{s}\ \mathsf{i},\ \langle \mathsf{comp'}\ \mathsf{e}\ \mathsf{c}\ ,\ \mathsf{s},\ \mathsf{i}\rangle \mid \mathsf{P}\ \mathsf{s}\ \mathsf{i}\} \Rightarrow\ & \{\mathsf{s}\ \mathsf{i}\ \mathsf{n},\ \langle \mathsf{c}\ ,\ \mathsf{VAL}\ \mathsf{n} :: \mathsf{s},\ \mathsf{i}\rangle \mid \mathsf{e} \Downarrow_{\mathsf{i}} \mathsf{Some}\ \mathsf{n} \wedge \mathsf{P}\ \mathsf{s}\ \mathsf{i}\} \\ \cup\ & \{\mathsf{s}\ \mathsf{i},\ \langle\!\langle \mathsf{s},\ \mathsf{i}\rangle\!\rangle \mid \mathsf{e} \Downarrow_{\mathsf{i}} \mathsf{None} \wedge \mathsf{P}\ \mathsf{s}\ \mathsf{i}\}\end{aligned}$$

Note that the predicate P is now applied to s *and* i. We could have started the calculation with P only covering s, but we would soon realise that we need the more general version of P in order to apply the induction hypothesis.

An important difference between the above correctness property and the correctness property we have considered in Sect. 3 is that it involves set union. As a consequence, we need to use the UNION proof rule from Fig. 3. We will always use the UNION rule together with the REFL rule, such that we only change one component of a union while the others remain constant. For example, given $\mathsf{X} \Rightarrow \mathsf{X'}$, we may derive $\mathsf{W} \cup \mathsf{X} \cup \mathsf{Y} \Rightarrow \mathsf{W} \cup \mathsf{X'} \cup \mathsf{Y}$.

Figure 7 shows the calculation for the two cases $\mathsf{e} = \mathsf{Val}\ \mathsf{n}$ and $\mathsf{e} = \mathsf{Catch}\ \mathsf{e}_1\ \mathsf{e}_2$. We focus on these two cases to illustrate the calculation. The complete calculation covering the remaining cases as well can be found in the accompanying Coq code.

We begin with the case $\mathsf{e} = \mathsf{Val}\ \mathsf{n}$. First, we use the definition of $\Downarrow_{\mathsf{i}}$: we can derive $\mathsf{Val}\ \mathsf{n} \Downarrow_{\mathsf{i}} \mathsf{Some}\ \mathsf{n'}$ iff $\mathsf{n} = \mathsf{n'}$; and we can derive $\mathsf{Val}\ \mathsf{n} \Downarrow_{\mathsf{i}} \mathsf{None}$ iff $\mathsf{i} = \mathsf{U}$. We can then proceed to transform the configuration set such that it matches the left-hand side of the correctness statement, i.e. it is of the form $\{\mathsf{s}\ \mathsf{i},\ \langle \mathsf{c'},\ \mathsf{s},\ \mathsf{i}\rangle \mid \mathsf{P}\ \mathsf{s}\ \mathsf{i}\}$. In particular, we must get rid of the union construction.

[4] Bahr and Hutton [4] use tail-recursive functions to represent virtual machines. In their approach, one has to introduce an additional tail-recursive function fail. In our approach, this corresponds to a new constructor for the type of configurations.

– <u>Val n:</u>

$$\{\mathsf{s\ i\ n'},\ \langle \mathsf{c},\ \mathsf{VAL\ n' :: s, i}\rangle \mid \mathsf{Val\ n} \Downarrow_{\mathsf{i}} \mathsf{Some\ n'} \wedge \mathsf{P\ s\ i}\}$$
$$\cup\ \{\mathsf{s\ i},\ \langle\langle \mathsf{s, i}\rangle\rangle \mid \mathsf{Val\ n} \Downarrow_{\mathsf{i}} \mathsf{None} \wedge \mathsf{P\ s\ i}\}$$
$$\equiv \quad \{ \text{ by definition of } \Downarrow_{\mathsf{i}} \}$$
$$\{\mathsf{s\ i},\ \langle \mathsf{c},\ \mathsf{VAL\ n :: s, i}\rangle \mid \mathsf{P\ s\ i}\} \cup \{\mathsf{s},\ \langle\langle \mathsf{s, U}\rangle\rangle \mid \mathsf{P\ s\ U}\}$$
$$\Longleftarrow \quad \{ \text{ define } \langle \mathsf{PUSH\ n :: c, s, i}\rangle \Longrightarrow \langle \mathsf{c},\ \mathsf{VAL\ n :: s, i}\rangle \}$$
$$\{\mathsf{s\ i},\ \langle \mathsf{PUSH\ n :: c, s, i}\rangle \mid \mathsf{P\ s\ i}\} \cup \{\mathsf{s},\ \langle\langle \mathsf{s, U}\rangle\rangle \mid \mathsf{P\ s\ U}\}$$
$$\Longleftarrow \quad \{ \text{ define } \langle \mathsf{PUSH\ n :: c, s, U}\rangle \Longrightarrow \langle\langle \mathsf{s, U}\rangle\rangle \}$$
$$\{\mathsf{s\ i},\ \langle \mathsf{PUSH\ n :: c, s, i}\rangle \mid \mathsf{P\ s\ i}\} \cup \{\mathsf{s},\ \langle \mathsf{PUSH\ n :: c, s, U}\rangle \mid \mathsf{P\ s\ U}\}$$
$$\equiv \quad \{ \text{ second set of the union is contained in first set } \}$$
$$\{\mathsf{s\ i},\ \langle \mathsf{PUSH\ n :: c, s, i}\rangle \mid \mathsf{P\ s\ i}\}$$

– <u>Catch e_1 e_2:</u>

$$\{\mathsf{s\ i\ n},\ \langle \mathsf{c},\ \mathsf{VAL\ n :: s, i}\rangle \mid \mathsf{Catch}\ e_1\ e_2 \Downarrow_{\mathsf{i}} \mathsf{Some\ n} \wedge \mathsf{P\ s\ i}\}$$
$$\cup\ \{\mathsf{s\ i},\ \langle\langle \mathsf{s, i}\rangle\rangle \mid \mathsf{Catch}\ e_1\ e_2 \Downarrow_{\mathsf{i}} \mathsf{None} \wedge \mathsf{P\ s\ i}\}$$
$$\equiv \quad \{ \text{ by definition of } \Downarrow_{\mathsf{i}} \}$$
$$\{\mathsf{s\ i\ n},\ \langle \mathsf{c},\ \mathsf{VAL\ n :: s, i}\rangle \mid e_1 \Downarrow_{\mathsf{i}} \mathsf{Some\ n} \wedge \mathsf{P\ s\ i}\}$$
$$\cup\ \{\mathsf{s\ i\ n},\ \langle \mathsf{c},\ \mathsf{VAL\ n :: s, i}\rangle \mid e_2 \Downarrow_{\mathsf{i}} \mathsf{Some\ n} \wedge (e_1 \Downarrow_{\mathsf{i}} \mathsf{None} \wedge \mathsf{P\ s\ i})\}$$
$$\cup\ \{\mathsf{s\ i},\ \langle\langle \mathsf{s, i}\rangle\rangle \mid e_2 \Downarrow_{\mathsf{i}} \mathsf{None} \wedge (e_1 \Downarrow_{\mathsf{i}} \mathsf{None} \wedge \mathsf{P\ s\ i})\}$$
$$\Leftarrow \quad \{ \text{ induction hypothesis for } e_2 \}$$
$$\{\mathsf{s\ i\ n},\ \langle \mathsf{c},\ \mathsf{VAL\ n :: s, i}\rangle \mid e_1 \Downarrow_{\mathsf{i}} \mathsf{Some\ n} \wedge \mathsf{P\ s\ i}\}$$
$$\cup\ \{\mathsf{s\ i},\ \langle \mathsf{comp'}\ e_2\ \mathsf{c, s, i}\rangle \mid e_1 \Downarrow_{\mathsf{i}} \mathsf{None} \wedge \mathsf{P\ s\ i}\}$$
$$\Longleftarrow \quad \{ \text{ define } \langle\langle \mathsf{HAN\ h :: s, i}\rangle\rangle \Longrightarrow \langle \mathsf{h,s,i}\rangle \}$$
$$\{\mathsf{s\ i\ n},\ \langle \mathsf{c},\ \mathsf{VAL\ n :: s, i}\rangle \mid e_1 \Downarrow_{\mathsf{i}} \mathsf{Some\ n} \wedge \mathsf{P\ s\ i}\}$$
$$\cup\ \{\mathsf{s\ i},\ \langle\langle \mathsf{HAN\ (comp'}\ e_2\ \mathsf{c) :: s, i}\rangle\rangle \mid e_1 \Downarrow_{\mathsf{i}} \mathsf{None} \wedge \mathsf{P\ s\ i}\}$$
$$\Longleftarrow \quad \{ \text{ define } \langle \mathsf{UNMARK :: c,\ VAL\ n :: HAN\ h :: s, i}\rangle \Longrightarrow \langle \mathsf{c},\ \mathsf{VAL\ n :: s, i}\rangle \}$$
$$\{\mathsf{s\ i\ n},\ \langle \mathsf{UNMARK :: c,\ VAL\ n :: HAN\ (comp'}\ e_2\ \mathsf{c) :: s, i}\rangle \mid e_1 \Downarrow_{\mathsf{i}} \mathsf{Some\ n} \wedge \mathsf{P\ s\ i}\}$$
$$\cup\ \{\mathsf{s\ i},\ \langle\langle \mathsf{HAN\ (comp'}\ e_2\ \mathsf{c) :: s, i}\rangle\rangle \mid e_1 \Downarrow_{\mathsf{i}} \mathsf{None} \wedge \mathsf{P\ s\ i}\}$$
$$\equiv \quad \{ \text{ move stack element } \mathsf{HAN\ (comp'}\ e_2\ \mathsf{c)} \text{ into the predicate } \}$$
$$\{\mathsf{s'\ i\ n},\ \langle \mathsf{UNMARK :: c,\ VAL\ n :: s', i}\rangle \mid e_1 \Downarrow_{\mathsf{i}} \mathsf{Some\ n} \wedge (\exists\, \mathsf{s},\ \mathsf{s'} = \mathsf{HAN\ (comp'}\ e_2\ \mathsf{c) :: s} \wedge \mathsf{P\ s\ i})\}$$
$$\cup\ \{\mathsf{s'\ i},\ \langle\langle \mathsf{s', i}\rangle\rangle \mid e_1 \Downarrow_{\mathsf{i}} \mathsf{None} \wedge (\exists\, \mathsf{s},\ \mathsf{s'} = \mathsf{HAN\ (comp'}\ e_2\ \mathsf{c) :: s} \wedge \mathsf{P\ s\ i})\}$$
$$\Leftarrow \quad \{ \text{ induction hypothesis for } e_1 \}$$
$$\{\mathsf{s'\ i},\ \langle \mathsf{comp'}\ e_1\ \mathsf{(UNMARK :: c)},\ \mathsf{s', i}\rangle \mid (\exists\, \mathsf{s},\ \mathsf{s'} = \mathsf{HAN\ (comp'}\ e_2\ \mathsf{c) :: s} \wedge \mathsf{P\ s\ i})\}$$
$$\equiv \quad \{ \text{ extract stack element } \mathsf{HAN\ (comp'}\ e_2\ \mathsf{c)} \text{ from the predicate } \}$$
$$\{\mathsf{s\ i},\ \langle \mathsf{comp'}\ e_1\ \mathsf{(UNMARK :: c)},\ \mathsf{HAN\ (comp'}\ e_2\ \mathsf{c) :: s, i}\rangle \mid \mathsf{P\ s\ i}\}.$$
$$\Longleftarrow \quad \{ \text{ define } \langle \mathsf{MARK\ h :: c, s, i}\rangle \Longrightarrow \langle \mathsf{c},\ \mathsf{HAN\ h :: s, i}\rangle \}$$
$$\{\mathsf{s\ i},\ \langle \mathsf{MARK\ (comp'}\ e_2\ \mathsf{c) :: comp'}\ e_1\ \mathsf{(UNMARK :: c)},\ \mathsf{s, i}\rangle \mid \mathsf{P\ s\ i}\}$$

Fig. 7. Calculation for cases Val and Catch.

The general strategy to achieve this is to transform the union into a form

$$\{\mathsf{s\ i},\ \langle \mathsf{c'},\ \mathsf{s, i}\rangle \mid \mathsf{P_1\ s\ i}\} \cup \{\mathsf{s\ i},\ \langle \mathsf{c'},\ \mathsf{s, i}\rangle \mid \mathsf{P_2\ s\ i}\}$$

such that we can replace it with a single set comprehension

$$\{\mathsf{s\ i},\ \langle \mathsf{c'},\ \mathsf{s, i}\rangle \mid \mathsf{P_1\ s\ i} \vee \mathsf{P_2\ s\ i}\}$$

In most cases, we have that $\mathsf{P_1}$ implies $\mathsf{P_2}$, or vice versa such that we can replace the union with the right or the left component of the union, respectively.

We first consider the left component of the union. Similarly to the simple language of Sect. 2, we introduce an instruction PUSH in order to get rid of the topmost stack element VAL n. In order to apply the above strategy to transform a union of set comprehensions into a single set comprehension, we must be able to transform the configuration ⟨⟨s, U⟩⟩ into the form ⟨PUSH n :: c, s', i'⟩ such that P s U implies P s' i'. Hence, we must have s' = s and i' = U. We thus decide to add the rule

$$\langle \mathsf{PUSH\ n :: c, s, U} \rangle \implies \langle\!\langle \mathsf{s, U} \rangle\!\rangle$$

We could have also added the more general rule

$$\langle \mathsf{PUSH\ n :: c, s, i} \rangle \implies \langle\!\langle \mathsf{s, i} \rangle\!\rangle$$

However, by adding this rule, the calculation step that we did before would not be correct anymore since we would not have that P s i implies that

$$\langle \mathsf{PUSH\ n :: c, s, i} \rangle \lhd \{\mathsf{s\,i}\,,\ \langle \mathsf{c, VAL\ n :: s, i} \rangle \mid \mathsf{P\,s\,i}\} \cup \{\mathsf{s}\,,\ \langle\!\langle \mathsf{s, U} \rangle\!\rangle \mid \mathsf{P\,s\,U}\}$$

This phenomenon illustrates the utility of the proof automation of our calculation framework that discharges side conditions as the one above: by adding new rules to the definition of the virtual machine, applications of the STEP proof rule have to be reconsidered and may fail as the side condition is not fulfilled anymore.

The union is almost in the right form now. The right component can be equivalently written as

$$\{\mathsf{s\,i}, \langle \mathsf{PUSH\ n :: c, s, i} \rangle \mid \mathsf{i} = \mathsf{U} \wedge \mathsf{P\,s\,i}\}$$

Since $\mathsf{i} = \mathsf{U} \wedge \mathsf{P\,s\,i}$ obviously implies P s i, we can replace the union with its left component.

Finally, we can observe that comp' (Val n) c must be equal to PUSH n :: c and we thus add the clause Val n ⇒ PUSH n :: c to the definition of comp'.

The calculation for the case e = Catch $\mathsf{e_1}$ $\mathsf{e_2}$ may use the induction hypotheses for $\mathsf{e_1}$ and $\mathsf{e_2}$, which reads as follows:

$$\begin{aligned} \forall\, \mathsf{c'\ P'},\quad & \{\mathsf{s\,i}, \langle \mathsf{comp'\ e_j\ c'\ , s, i} \rangle \mid \mathsf{P'\,s\,i}\} \Rightarrow \\ & \qquad \{\mathsf{s\,i\,n}, \langle \mathsf{c'\ , VAL\ n :: s, i} \rangle \mid \mathsf{e_j} \Downarrow_{\mathsf{i}} \mathsf{Some\ n} \wedge \mathsf{P'\,s\,i}\} \\ & \qquad \cup \{\mathsf{s\,i}, \quad \langle\!\langle \mathsf{s, i} \rangle\!\rangle \qquad\qquad\quad \mid \mathsf{e_j} \Downarrow_{\mathsf{i}} \mathsf{None} \quad \wedge \mathsf{P'\,s\,i}\} \end{aligned}$$

The calculation is driven by the desire to apply these induction hypotheses, which means we want to transform the configuration set such that it matches the right-hand side of one of the induction hypotheses. Achieving this for $\mathsf{e_2}$ is easy: we use the definition of $\Downarrow_{\mathsf{i}}$ to reformulate Catch $\mathsf{e_1}$ $\mathsf{e_2}$ $\Downarrow_{\mathsf{i}}$ ... in terms of $\mathsf{e_1} \Downarrow_{\mathsf{i}} \ldots$ and $\mathsf{e_2} \Downarrow_{\mathsf{i}} \ldots$. The second set comprehension together with the third one then already have the right shape for the induction hypothesis for $\mathsf{e_2}$. We instantiate c' with c and P' s i with $\mathsf{e_1} \Downarrow_{\mathsf{i}}$ None ∧ P s i.

We then have to transform the resulting union of two set comprehensions into the right shape for the induction hypothesis for e_1. At first we notice that the second set comprehension has the condition $e_1 \Downarrow_i$ None, and thus we have to transform it such that the configuration is of the shape ⟪s', i'⟫ and not ⟨c' , s', i'⟩. That means that we need to have a rule for $\Longrightarrow$ that transforms something of the form ⟪s', i'⟫ into ⟨comp' e_2 c, s, i⟩. To do so, we need to be able to draw the code component comp' e_2 c of the target configuration from s' (or i', but that is not possible). Hence, we introduce a new stack element constructor HAN : Code → Elem, which is able to store code on the stack. Thus the only reasonable choice for the new rule is ⟪HAN h :: s, i⟫ $\Longrightarrow$ ⟨h,s,i⟩.

Next, in order to be able to apply the induction hypothesis, we must manipulate the stack in the first set comprehension. It is of the form VAL n :: s, but it needs to be of the form VAL n :: HAN (comp' e_2 c) :: s. To this end, we introduce an instruction UNMARK, which removes the HAN element from the stack. It is then a simple matter of moving the constraint on the shape of the stack into the predicate of the set comprehension such that we can apply the induction hypothesis. After we have applied the induction hypothesis for e_1, we reverse this encoding.

Finally, we need to get rid of the top element of the stack such that the stack becomes just s. As in previous cases, we introduce an instruction that does this job. This completes the calculation for this case, and we can now read off the clause that we need to add to the definition of comp':

```
Catch e₁ e₂ ⇒ MARK (comp' e₂ c) :: comp' e₁ (UNMARK :: c)
```

The other cases of the calculation can be completed using the same strategies that we have used above. As the result of the calculation, we derive the following compiler definition:

```
Fixpoint comp' (e : Expr) (c : Code) : Code :=
  match e with
    | Val n       ⇒ PUSH n :: c
    | Add x y     ⇒ comp' x (comp' y (ADD :: c))
    | Throw       ⇒ [THROW]
    | Catch e₁ e₂ ⇒ MARK (comp' e₂ c) :: comp' e₁ (UNMARK :: c)
    | Seqn e₁ e₂  ⇒ comp' e₁ (POP :: comp' e₂ c)
    | Block e     ⇒ BLOCK :: comp' e (RESET :: c)
    | Unblock e   ⇒ UNBLOCK :: comp' e (RESET :: c)
  end.
```

The derived target language is the following:

```
Inductive Instr : Set := PUSH (n : ℤ) | ADD | THROW
                       | UNMARK | MARK (h : list Instr)
                       | POP | RESET | BLOCK | UNBLOCK.
```

Moreover, during the calculation we needed to extend the type of stack elements such that we can store exception handlers and interrupt status on the stack:

Inductive Elem : Set := VAL (n : $\mathbb{Z}$) | HAN (c : Code) | INT (s : Status).

This is almost the same compiler as the one given by Hutton and Wright [7]. There are only two minor differences: Hutton and Wright's compiler compiles Throw to THROW :: c instead of [THROW]; and instead of the two instructions BLOCK and UNBLOCK they use a single instruction SET, which takes an argument of type Status such that SET B and SET U correspond to BLOCK and UNBLOCK, respectively.

However, these differences are rather superficial. More interesting differences can be found in the virtual machine that we derived from the calculation. The definition of the virtual machine is shown in Fig. 8. The virtual machine used by Hutton and Wright [7] may go into a fail configuration from any unblocked configuration, no matter what the current instruction is. In our virtual machine only the instructions PUSH, THROW and BLOCK may be interrupted. It turns out that there is some room of freedom in choosing appropriate rules for $\Longrightarrow$.

$$
\begin{aligned}
\langle \mathsf{PUSH\ n :: c, s, i} \rangle &\Longrightarrow \langle \mathsf{c\ ,\ VAL\ n :: s, i} \rangle \\
\langle \mathsf{PUSH\ n :: c, s, U} \rangle &\Longrightarrow \langle\!\langle \mathsf{s, U} \rangle\!\rangle \\
\langle \mathsf{ADD :: c, VAL\ m :: VAL\ n :: s, i} \rangle &\Longrightarrow \langle \mathsf{c, VAL\ (n + m) :: s, i} \rangle \\
\langle\!\langle \mathsf{VAL\ m :: s, i} \rangle\!\rangle &\Longrightarrow \langle\!\langle \mathsf{s, i} \rangle\!\rangle \\
\langle \mathsf{THROW :: c, s, i} \rangle &\Longrightarrow \langle\!\langle \mathsf{s, i} \rangle\!\rangle \\
\langle\!\langle \mathsf{HAN\ h :: s, i} \rangle\!\rangle &\Longrightarrow \langle \mathsf{h,s,i} \rangle \\
\langle \mathsf{UNMARK :: c, VAL\ n :: HAN\ h :: s, i} \rangle &\Longrightarrow \langle \mathsf{c, VAL\ n :: s, i} \rangle \\
\langle \mathsf{MARK\ h :: c, s, i} \rangle &\Longrightarrow \langle \mathsf{c, HAN\ h :: s, i} \rangle \\
\langle \mathsf{POP :: c, VAL\ n :: s, i} \rangle &\Longrightarrow \langle \mathsf{c, s, i} \rangle \\
\langle \mathsf{RESET :: c, VAL\ n :: INT\ i :: s, j} \rangle &\Longrightarrow \langle \mathsf{c, VAL\ n :: s, i} \rangle \\
\langle \mathsf{BLOCK :: c, s, i} \rangle &\Longrightarrow \langle \mathsf{c, INT\ i :: s, B} \rangle \\
\langle \mathsf{BLOCK :: c, s, U} \rangle &\Longrightarrow \langle\!\langle \mathsf{s, U} \rangle\!\rangle \\
\langle \mathsf{UNBLOCK :: c, s, i} \rangle &\Longrightarrow \langle \mathsf{c, INT\ i :: s, U} \rangle \\
\langle\!\langle \mathsf{INT\ i :: s, j} \rangle\!\rangle &\Longrightarrow \langle\!\langle \mathsf{s,i} \rangle\!\rangle
\end{aligned}
$$

Fig. 8. Definition of the virtual machine.

We could have equally well made different choices during the calculation process, which would have resulted in a virtual machine equivalent to Hutton and Wright's.[5] For example, for the case e = Val n, we have introduced the rule

$$\langle \mathsf{PUSH\ n :: c, s, U} \rangle \Longrightarrow \langle\!\langle \mathsf{s, U} \rangle\!\rangle$$

[5] Similarly, we could have chosen to use a single instruction SET instead of BLOCK and UNBLOCK.

i.e. the PUSH instruction may be interrupted. However, we could have instead introduced the more general rule

$$\langle \mathsf{op} :: \mathsf{c}, \mathsf{s}, \mathsf{U} \rangle \implies \langle\!\langle \mathsf{s}, \mathsf{U} \rangle\!\rangle$$

i.e. any instruction may be interrupted, which is the semantics Hutton and Wright chose for their virtual machine. The fact that the calculation is performed in one go makes this flexibility in the semantics of the virtual machine immediately apparent. The calculation that yields the compiler and virtual machine of Hutton and Wright can be found in the accompanying Coq code.

6 Representation in Coq

In this section, we briefly outline some of the technical setup for our calculation framework in Coq. In addition to the proof rules that we discussed in this paper, our calculation framework consists of three essential components: a representation of sets of configurations suitable for proof automation, a syntax for calculation proofs, and proof tactic that is able to prove the "barred" side condition of the STEP proof rule automatically. This setup – including the proof rules – is independent of the specific definition of the virtual machine and the compiler as well as the source and the target language. In particular, the framework is defined as a functor (i.e. a parametrised module) that takes the definition of the virtual machine (given by Conf and $\implies$) as a parameter.

6.1 Sets and Set Comprehensions

In our calculations, we need to reason over sets of configurations. The most straightforward and general representation of such sets is provided by the type Conf $\rightarrow$ Prop, i.e. the type of predicates over configurations. However, in order to prove properties over such sets and – most importantly – to automate proofs, it is better to define an inductive type ConfSet together with an interpretation function that maps each such set to a predicate of type Conf $\rightarrow$ Prop.

At first we define a type of set comprehensions, which is indexed by the list of types that are available for existential quantification:

```
Inductive SetCom : list Type → Type :=
  | BaseSet      : Conf → Prop → SetCom []
  | ExSet {t ts} : (t → SetCom ts) → SetCom (t :: ts).
```

The first constructor defines a singleton set that consists of a single configuration that is subject to a predicate[6]. The second constructor adds an existential quantifier. The meaning of these constructors is best explained by the interpretation function, which defines the membership predicate for each set comprehension:

[6] Thus BaseSet may in fact represent the empty set if the predicate is false.

```
Fixpoint SetComElem {ts} (C : Conf) (S : SetCom ts) : Prop :=
  match S with
    | BaseSet C' P   ⇒ C' = C ∧ P
    | ExSet _ _ e    ⇒ ∃ x, SetComElem C (e x)
  end.
```

For the base case we check whether the configuration equals the configuration in the set comprehension and whether the predicate is fulfilled. The second constructor is simply interpreted as existential quantification.

The type ConfSet extends set comprehensions with a union operator. The corresponding interpretation function ConfElem is straightforward:

```
Inductive ConfSet : Type :=
  | Sing {ts} : SetCom ts → ConfSet
  | Union     : ConfSet → ConfSet → ConfSet.
Fixpoint ConfElem (C : Conf) (S : ConfSet) : Prop :=
  match S with
    | Sing _ s      ⇒ SetComElem C s
    | Union S1 S2   ⇒ ConfElem C S1 ∨ ConfElem C S2
  end.
```

The equivalence of sets is then simply defined in terms of the above interpretation function, and the union and set comprehension notation is mapped to the constructors of ConfSet and SetCom:

```
Notation "S ≡ T" := (∀ x, ConfElem x S ↔ ConfElem x T)
                    (at level 80 , no associativity).
Infix "∪" := Union (at level 76 , left associativity).
Notation "{ x .. y , C | P }" :=
  (Sing (ExSet (fun x ⇒ ... (ExSet (fun y ⇒ BaseSet C P)) ...)))
  (at level 70 , x binder, y binder, no associativity).
```

Using this setup, we can then formulate and prove the proof rules that we listed in Fig. 3. However, some care has to be taken in formulating the proof rules in a way that they can be readily applied to a proof goal in a calculation proof. For instance the STEP proof rule is formulated as follows:

```
Theorem step : ∀ ts (S S' : SetCom ts) (T : ConfSet) ,
  (∀ x, getProp S' x → getConf S x ⟹ getConf S' x) →
  (∀ x, getProp S' x → getProp S x) →
  (∀ x, getProp S x → getConf S x ◁ Sing S' ∪ T) →
  Sing S ∪ T ⇒▷ Sing S' ∪ T.
```

In the above theorem, we use getProp and getConf to extract the configuration and the predicate of a given set comprehension. We define getConf as follows:

```
Fixpoint getConf {ts} (S : SetCom ts) : tuple ts → Conf :=
  match S with
    | BaseSet C P ⇒ fun xs ⇒ C
    | ExSet _ _ ex ⇒ fun xs ⇒ let (x, xs') := xs in getConf (ex x) xs'
  end.
```

The function getProp is defined analogously. The definition uses the type tuple ts, which is a nested product type of all types in the list ts defined as follows:

```
Fixpoint tuple (ts : list Type) : Type :=
  match ts with
    | []       ⇒ unit
    | t :: ts' ⇒ t * tuple ts'
  end.
```

We would not have been able to define getConf and getProp, if we had represented configuration sets using simply the type Conf → Prop. The ability to define these functions is crucial in order to formulate the STEP rule in such a way that the Coq system can readily apply it to a given proof goal.

6.2 Calculation Syntax and Proof Automation

The calculation syntax is quite easy to achieve using Coq's Tactic Notation command to define custom tactics. The implementation of our calculation syntax closely follows the work by Tesson et al. [12]. But we use a somewhat simpler setup. The details can be found in the accompanying Coq source code.

In our pen-and-paper proofs (e.g. in Fig. 4) we use the notations "$\Leftarrow$=", "$\equiv$", and "$\Longleftarrow$" to indicate the proof rules that we use. This notation is reflected in the Coq proofs (cf. Fig. 5), where we have the corresponding tactic notations "<|= { t } S", "= { t } S", and "<== { t } S", which refer to a tactic t and a configuration set S.

Given a proof goal of the form T $\Longrightarrow$ U, all three tactic notations try to prove the proof goal S $\Longrightarrow$ U. If successful, the original proof goal T $\Longrightarrow$ U is replaced by T $\Longrightarrow$ S using the TRANS proof rule. Once we have transformed the proof goal into the form S $\Longrightarrow$ S, we can use the tactic [], which applies the REFL proof rule to complete the calculation proof.

The basic tactic <|= { t } S does little proof automation: it tries to prove the goal S $\Longrightarrow$ U using tactic t using the UNION and REFL proof rule, which allows us to prove goals of the form

$$S_1 \cup \ldots \cup S_i \cup \ldots S_n \Longrightarrow S_1 \cup \ldots \cup S'_i \cup \ldots S_n$$

using tactic t to prove $S_i \Longrightarrow S'_i$.

The tactic = { t } S is even simpler: it tries to prove S $\Longrightarrow$ U using the IFF proof rule, and it uses t to prove S $\equiv$ U.

Finally, $\texttt{<==}\ \{\ \texttt{t}\ \}\ \texttt{S}$ applies the proof rule Step to prove $\mathsf{S} \Rightarrow \mathsf{U}$. That is, the proof goal has to be of the shape

$$\{\overline{\mathsf{x}}\,,\,\mathsf{C}\ \mid\ \mathsf{P}\} \cup \mathsf{S}' \Rightarrow \{\overline{\mathsf{x}}\,,\,\mathsf{C}'\ \mid\ \mathsf{P}\} \cup \mathsf{S}'.$$

The tactic $\mathtt{t}$ is then used to discharge the proof obligation for $\forall \overline{\mathsf{x}}, \mathsf{P} \rightarrow \mathsf{C} \Longrightarrow \mathsf{C}'$, whereas it tries to prove the side condition $\forall\,\overline{\mathsf{x}}\,,\,\mathsf{P}\ \rightarrow\ \mathsf{C} \lhd \{\overline{\mathsf{x}}\,,\,\mathsf{C}'\ \mid\ \mathsf{P}\} \cup \mathsf{S}'$ fully generically. It does so by applying the two rules Here-$\lhd$ and Step-$\lhd$ that define $\lhd$ (cf. Sect. 3). Successfully applying Step-$\lhd$ and then Here-$\lhd$ to all subsequent subgoals, means that all single step $\Longrightarrow$-derivations from C reach a configuration in $\{\overline{\mathsf{x}}\,,\,\mathsf{C}'\ \mid\ \mathsf{P}\} \cup \mathsf{S}'$. This combination proves most barred side conditions and is thus tried first. If that fails, our tactic tries to prove the side condition by systematically trying all $\overset{*}{\Longrightarrow}$-derivations of a bounded length.

7 Concluding Remarks

We presented a framework for deriving correct-by-construction compilers from formal specifications by means of calculations. The distinguishing feature of our calculation framework is that it accommodates non-deterministic semantics. The key ingredient of this framework is the set of proof rules that allows us to prove the compiler correctness property in one go despite the non-deterministic semantics. The mechanisation of the framework in Coq helps to scale our approach to more intricate languages. Moreover, this mechanisation allows us to utilise proof search to discharge side conditions that are subject to the changing virtual machine semantics.

We conclude this paper with a brief discussion on related work, limitations of our approach, and possible further work.

Related work. To the best of our knowledge, non-deterministic languages have not been considered in the literature on calculating compilers. Meijer [10] does consider a language with backtracking semantics, called $\mathcal{B}$, but strictly speaking it is not a non-deterministic language. The language $\mathcal{B}$ has a set-valued semantics that describes the search space, which the language is able to traverse. However, non-deterministic languages can be represented using such a set-valued semantics as well, and we believe Meijer's approach can be used to calculate a compiler for a non-deterministic language. Unfortunately, this setup requires a detour: from the big-step semantics of the target language we need to calculate an equivalent set-valued semantics. From that semantics we calculate a compiler and a virtual machine, which is a tail recursive function on sets of machine configurations. From this representation of the virtual machine a small-step semantics representation can be calculated. Bahr and Hutton [4] describe this technique and argue that it can be used to extend their calculation approach to non-deterministic languages. In contrast, the calculation framework presented in this paper allows us to calculate a compiler and a virtual machine directly without pre-processing the input semantics or post-processing the output semantics.

We also briefly remark on the importance of the antecedent $\exists\, \mathsf{D}, \mathsf{C} \Longrightarrow \mathsf{D}$ of the STEP-$\lhd$ rule, which is missing in the corresponding definition by Hutton and Wright [7]. If we removed it, we would be able to prove "correctness" of compilers that are in fact not correct. For example, assume that we extended the definition of the virtual machine from Sect. 2 by the following rule

$$\langle \mathsf{PUSH\ n :: c\ ,\ s} \rangle \Longrightarrow \langle [], [42] \rangle$$

Then the calculation in Fig. 4 would still be valid even though the compiler is not correct for this virtual machine. The virtual machine admits the following single step execution, which computes the result 42 for the expression Val 0:

$$\langle \mathsf{comp\ (Val\ 0)}, [] \rangle = \langle [\mathsf{PUSH\ 0}], [] \rangle \Longrightarrow \langle [], [42] \rangle$$

The underlying problem is that the definition of $\lhd$ by Hutton and Wright [7] does not satisfy Proposition 1. In particular, their definition allows us to prove that $\langle \mathsf{comp\ (Val\ 0)}, [] \rangle$ is barred by the singleton set $\{\langle [], [0] \rangle\}$. However, we cannot extend the above execution of the virtual machine such that it ends in the configuration $\langle [], [0] \rangle$, which contradicts Proposition 1.

Our implementation of the calculation framework in Coq is derived from the work of Tesson et al. [12]. A similar framework for writing calculation proofs in Agda has been developed by Mu et al. [11]. However, since our approach relies on proof automation, we prefer the Coq system over Agda.

Future work. The calculation proofs that we presented here proceed by induction on the structure of the source language. This may be a problem if the semantics is not given in a compositional manner. For instance, in their treatment of lambda calculi Bahr and Hutton [4] start with a compositional semantics, but then transform it using defunctionalisation. The resulting big-step operational semantics is not compositional anymore and the calculation proof proceeds by induction on the big-step operational semantics.

The use of induction on the source language (as opposed to induction on the semantics) appears to be unavoidable for proving the soundness property for a non-deterministic language: The very goal of proving soundness is to prove that $\mathsf{e} \Downarrow \mathsf{n}$ holds, given that n is a possible result that the compiled program yields. Hence, we cannot perform a proof by induction on $\mathsf{e} \Downarrow \mathsf{n}$.

Another computational feature worth considering is concurrency. We did consider a language with interrupts – arguably a concurrency feature, albeit a simple one. However, if we want to deal with "proper" concurrency features, the calculation framework needs to be able to deal with a semantics for the source language that is able to properly capture concurrent behaviour, e.g. small-step operational semantics. Moreover, in a setting of concurrency we need to be able to reason not only about the result of a computation but also its I/O behaviour.

References

1. Ager, M.S., Biernacki, D., Danvy, O., Midtgaard., J.: From interpreter to compiler and virtual machine: a functional derivation. Technical report RS-03-14, Department of Computer Science, University of Aarhus (2003)

2. Ager, M.S., Biernacki, D., Danvy, O., Midtgaard, J.: A functional correspondence between evaluators and abstract machines. In: Proceedings of the 5th ACM SIGPLAN International Conference on Principles and Practice of Declaritive Programming, pp. 8–19 (2003)
3. Backhouse, R.: Program Construction: Calculating Implementations from Specifications. Wiley, UK (2003)
4. Bahr, P., Hutton, G.: Calculating correct compilers, July 2014. submitted to J. Funct. Program
5. Danvy, O., Millikin, K.: On the equivalence between small-step and big-step abstract machines: a simple application of lightweight fusion. Inf. Process. Lett. **106**(3), 100–109 (2008)
6. Hutton, G.: Programming in Haskell, vol. 2. Cambridge University Press, Cambridge (2007)
7. Hutton, G., Wright, J.: What is the meaning of these constant interruptions? J. Funct. Program. **17**(06), 777–792 (2007)
8. Kahn, G.: Natural semantics. In: Proceedings of the 4th Annual Symposium on Theoretical Aspects of Computer Science, pp. 22–39 (1987)
9. Leroy, X.: Formal certification of a compiler back-end or: programming a compiler with a proof assistant. In: Proceedings of the 33rd ACM SIGPLAN-SIGACT Symposium on Principles of Programming Languages, pp. 42–54 (2006)
10. Meijer, E.: Calculating compilers. Ph.D. thesis, Katholieke Universiteit Nijmegen (1992)
11. Mu, S.C., Ko, H.S., Jansson, P.: Algebra of programming in Agda: dependent types for relational program derivation. J. Funct. Program. **19**, 545–579 (2009)
12. Tesson, J., Hashimoto, H., Hu, Z., Loulergue, F., Takeichi, M.: Program calculation in Coq. In: Johnson, M., Pavlovic, D. (eds.) AMAST 2010. LNCS, vol. 6486, pp. 163–179. Springer, Heidelberg (2011)
13. Troelstra, A.S., van Dalen, D.: Constructivism in Mathematics: An Introduction, vol. 1. Elsevier, USA (1988)
14. Wand, M.: Deriving target code as a representation of continuation semantics. ACM Trans. Program. Lang. Syst. **4**(3), 496–517 (1982)

Notions of Bidirectional Computation and Entangled State Monads

Faris Abou-Saleh[1], James Cheney[2], Jeremy Gibbons[1(✉)], James McKinna[2], and Perdita Stevens[2]

[1] Department of Computer Science, University of Oxford, Oxford, UK
{faris.abou-saleh,jeremy.gibbons}@cs.ox.ac.uk
[2] School of Informatics, University of Edinburgh, Edinburgh, UK
{james.cheney,james.mckinna,perdita.stevens}@ed.ac.uk

Abstract. Bidirectional transformations (bx) support principled consistency maintenance between data sources. Each data source corresponds to one perspective on a composite system, manifested by operations to 'get' and 'set' a view of the whole from that particular perspective. Bx are important in a wide range of settings, including databases, interactive applications, and model-driven development. We show that bx are naturally modelled in terms of mutable state; in particular, the 'set' operations are stateful functions. This leads naturally to considering bx that exploit other computational effects too, such as I/O, nondeterminism, and failure, all largely ignored in the bx literature to date. We present a semantic foundation for symmetric bidirectional transformations with effects. We build on the mature theory of monadic encapsulation of effects in functional programming, develop the equational theory and important combinators for effectful bx, and provide a prototype implementation in Haskell along with several illustrative examples.

1 Introduction

Bidirectional transformations (bx) arise when synchronising data in different data sources: updates to one source entail corresponding updates to the others, in order to maintain consistency. When a data source represents the complete information, this is a straightforward task; an update can be matched by discarding and regenerating the other sources. It becomes more interesting when one data representation lacks some information that is recorded by another; then the corresponding update has to merge new information on one side with old information on the other side. Such bidirectional transformations have been the focus of a flurry of recent activity—in databases, in programming languages, and in software engineering, among other fields—giving rise to a flourishing series of BX Workshops (see http://bx-community.wikidot.com/) and BX Seminars (in Japan, Germany, and Canada so far: see [6] for an early report on the state of the art).

The different branches of the bx community have come up with a variety of different formalisations of bx with conflicting definitions and incompatible

R. Hinze and J. Voigtländer (Eds.): MPC 2015, LNCS 9129, pp. 187–214, 2015.
DOI:10.1007/978-3-319-19797-5_9

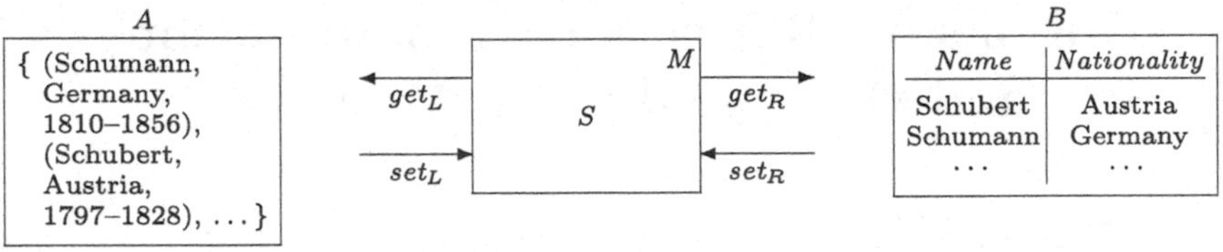

Fig. 1. Effectful bx between sources A and B and with effects in monad M, with hidden state S, illustrated using the Composers example

extensions, such as *lenses* [10], *relational bx* [31], *symmetric lenses* [13], *putback-based lenses* [27], and *profunctors* [20]. We have been seeking a unification of the varying approaches. It turns out that quite a satisfying unifying formalism can be obtained from the perspective of the state monad. More specifically, we are thinking about two data sources, and stateful computations acting on pairs representing these two sources. However, the two components of the pair are not independent, as two distinct memory cells would be, but are *entangled*—a change to one component generally entails a consequent change to the other.

This stateful perspective suggests using monads for bx, much as Moggi showed that monads unify many computational effects [24]. But not only that; it suggests a way to *generalise* bx to encompass other features that monads can handle. In fact, several approaches to lenses do in practice allow for monadic operations [20,27]. But there are natural concerns about such an extension: Are "monadic lenses" legitimate bidirectional transformations? Do they satisfy laws analogous to the roundtripping ('GetPut' and 'PutGet') laws of traditional lenses? Can we compose such transformations? We show that bidirectional computations can be encapsulated in monads, and be combined with other standard monads to accommodate effects, while still satisfying appropriate equational laws and supporting composition.

To illustrate our approach informally, consider Fig. 1, which is based on the *Composers* example [32]. This example relates two data sources A and B: on the left, an $a :: A$ consists of a set of triples (*name*, *nationality*, *dates*), and on the right, a $b :: B$ consists of an ordered list of pairs (*name*, *nationality*). The two sources a and b are consistent when they contain the same name–nationality pairs, ignoring dates and ordering. The centre of the figure illustrates the interface and typical operations of our (effectful) bx, including a monad M, operations $get_L :: M\ A$ and $get_R :: M\ B$ that return the current values of the left and right sides, and operations $set_L :: A \to M\ ()$ and $set_R :: B \to M\ ()$ that accept a new value for the left-hand or right-hand side, possibly performing side-effects in M.

In this example, neither A nor B is obtainable from the other; A omits the ordering information from B, while B omits the date information from A. This means that there may be multiple ways to change one side to match a change to the other. For example, the monadic computation **do** $\{b \leftarrow get_R; set_R\ (b +\!\!+ [(\texttt{"Bach"}, \texttt{"Germany"})])\}$ looks at the current value of the B-side, and modifies it to include a new pair at the end. Of course, this addition is ambiguous: do we mean J. S. Bach (1685–1750), J. C. Bach (1735–1782), or another Bach? There is no way to give the dates through the B interface; typically, pure bx would initialise the unspecified part to some default value such as `"????-????"`. Conversely, had we inserted the triple (`"Bach"`, `"Germany"`, `"1685-1750"`) on the left, then there may be several consistent ways to change the right-hand side to match: for example, inserting at the beginning, the end, or somewhere in the middle of the list. Again, conventional pure bx must fix some strategy in advance.

A conventional (pure) bx corresponds (roughly) to taking $M = State\ S$, where S is some type of states from which we can obtain both A and B; for example, S could consist of *lists* of triples (*name*, *nationality*, *dates*) from which we can easily extract both A (by forgetting the order) and B (by forgetting the dates). However, our approach to effectful bx allows many other choices for M that allow us to use side-effects when restoring consistency. Consider the following two scenarios, taken from the model-driven development domain, where the entities being synchronised are 'model states' a, b such as UML models or RDBMS schemas, drawn from suitable 'model spaces' A, B. We will revisit them among the concrete examples in Sect. 5.

Scenario 1.1 (nondeterminism). As mentioned above, most formal notions of bx require that the transformation (or programmer) decide on a consistency-restoration strategy in advance. By contrast, in the Janus Transformation Language (JTL) [5], programmers need only specify a consistency relation, allowing the bx engine to resolve the underspecification nondeterministically. Given a change to one source, JTL uses an external constraint solver to find a consistent choice for the other source; there might be multiple choices.

Our effectful bx can handle this by combining state with a nondeterministic choice monad, taking $M = StateT\ S\ [\,]$. For example, an attempt to add Bach to the right-hand side could result in several possibilities for the left-hand side (perhaps using an external source such as Wikipedia to find the dates for candidate matches). Conversely, adding Bach to the left-hand side could result in a nondeterministic choice of all possible positions to add the matching record in the right-hand list. No previous formalism permits such *nondeterministic bx* to be composed with conventional deterministic transformations, or characterises the laws that such transformations ought to satisfy. ◇

Scenario 1.2 (interaction). Alternatively, instead of automatically trying to find a single (or all possible) dates for Bach, why not ask the user for help? In unidirectional model transformation settings, Varró [35] proposed "model transformation by example", where the transformation system 'learns' gradually, by prompting its users over time, the desired way to restore consistency in various situations. One can see this as an (interactive) instance of *memoisation*.

Our effectful bx can also handle this, using the IO monad in concert with some additional state to remember past questions and their answers. For example, when Bach is added to the right-hand side, the bx can check to see whether it already knows how to restore consistency. If so, it does so without further ado. If not, it queries the user to determine how to fill in the missing values needed for the right-hand side, perhaps having first appealed to some external knowledge base to generate helpful suggestions. It then records the new version of the right-hand side and updates its state so that the same question is not asked again. No previous formalism allows for I/O during consistency restoration. $\diamondsuit$

The paper is structured as follows. Section 2 reviews monads as a foundation for effectful programming, fixing (idealised) Haskell notation used throughout the paper, and recaps definitions of lenses. Our contributions start in Sect. 3, with a presentation of our monadic approach to bx. Section 4 considers a definition of composition for effectful bx. In Sect. 5 we discuss initialisation, and formalise the motivating examples above, along with other combinators and examples of effectful bx. These examples are the first formal treatments of effects such as nondeterminism or interaction for symmetric bidirectional transformations, and they illustrate the generality of our approach. Finally we discuss related work and conclude. All proofs and code for examples can be found in an extended version of this paper [1].

2 Background

Our approach to bx is semantics-driven, so we here provide some preliminaries on semantics of effectful computation – focusing on monads, Haskell's use of type classes for them, and some key instances that we exploit heavily in what follows. We also briefly recap the definitions of asymmetric and symmetric lenses.

2.1 Effectful Computation

Moggi's seminal work on the semantics of effectful computation [24], and much continued investigation, shows how computational effects can be described using monads. Building on this, we assume that computations are represented as Kleisli arrows for a strong monad T defined on a cartesian closed category $\mathbb{C}$ of 'value' types and 'pure' functions. The reader uncomfortable with such generality can safely consider our definitions in terms of the category of sets and total functions, with T encapsulating the 'ambient' programming language effects: none in a total functional programming language like Agda, partiality in Haskell, global state in Pascal, network access in Java, etc.

2.2 Notational Conventions

We write in Haskell notation, except for the following few idealisations. We assume a cartesian closed category $\mathbb{C}$, avoiding niceties about lifted types and undefined values in Haskell; we further restrict attention to terminating programs. We use lowercase (Greek) letters for polymorphic type variables in code, and uppercase (Roman) letters for monomorphic instantiations of those variables in accompanying prose. We elide constructors and destructors for a **newtype**, the explicit witnesses to the isomorphism between the defined type and its structure, and use instead a **type** synonym that equates the defined type and its structure; e.g., in Sect. 2.4 we omit the function *runStateT* from *StateT S T A* to $S \to T\ (A, S)$. Except where expressly noted, we assume a kind of Barendregt convention, that bound variables are chosen not to clash with free variables; for example, in the definition below of a commutative monad, we elide the explicit proviso "for x, y distinct variables not free in m, n" (one might take the view that m, n are themselves variables, rather than possibly open terms that might capture x, y). We use a tightest-binding lowered dot for field access in records; e.g., in Definition 3.8 we write $bx.get_L$ rather than $get_L\ bx$; we therefore write function composition using a centred dot, $f \cdot g$. The code online expands these conventions into real Haskell. We also make extensive use of equational reasoning over monads in **do** notation [11]. Different branches of the bx community have conflicting naming conventions for various operations, so we have renamed some of them, favouring internal over external consistency.

2.3 Monads

Definition 2.1 (monad type class). Type constructors representing notions of effectful computation are represented as instances of the Haskell type class *Monad*:

$$
\begin{array}{l}
\mathbf{class}\ \mathit{Monad}\ \tau\ \mathbf{where} \\
\quad \mathit{return} :: \alpha \to \tau\ \alpha \\
\quad (\gg\!=) \ :: \tau\ \alpha \to (\alpha \to \tau\ \beta) \to \tau\ \beta \quad \text{-- pronounced `bind'}
\end{array}
$$

A *Monad* instance should satisfy the following laws:

$$
\begin{array}{ll}
\mathit{return}\ x \gg\!= f & = f\ x \\
m \gg\!= \mathit{return} & = m \\
(m \gg\!= f) \gg\!= g & = m \gg\!= \lambda x.\,(f\ x \gg\!= g)
\end{array}
$$

◇

Common examples in Haskell (with which we assume familiarity) include:

$$
\begin{array}{lll}
\mathbf{type}\ \mathit{Id}\ \alpha & = \alpha & \text{-- no effects} \\
\mathbf{data}\ \mathit{Maybe}\ \alpha & = \mathit{Just}\ \alpha \mid \mathit{Nothing} & \text{-- failure/exceptions} \\
\mathbf{data}\ [\alpha] & = [\,] \mid \alpha : [\alpha] & \text{-- choice} \\
\mathbf{type}\ \mathit{State}\ \sigma\ \alpha & = \sigma \to (\alpha, \sigma) & \text{-- state} \\
\mathbf{type}\ \mathit{Reader}\ \sigma\ \alpha & = \sigma \to \alpha & \text{-- environment} \\
\mathbf{type}\ \mathit{Writer}\ \sigma\ \alpha & = (\alpha, \sigma) & \text{-- logging}
\end{array}
$$

as well as the (in)famous *IO* monad, which encapsulates interaction with the outside world. We need a *Monoid* σ instance for the *Writer* σ monad, in order to support empty and composite logs.

Definition 2.2. In Haskell, monadic expressions may be written using **do** notation, which is defined by translation into applications of bind:

$$\begin{aligned}
&\mathbf{do}\,\{\,\mathbf{let}\ \mathit{decls}; \mathit{ms}\,\} = \mathbf{let}\ \mathit{decls}\ \mathbf{in}\ \mathbf{do}\,\{\,\mathit{ms}\,\}\\
&\mathbf{do}\,\{\,a \leftarrow m; \mathit{ms}\,\} \quad = m \gg\!\!= \lambda a.\,\mathbf{do}\,\{\,\mathit{ms}\,\}\\
&\mathbf{do}\,\{\,m\,\} \qquad\qquad\quad = m
\end{aligned}$$

The *body* ms of a **do** expression consists of zero or more 'qualifiers', and a final expression m of monadic type; qualifiers are either 'declarations' **let** *decls* (with *decls* a collection of bindings $a = e$ of patterns a to expressions e) or 'generators' $a \leftarrow m$ (with pattern a and monadic expression m). Variables bound in pattern a may appear free in the subsequent body ms; in contrast to Haskell, we assume that the pattern cannot fail to match. When the return value of m is not used – e.g., when void – we write $\mathbf{do}\,\{\,m; \mathit{ms}\,\}$ as shorthand for $\mathbf{do}\,\{\,\_ \leftarrow m; \mathit{ms}\,\}$ with its wildcard pattern. ♢

Definition 2.3 (commutative monad). We say that $m :: T\ A$ *commutes in* T if the following holds for all $n :: T\ B$:

$$\mathbf{do}\,\{x \leftarrow m; y \leftarrow n; \mathit{return}\ (x, y)\} = \mathbf{do}\,\{\,y \leftarrow n; x \leftarrow m; \mathit{return}\ (x, y)\}$$

A monad T is *commutative* if all $m :: T\ A$ commute, for all A. ♢

Definition 2.4. An element z of a monad is called a *zero element* if it satisfies:

$$\mathbf{do}\,\{x \leftarrow z; f\ x\} = z = \mathbf{do}\,\{x \leftarrow m; z\}$$ ♢

Among monads discussed so far, *Id*, *Reader* and *Maybe* are commutative; if σ is a commutative monoid, *Writer* σ is commutative; but many interesting monads, such as *IO* and *State*, are not. The *Maybe* monad has zero element *Nothing*, and *List* has zero *Nil*; the zero element is unique if it exists.

Definition 2.5 (monad morphism). Given monads T and T', a *monad morphism* is a polymorphic function $\varphi :: \forall \alpha. T\ \alpha \to T'\ \alpha$ satisfying

$$\begin{aligned}
&\varphi\,(\mathbf{do}_T\,\{\,\mathit{return}\ a\}) \qquad = \mathbf{do}_{T'}\,\{\,\mathit{return}\ a\}\\
&\varphi\,(\mathbf{do}_T\,\{a \leftarrow m; k\ a\}) = \mathbf{do}_{T'}\,\{a \leftarrow \varphi\ m; \varphi\ (k\ a)\}
\end{aligned}$$

(subscripting to make clear which monad is used where). ♢

2.4 Combining State and Other Effects

We recall the *state monad transformer* (see e.g. Liang *et al.* [21]).

Definition 2.6 (state monad transformer). State can be combined with effects arising from an arbitrary monad T using the *StateT* monad transformer:

$$\begin{aligned}
&\mathbf{type}\ \mathit{StateT}\ \sigma\ \tau\ \alpha = \sigma \to \tau\ (\alpha, \sigma)\\
&\mathbf{instance}\ \mathit{Monad}\ \tau \Rightarrow \mathit{Monad}\ (\mathit{StateT}\ \sigma\ \tau)\ \mathbf{where}\\
&\quad \mathit{return}\ a = \lambda s.\,\mathit{return}\ (a, s)\\
&\quad m \gg\!\!= k \ \ = \lambda s.\,\mathbf{do}\,\{(a, s') \leftarrow m\ s; k\ a\ s'\}
\end{aligned}$$

This provides *get* and *set* operations for the state type:

$$
\begin{aligned}
&\mathit{get} :: \mathit{Monad}\ \tau \Rightarrow \mathit{StateT}\ \sigma\ \tau\ \sigma \\
&\mathit{get} = \lambda s.\ \mathit{return}\ (s, s) \\
&\mathit{set} :: \mathit{Monad}\ \tau \Rightarrow \sigma \rightarrow \mathit{StateT}\ \sigma\ \tau\ () \\
&\mathit{set}\ s' = \lambda s.\ \mathit{return}\ ((), s')
\end{aligned}
$$

which satisfy the following four laws [28]:

$$
\begin{array}{lll}
\text{(GG)} & \mathbf{do}\,\{\, s \leftarrow \mathit{get}; s' \leftarrow \mathit{get}; \mathit{return}\ (s, s') \,\} & = \mathbf{do}\,\{\, s \leftarrow \mathit{get}; \mathit{return}\ (s, s) \,\} \\
\text{(SG)} & \mathbf{do}\,\{\, \mathit{set}\ s; \mathit{get} \,\} & = \mathbf{do}\,\{\, \mathit{set}\ s; \mathit{return}\ s \,\} \\
\text{(GS)} & \mathbf{do}\,\{\, s \leftarrow \mathit{get}; \mathit{set}\ s \,\} & = \mathbf{do}\,\{\, \mathit{return}\ () \,\} \\
\text{(SS)} & \mathbf{do}\,\{\, \mathit{set}\ s; \mathit{set}\ s' \,\} & = \mathbf{do}\,\{\, \mathit{set}\ s' \,\}
\end{array}
$$

Computations in T embed into $\mathit{StateT}\ S\ T$ via the monad morphism *lift*:

$$
\begin{aligned}
&\mathit{lift} :: \mathit{Monad}\ \tau \Rightarrow \tau\ \alpha \rightarrow \mathit{StateT}\ \sigma\ \tau\ \alpha \\
&\mathit{lift}\ m = \lambda s.\ \mathbf{do}\,\{\, a \leftarrow m; \mathit{return}\ (a, s) \,\}
\end{aligned}
$$

◇

Lemma 2.7. Unused *get*s are discardable:

$$\mathbf{do}\,\{\, \_ \leftarrow \mathit{get}; m \,\} = \mathbf{do}\,\{\, m \,\}$$

◇

Lemma 2.8 (liftings commute with *get* and *set*). We have:

$$
\begin{aligned}
&\mathbf{do}\,\{\, a \leftarrow \mathit{get}; b \leftarrow \mathit{lift}\ m; \mathit{return}\ (a, b) \,\} \\
&\qquad\qquad = \mathbf{do}\,\{\, b \leftarrow \mathit{lift}\ m; a \leftarrow \mathit{get}; \mathit{return}\ (a, b) \,\} \\
&\mathbf{do}\,\{\, \mathit{set}\ a; b \leftarrow \mathit{lift}\ m; \mathit{return}\ b \,\} = \mathbf{do}\,\{\, b \leftarrow \mathit{lift}\ m; \mathit{set}\ a; \mathit{return}\ b \,\}
\end{aligned}
$$

◇

Definition 2.9. Some convenient shorthands:

$$
\begin{aligned}
&\mathit{gets} :: \mathit{Monad}\ \tau \Rightarrow (\sigma \rightarrow \alpha) \rightarrow \mathit{StateT}\ \sigma\ \tau\ \alpha \\
&\mathit{gets}\ f = \mathbf{do}\,\{\, s \leftarrow \mathit{get}; \mathit{return}\ (f\ s) \,\} \\
&\mathit{eval} :: \mathit{Monad}\ \tau \Rightarrow \mathit{StateT}\ \sigma\ \tau\ \alpha \rightarrow \sigma \rightarrow \tau\ \alpha \\
&\mathit{eval}\ m\ s = \mathbf{do}\,\{\, (a, s') \leftarrow m\ s; \mathit{return}\ a \,\} \\
&\mathit{exec} :: \mathit{Monad}\ \tau \Rightarrow \mathit{StateT}\ \sigma\ \tau\ \alpha \rightarrow \sigma \rightarrow \tau\ \sigma \\
&\mathit{exec}\ m\ s = \mathbf{do}\,\{\, (a, s') \leftarrow m\ s; \mathit{return}\ s' \,\}
\end{aligned}
$$

◇

Definition 2.10. We say that a computation $m :: \mathit{StateT}\ S\ T\ A$ is a *T-pure query* if it cannot change the state, and is pure with respect to the base monad T; that is, $m = \mathit{gets}\ h$ for some $h :: S \rightarrow A$. Note that a T-pure query need not be pure with respect to $\mathit{StateT}\ S\ T$; in particular, it will typically read the state. ◇

Definition 2.11 (data refinement). Given monads M of 'abstract computations' and M' of 'concrete computations', various 'abstract operations' $\mathit{op} :: A \rightarrow M\ B$ with corresponding 'concrete operations' $\mathit{op}' :: A \rightarrow M'\ B$, an 'abstraction function' $\mathit{abs} :: M'\ \alpha \rightarrow M\ \alpha$ and a 'reification function' $\mathit{conc} :: M\ \alpha \rightarrow M'\ \alpha$, we say that *conc* is a *data refinement* from (M, op) to (M', op') if:

- *conc* distributes over $(\gg\!=)$
- $abs \cdot conc = id$, and
- $op' = conc \cdot op$ for each of the operations. ◇

Remark 2.12. Given such a data refinement, a composite abstract computation can be faithfully simulated by a concrete one:

$$\begin{array}{ll} & \mathbf{do}\,\{\,a \leftarrow op_1\,();b \leftarrow op_2\,(a);op_3\,(a,b)\,\} \\ = & \quad [\![\; abs \cdot conc = id \;]\!] \\ & abs\,(conc\,(\mathbf{do}\,\{\,a \leftarrow op_1\,();b \leftarrow op_2\,(a);op_3\,(a,b)\}))\\ = & \quad [\![\; conc \text{ distributes over } (\gg\!=) \;]\!] \\ & abs\,(\mathbf{do}\,\{\,a \leftarrow conc\,(op_1\,());b \leftarrow conc\,(op_2\,(a));conc\,(op_3\,(a,b))\})\\ = & \quad [\![\; \text{concrete operations} \;]\!] \\ & abs\,(\mathbf{do}\,\{\,a \leftarrow op'_1\,();b \leftarrow op'_2\,(a);op'_3\,(a,b)\,\}) \end{array}$$

If *conc* also preserves *return* (so *conc* is a monad morphism), then we would have a similar result for 'empty' abstract computations too; but we don't need that stronger property in this paper, and it does not hold for our main example (Remark 3.7). ◇

Lemma 2.13. Given an arbitrary monad T, not assumed to be an instance of $StateT$, with operations $get_T :: T\ S$ and $set_T :: S \rightarrow T\ ()$ for a type S, such that get_T and set_T satisfy the laws (GG), (GS), and (SG) of Definition 2.6, then there is a data refinement from T to $StateT\ S\ T$. ◇

Proof (sketch). The abstraction function *abs* from $StateT\ S\ T$ to T and the reification function *conc* in the opposite direction are given by

$$\begin{array}{ll} abs\ m & = \mathbf{do}\,\{\,s \leftarrow get_T;(a,s') \leftarrow m\ s;set_T\ s';return\ a\,\} \\ conc\ m & = \lambda s.\,\mathbf{do}\,\{\,a \leftarrow m;s' \leftarrow get_T;return\ (a,s')\,\} \\ & = \mathbf{do}\,\{\,a \leftarrow lift\ m;s' \leftarrow lift\ get_T;set\ s';return\ a\,\} \end{array}$$ □

Remark 2.14. Informally, if T provides suitable get and set operations, we can without loss of generality assume it to be an instance of $StateT$. The essence of the data refinement is for concrete computations to maintain a shadow copy of the implicit state; *conc* m synchronises the outer copy of the state with the inner copy after executing m, and *abs* m runs the $StateT$ computation m on an initial state extracted from T, and stores the final state back there. ◇

2.5 Lenses

The notion of an (asymmetric) 'lens' between a source and a view was introduced by Foster *et al.* [10]. We adapt their notation, as follows.

Definition 2.15. A lens $l :: Lens\ S\ V$ from source type S to view type V consists of a pair of functions which get a *view* of the source, and *update* an old source with a modified view:

$$\mathbf{data}\ \mathit{Lens}\ \alpha\ \beta = \mathit{Lens}\ \{\ \mathit{view} :: \alpha \to \beta, \mathit{update} :: \alpha \to \beta \to \alpha\}$$

We say that a lens $l :: \mathit{Lens}\ S\ V$ is *well behaved* if it satisfies the two round-tripping laws

$$\begin{array}{ll} \text{(UV)} & l.\mathit{view}\ (l.\mathit{update}\ s\ v) = v \\ \text{(VU)} & l.\mathit{update}\ s\ (l.\mathit{view}\ s) = s \end{array}$$

and *very well-behaved* or *overwritable* if in addition

$$\text{(UU)} \quad l.\mathit{update}\ (l.\mathit{update}\ s\ v)\ v' = l.\mathit{update}\ s\ v'$$

◇

Remark 2.16. Very well-behavedness captures the idea that, after two successive updates, the second update completely *overwrites* the first. It turns out to be a rather strong condition, and many natural lenses do not satisfy it. Those that do generally have the special property that source S factorises cleanly into $V \times C$ for some type C of 'complements' independent of V, and so the view is a projection. For example:

$$\begin{array}{l} \mathit{fstLens} :: \mathit{Lens}\ (a, b)\ a \\ \mathit{fstLens} = \mathit{Lens}\ \mathit{fst}\ u\ \mathbf{where}\ u\ (a, b)\ a' = (a', b) \\ \mathit{sndLens} :: \mathit{Lens}\ (a, b)\ b \\ \mathit{sndLens} = \mathit{Lens}\ \mathit{snd}\ u\ \mathbf{where}\ u\ (a, b)\ b' = (a, b') \end{array}$$

But in general, the V may be computed from and therefore depend on all of the S value, and there is no clean factorisation of S into $V \times C$. ◇

Asymmetric lenses are constrained, in the sense that they relate two types S and V in which the view V is completely determined by the source S. Hofmann *et al.* [13] relaxed this constraint, introducing *symmetric lenses* between two types A and B, neither of which need determine the other:

Definition 2.17. A *symmetric lens* from A to B with complement type C consists of two functions converting to and from A and B, each also operating on C.

$$\mathbf{data}\ \mathit{SLens}\ \gamma\ \alpha\ \beta = \mathit{SLens}\ \{\mathit{put}_R :: (\alpha, \gamma) \to (\beta, \gamma),\ \mathit{put}_L :: (\beta, \gamma) \to (\alpha, \gamma)\}$$

We say that symmetric lens l is *well-behaved* if it satisfies the following two laws:

$$\begin{array}{llll} \text{(PutRL)} & l.\mathit{put}_R\ (a, c) = (b, c') & \Rightarrow & l.\mathit{put}_L\ (b, c') = (a, c') \\ \text{(PutLR)} & l.\mathit{put}_L\ (b, c) = (a, c') & \Rightarrow & l.\mathit{put}_R\ (a, c') = (b, c') \end{array}$$

(There is also a stronger notion of very well-behavedness, but we do not need it for this paper.) ◇

Remark 2.18. The idea is that A and B represent two overlapping but distinct views of some common underlying data, and the so-called complement C represents their amalgamation (*not* necessarily containing all the information from both: rather, one view plus the complement together contain enough information to reconstruct the other view). Each function takes a new view and the old complement, and returns a new opposite view and a new complement. The two well-behavedness properties each say that after one update operation, the complement c' is fully consistent with the current views, and so a subsequent opposite update with the same view has no further effect on the complement. ◇

3 Monadic Bidirectional Transformations

We have seen that the state monad provides a pair *get*, *set* of operations on the state. A symmetric bx should provide two such pairs, one for each data source; these four operations should be effectful, not least because they should operate on some shared consistent state. We therefore introduce the following general notion of *monadic bx* (which we sometimes call 'mbx', for short).

Definition 3.1. We say that a data structure $t :: BX\ T\ A\ B$ is a *bx between A and B in monad T* when it provides appropriately typed functions:

$$\begin{aligned}\mathbf{data}\ BX\ \tau\ \alpha\ \beta = BX\ \{ & get_L :: \tau\ \alpha, & set_L :: \alpha \to \tau\ (), \\ & get_R :: \tau\ \beta, & set_R :: \beta \to \tau\ ()\}\end{aligned}$$

◇

3.1 Entangled State

The *get* and *set* operations of the state monad satisfy the four laws (GG), (SG), (GS), (SS) of Definition 2.6. More generally, one can give an equational theory of state with multiple memory locations; in particular, with just two locations 'left' (L) and 'right' (R), the equational theory has four operations get_L, set_L, get_R, set_R that match the BX interface. This theory has four laws for L analogous to those of Definition 2.6, another four such laws for R, and a final four laws stating that the L-operations commute with the R-operations. But this equational theory of two memory locations is too strong for interesting bx, because of the commutativity requirement: the whole point of the exercise is that invoking set_L should indeed affect the behaviour of a subsequent get_R, and symmetrically. We therefore impose only a subset of those twelve laws on the BX interface.

Definition 3.2. A *well-behaved BX* is one satisfying the following seven laws:

$$\begin{aligned}
&(\mathrm{G_L G_L}) && \mathbf{do}\ \{a \leftarrow get_L; a' \leftarrow get_L; return\ (a, a')\} \\
& && \qquad = \mathbf{do}\ \{a \leftarrow get_L; return\ (a, a)\} \\
&(\mathrm{S_L G_L}) && \mathbf{do}\ \{set_L\ a; get_L\} = \mathbf{do}\ \{set_L\ a; return\ a\} \\
&(\mathrm{G_L S_L}) && \mathbf{do}\ \{a \leftarrow get_L; set_L\ a\} = \mathbf{do}\ \{return\ ()\}
\end{aligned}$$

$$\begin{array}{lll}
(\mathrm{G_RG_R}) & \mathbf{do}\,\{\, a \leftarrow get_R; a' \leftarrow get_R; return\ (a, a')\,\} & \\
 & & = \mathbf{do}\,\{\, a \leftarrow get_R; return\ (a, a)\,\} \\
(\mathrm{S_RG_R}) & \mathbf{do}\,\{\, set_R\ a; get_R \,\} & = \mathbf{do}\,\{\, set_R\ a; return\ a\,\} \\
(\mathrm{G_RS_R}) & \mathbf{do}\,\{\, a \leftarrow get_R; set_R\ a\,\} & = \mathbf{do}\,\{\, return\ ()\,\} \\
(\mathrm{G_LG_R}) & \mathbf{do}\,\{\, a \leftarrow get_L; b \leftarrow get_R; return\ (a, b)\,\} & \\
 & & = \mathbf{do}\,\{\, b \leftarrow get_R; a \leftarrow get_L; return\ (a, b)\,\}
\end{array}$$

We further say that a *BX* is *overwritable* if it satisfies

$$\begin{array}{lll}
(\mathrm{S_LS_L}) & \mathbf{do}\,\{\, set_L\ a; set_L\ a' \,\} & = \mathbf{do}\,\{\, set_L\ a' \,\} \\
(\mathrm{S_RS_R}) & \mathbf{do}\,\{\, set_R\ a; set_R\ a' \,\} & = \mathbf{do}\,\{\, set_R\ a' \,\}
\end{array}$$

◇

We might think of the A and B views as being *entangled*; in particular, we call the monad arising as the initial model of the theory with the four operations $get_L, set_L, get_R, set_R$ and the seven laws $(\mathrm{G_LG_L})\ldots(\mathrm{G_LG_R})$ the *entangled state monad.*

Remark 3.3. Overwritability is a strong condition, corresponding to very well-behavedness of lenses [10], history-ignorance of relational bx [31] etc.; many interesting bx fail to satisfy it. Indeed, in an effectful setting, a law such as $(\mathrm{S_LS_L})$ demands that $set_L\ a'$ be able to undo (or overwrite) any effects arising from $set_L\ a$; such behaviour is plausible in the pure state-based setting, but not in general. Consequently, we do not demand overwritability in what follows. ◇

Definition 3.4. A *bx morphism* from $bx_1 :: BX\ T_1\ A\ B$ to $bx_2 :: BX\ T_2\ A\ B$ is a monad morphism $\varphi : \forall \alpha. T_1\ \alpha \to T_2\ \alpha$ that preserves the bx operations, in the sense that $\varphi\ (bx_1.get_L) = bx_2.get_L$ and so on. A *bx isomorphism* is an invertible bx morphism, i.e. a pair of monad morphisms $\iota :: \forall \alpha. T_1\ \alpha \to T_2\ \alpha$ and $\iota^{-1} : \forall \alpha. T_2\ \alpha \to T_1\ \alpha$ which are mutually inverse, and which also preserve the operations. We say that bx_1 and bx_2 are *equivalent* (and write $bx_1 \equiv bx_2$) if there is a bx isomorphism between them. ◇

3.2 Stateful BX

The get and set operations of a *BX*, and the relationship via entanglement with the equational theory of the state monad, strongly suggest that there is something inherently stateful about bx; that will be a crucial observation in what follows. In particular, the get_L and get_R operations of a $BX\ T\ A\ B$ reveal that it is in some sense storing an $A \times B$ pair; conversely, the set_L and set_R operations allow that pair to be updated. We therefore focus on monads of the form $StateT\ S\ T$, where S is the 'state' of the bx itself, capable of recording an A and a B, and T is a monad encapsulating any other ambient effects that can be performed by the 'get' and 'set' operations.

Definition 3.5. We introduce the following instance of the *BX* signature (note the inverted argument order):

$$\mathbf{type}\ StateTBX\ \tau\ \sigma\ \alpha\ \beta = BX\ (StateT\ \sigma\ \tau)\ \alpha\ \beta$$

◇

Remark 3.6. In fact, we can say more about the pair inside a $bx :: BX\ T\ A\ B$: it will generally be the case that only certain such pairs are observable. Specifically, we can define the subset $R \subseteq A \times B$ of *consistent pairs* according to bx, namely those pairs (a, b) that may be returned by

$$\mathbf{do}\ \{a \leftarrow get_L; b \leftarrow get_R; return\ (a, b)\}$$

We can see this subset R as the *consistency relation* between A and B maintained by bx. We sometimes write $A \bowtie B$ for this relation, when the bx in question is clear from context. ◇

Remark 3.7. Note that restricting attention to instances of $StateT$ is not as great a loss of generality as might at first appear. Consider a well-behaved bx of type $BX\ T\ A\ B$, over some monad T not assumed to be an instance of $StateT$. We say that a consistent pair $(a, b) :: A \bowtie B$ is *stable* if, when setting the components in either order, the later one does not disturb the earlier:

$$\mathbf{do}\ \{set_L\ a; set_R\ b; get_L\} = \mathbf{do}\ \{set_L\ a; set_R\ b; return\ a\}$$
$$\mathbf{do}\ \{set_R\ b; set_L\ a; get_R\} = \mathbf{do}\ \{set_R\ b; set_L\ a; return\ b\}$$

We say that the bx itself is stable if all its consistent pairs are stable. Stability does not follow from the laws, but many bx do satisfy this stronger condition. And given a stable bx, we can construct get and set operations for $A \bowtie B$ pairs, satisfying the three laws (GG), (GS), (SG) of Definition 2.6. By Lemma 2.13, this gives a data refinement from T to $StateT\ S\ T$, and so we lose nothing by using $StateT\ S\ T$ instead of T. Despite this, we do not impose stability as a requirement in the following, because some interesting bx are not stable. ◇

We have not found convincing examples of $StateTBX$ in which the two get functions have effects from T, rather than being T-pure queries. When the get functions are T-pure queries, we obtain the get/get commutation laws ($\mathrm{G_L G_L}$), ($\mathrm{G_R G_R}$), ($\mathrm{G_L G_R}$) for free [11], motivating the following:

Definition 3.8. We say that a well-behaved $bx :: StateTBX\ T\ S\ A\ B$ in the monad $StateT\ S\ T$ is *transparent* if get_L, get_R are T-pure queries, *i.e.* there exist $read_L :: S \to A$ and $read_R :: S \to B$ such that $bx.get_L = gets\ read_L$ and $bx.get_R = gets\ read_R$. ◇

Remark 3.9. Under the mild condition (Moggi's *monomorphism condition* [24]) on T that *return* be injective, $read_L$ and $read_R$ are *uniquely* determined for a transparent bx; so informally, we refer to $bx.read_L$ and $bx.read_R$, regarding them as part of the signature of bx. The monomorphism condition holds for the various monads we consider here (provided we have non-empty types σ for *State*, *Reader*, *Writer*). ◇

Now, transparent $StateTBX$ compose (Sect. 4), while general bx with effectful gets do not. So, in what follows, we confine our attention to transparent bx.

3.3 Subsuming Lenses

Asymmetric lenses, as in Definition 2.15, are subsumed by *StateTBX*. To simulate $l :: Lens\ A\ B$, one uses a *StateTBX* on state A and underlying monad *Id*:

$$\begin{array}{l} BX\ get\ set\ get_R\ set_R\ \mathbf{where} \\ \quad get_R \quad = \mathbf{do}\ \{\, a \leftarrow get; return\ (l.view\ a)\} \\ \quad set_R\ b' = \mathbf{do}\ \{\, a \leftarrow get; set\ (l.update\ a\ b')\} \end{array}$$

Symmetric lenses, as in Definition 2.17, are subsumed by our effectful bx too. In a nutshell, to simulate $sl :: SLens\ C\ A\ B$ one uses $StateTBX\ Id\ S$ where $S \subseteq A \times B \times C$ is the set of 'consistent triples' (a, b, c), in the sense that $sl.put_R\ (a, c) = (b, c)$ and $sl.put_L\ (b, c) = (a, c)$:

$$\begin{array}{l} BX\ get_L\ set_L\ get_L\ set_R\ \mathbf{where} \\ \quad get_L \quad = \mathbf{do}\ \{(a, b, c) \leftarrow get; return\ a\} \\ \quad get_R \quad = \mathbf{do}\ \{(a, b, c) \leftarrow get; return\ b\} \\ \quad set_L\ a' = \mathbf{do}\ \{(a, b, c) \leftarrow get; \mathbf{let}\ (b', c') = sl.put_R\ (a, c); set\ (a', b', c')\} \\ \quad set_R\ b' = \mathbf{do}\ \{(a, b, c) \leftarrow get; \mathbf{let}\ (a', c') = sl.put_L\ (b, c); set\ (a', b', c')\} \end{array}$$

Asymmetric lenses generalise straightforwardly to accommodate effects in an underlying monad too. One can define

$$\begin{array}{l} \mathbf{data}\ MLens\ \tau\ \alpha\ \beta = MLens\ \{mview \quad :: \alpha \rightarrow \beta, \\ \qquad\qquad\qquad\qquad\qquad\qquad mupdate :: \alpha \rightarrow \beta \rightarrow \tau\ \alpha\} \end{array}$$

with corresponding notions of well-behaved and very-well-behaved monadic lens. (Diviánszky [8] and Pacheco et al. [27], among others, have proposed similar notions.) However, it turns out not to be straightforward to establish a corresponding notion of 'monadic symmetric lens' incorporating other effects. In this paper, we take a different approach to combining symmetry and effects; we defer further discussion of the different approaches to *MLens*es and the complications involved in extending symmetric lenses with effects to a future paper.

4 Composition

An obviously crucial question is whether well-behaved monadic bx compose. They do, but the issue is more delicate than might at first be expected. Of course, we cannot expect arbitrary *BX* to compose, because arbitrary monads do not. Here, we present one successful approach for *StateTBX*, based on lifting the component operations on different state types (but the same underlying monad of effects) into a common compound state.

Definition 4.1 (*StateT* embeddings from lenses). Given a lens from A to B, we can embed a *StateT* computation on the narrower type B into another computation on the wider type A, wrt the same underlying monad T:

$$\begin{array}{l}\vartheta :: \mathit{Monad}\ \tau \Rightarrow \mathit{Lens}\ \alpha\ \beta \to \mathit{StateT}\ \beta\ \tau\ \gamma \to \mathit{StateT}\ \alpha\ \tau\ \gamma \\ \vartheta\ l\ m = \mathbf{do}\ a \leftarrow \mathit{get}; \mathbf{let}\ b = l.\mathit{view}\ a; \\ \qquad\qquad (c, b') \leftarrow \mathit{lift}\ (m\ b); \\ \qquad\qquad \mathbf{let}\ a' = l.\mathit{update}\ a\ b'; \\ \qquad\qquad \mathit{set}\ a'; \mathit{return}\ c\end{array}$$

◇

Essentially, $\vartheta\ l\ m$ uses l to get a view b of the source a, runs m to get a return value c and updated view b', uses l to update the view yielding an updated source a', and returns c.

Lemma 4.2. If $l :: \mathit{Lens}\ A\ B$ is very well-behaved, then $\vartheta\ l$ is a monad morphism. ◇

Definition 4.3. By Lemma 4.2, and since *fstLens* and *sndLens* are very well-behaved, we have the following monad morphisms lifting stateful computations to a product state space:

$$\begin{array}{l}\mathit{left}\ \ :: \mathit{Monad}\ \tau \Rightarrow \mathit{StateT}\ \sigma_1\ \tau\ \alpha \to \mathit{StateT}\ (\sigma_1, \sigma_2)\ \tau\ \alpha \\ \mathit{left}\ \ = \vartheta\ \mathit{fstLens} \\ \mathit{right} :: \mathit{Monad}\ \tau \Rightarrow \mathit{StateT}\ \sigma_2\ \tau\ \alpha \to \mathit{StateT}\ (\sigma_1, \sigma_2)\ \tau\ \alpha \\ \mathit{right} = \vartheta\ \mathit{sndLens}\end{array}$$

◇

Definition 4.4. For $bx_1 :: \mathit{StateTBX}\ T\ S_1\ A\ B$, $bx_2 :: \mathit{StateTBX}\ T\ S_2\ B\ C$, define the join $S_1\ {}_{bx_1}\!\bowtie_{bx_2}\ S_2$ informally as the subset of $S_1 \times S_2$ consisting of the pairs (s_1, s_2) in which observing the middle component of type B in state s_1 yields the same result as in state s_2. We might express this set-theoretically as follows:

$$S_1\ {}_{bx_1}\!\bowtie_{bx_2}\ S_2 = \{(s_1, s_2) \mid \mathit{eval}\ (bx_1.\mathit{get}_R)\ s_1 = \mathit{eval}\ (bx_2.\mathit{get}_L)\ s_2\}$$

More generally, one could state a categorical definition in terms of pullbacks. In Haskell, we can only work with the coarser type of raw pairs (S_1, S_2). Note that the equation in the set comprehension compares two computations of type $T\ B$; but if the bx are transparent, and *return* injective as per Remark 3.9, then the definition simplifies to:

$$S_1\ {}_{bx_1}\!\bowtie_{bx_2}\ S_2 = \{(s_1, s_2) \mid bx_1.\mathit{read}_R\ s_1 = bx_2.\mathit{read}_L\ s_2\}$$

The notation $S_1\ {}_{bx_1}\!\bowtie_{bx_2}\ S_2$ explicitly mentions bx_1 and bx_2, but we usually just write $S_1 \bowtie S_2$. No confusion should arise from using the same symbol to denote the consistent pairs of a single bx, as we did in Remark 3.6. ◇

Definition 4.5. Using *left* and *right*, we can define composition by:

$$\begin{array}{l}(\fatsemi) :: \mathit{Monad}\ \tau \Rightarrow \\ \qquad \mathit{StateTBX}\ \sigma_1\ \tau\ \alpha\ \beta \to \mathit{StateTBX}\ \sigma_2\ \tau\ \beta\ \gamma \to \mathit{StateTBX}\ (\sigma_1 \bowtie \sigma_2)\ \tau\ \alpha\ \gamma \\ bx_1 \fatsemi bx_2 = \mathit{BX}\ \mathit{get}_L\ \mathit{set}_L\ \mathit{get}_R\ \mathit{set}_R\ \mathbf{where} \\ \quad \mathit{get}_L\ = \mathbf{do}\ \{\mathit{left}\ (bx_1.\mathit{get}_L)\} \\ \quad \mathit{get}_R\ = \mathbf{do}\ \{\mathit{right}\ (bx_2.\mathit{get}_R)\}\end{array}$$

$$\begin{array}{l} set_L\ a = \mathbf{do}\ \{ left\ (bx_1.set_L\ a); b \leftarrow left\ (bx_1.get_R); right\ (bx_2.set_L\ b) \} \\ set_R\ c = \mathbf{do}\ \{ right\ (bx_2.set_R\ c); b \leftarrow right\ (bx_2.get_L); left\ (bx_1.set_R\ b) \} \end{array}$$

Essentially, to set the left-hand side of the composed bx, we first set the left-hand side of the left component bx_1, then get bx_1's R-value b, and set the left-hand side of bx_2 to this value; and similarly on the right. Note that the composition maintains the invariant that the compound state is in the subset $\sigma_1 \bowtie \sigma_2$ of $\sigma_1 \times \sigma_2$. ◇

Theorem 4.6 (transparent composition). Given transparent (and hence well-behaved) $bx_1 :: StateTBX\ S_1\ T\ A\ B$ and $bx_2 :: StateTBX\ S_2\ T\ B\ C$, their composition $bx_1 \mathbin{⨾} bx_2 :: StateTBX\ (S_1 \bowtie S_2)\ T\ A\ C$ is also transparent. ◇

Remark 4.7. Unpacking and simplifying the definitions, we have:

$$\begin{array}{l} bx_1 \mathbin{⨾} bx_2 = BX\ get_L\ set_L\ get_R\ set_R\ \mathbf{where} \\ \quad get_L \quad = \mathbf{do}\ \{ (s_1, \_) \leftarrow get; return\ (bx_1.read_L\ s_1) \} \\ \quad get_R \quad = \mathbf{do}\ \{ (\_, s_2) \leftarrow get; return\ (bx_2.read_R\ s_2) \} \\ \quad set_L\ a' = \mathbf{do}\ \{ (s_1, s_2) \leftarrow get; \\ \qquad\qquad ((), s_1') \leftarrow lift\ (bx_1.set_L\ a'\ s_1); \\ \qquad\qquad \mathbf{let}\ b = bx_1.read_R\ s_1'; \\ \qquad\qquad ((), s_2') \leftarrow lift\ (bx_2.set_L\ b\ s_2); \\ \qquad\qquad set\ (s_1', s_2') \} \\ \quad set_R\ c' = \mathbf{do}\ \{ (s_1, s_2) \leftarrow get; \\ \qquad\qquad ((), s_2') \leftarrow lift\ (bx_2.set_R\ c'\ s_2); \\ \qquad\qquad \mathbf{let}\ b = bx_2.read_L\ s_2'; \\ \qquad\qquad ((), s_1') \leftarrow lift\ (bx_1.set_R\ b\ s_1); \\ \qquad\qquad set\ (s_1', s_2') \} \end{array}$$

◇

Remark 4.8. Allowing effectful *gets* turns out to impose appreciable extra technical difficulty. In particular, while it still appears possible to prove that composition preserves well-behavedness, the identity laws of composition do not appear to hold in general. At the same time, we currently lack compelling examples that motivate effectful *gets*; the only example we have considered that requires this capability is Example 5.11 in Sect. 5. This is why we mostly limit attention to transparent bx. ◇

Composition is usually expected to be associative and to satisfy identity laws. We can define a family of identity bx as follows:

Definition 4.9 (identity). For any underlying monad instance, we can form the *identity* bx as follows:

$$\begin{array}{l} identity :: Monad\ \tau \Rightarrow StateTBX\ \tau\ \alpha\ \alpha\ \alpha \\ identity = BX\ get\ set\ get\ set \end{array}$$

Clearly, this bx is well-behaved, indeed transparent, and overwritable. ◇

However, if we ask whether $bx = identity \mathbin{;} bx$, we are immediately faced with a problem: the two bx do not even have the same state types. Similarly when we ask about associativity of composition. Apparently, therefore, as for symmetric lenses [13], we must satisfy ourselves with equality only up to some notion of equivalence, such as the one introduced in Definition 3.4.

Theorem 4.10. Composition of transparent bx satisfies the identity and associativity laws, modulo $\equiv$.

$$\begin{array}{ll} \text{(Identity)} & identity \mathbin{;} bx \equiv bx \equiv bx \mathbin{;} identity \\ \text{(Assoc)} & bx_1 \mathbin{;} (bx_2 \mathbin{;} bx_3) \equiv (bx_1 \mathbin{;} bx_2) \mathbin{;} bx_3 \end{array}$$

◇

5 Examples

We now show how to use and combine bx, and discuss how to extend our approach to support initialisation. We adapt some standard constructions on symmetric lenses, involving pairs, sums and lists. Finally we investigate some effectful bx primitives and combinators, culminating with the two examples from Sect. 1.

5.1 Initialisation

Readers familiar with bx will have noticed that so far we have not mentioned mechanisms for initialisation, e.g. 'create' for asymmetric lenses [10], 'missing' in symmetric lenses [13], or Ω in relational bx terminology [31]. Moreover, as we shall see in Sect. 5.2, initialisation is also needed for certain combinators.

Definition 5.1. An *initialisable StateTBX* is a *StateTBX* with two additional operations for initialisation:

$$\begin{array}{lll} \multicolumn{3}{l}{\mathbf{data}\ \mathit{InitStateTBX}\ \tau\ \sigma\ \alpha\ \beta = \mathit{InitStateTBX}\ \{} \\ \quad get_L :: \mathit{StateT}\ \sigma\ \tau\ \alpha, & set_L :: \alpha \rightarrow \mathit{StateT}\ \sigma\ \tau\ (), & init_L :: \alpha \rightarrow \tau\ \sigma, \\ \quad get_R :: \mathit{StateT}\ \sigma\ \tau\ \beta, & set_R :: \beta \rightarrow \mathit{StateT}\ \sigma\ \tau\ (), & init_R :: \beta \rightarrow \tau\ \sigma\} \end{array}$$

The $init_L$ and $init_R$ operations build an initial state from one view or the other, possibly incurring effects in the underlying monad. Well-behavedness of the bx requires in addition:

$$\begin{array}{ll} (\mathrm{I_L G_L}) & \mathbf{do}\ \{s \leftarrow bx.init_L\ a; bx.get_L\ s\} \\ & \quad = \mathbf{do}\ \{s \leftarrow bx.init_L\ a; return\ (a, s)\} \\ (\mathrm{I_R G_R}) & \mathbf{do}\ \{s \leftarrow bx.init_R\ b; bx.get_R\ s\} \\ & \quad = \mathbf{do}\ \{s \leftarrow bx.init_R\ b; return\ (b, s)\} \end{array}$$

stating informally that initialising then getting yields the initialised value. There are no laws concerning initialising then setting. ◇

We can extend composition to handle initialisation as follows:

$$(bx_1 \fatsemi bx_2).init_L\ a = \mathbf{do}\ \{ s_1 \leftarrow bx_1.init_L\ a; b \leftarrow bx_1.get_R\ s_1; \\ s_2 \leftarrow bx_2.init_L\ b; return\ (s_1, s_2) \}$$

and symmetrically for $init_R$. We refine the notions of bx isomorphism and equivalence to *InitStateTBX* as follows. A monad isomorphism $\iota :: StateT\ S_1\ T \to StateT\ S_2\ T$ amounts to a bijection $h :: S_1 \to S_2$ on the state spaces. An isomorphism of *InitStateTBX*s consists of such an ι and h satisfying the following equations (and their duals):

$$\begin{aligned}
\iota\ (bx_1.get_L) &= bx_2.get_L \\
\iota\ (bx_1.set_L\ a) &= bx_2.set_L\ a \\
\mathbf{do}\ \{ s \leftarrow bx_1.init_L\ a; return\ (h\ s) \} &= bx_2.init_L\ a
\end{aligned}$$

Note that the first two equations (and their duals) imply that ι is a conventional isomorphism between the underlying bx structures of bx_1 and bx_2 if we ignore the initialisation operations. The third equation simply says that h maps the state obtained by initialising bx_1 with a to the state obtained by initialising bx_2 with a. Equivalence of *InitStateTBX*s amounts to the existence of such an isomorphism.

Remark 5.2. Of course, there may be situations where these operations are not what is desired. We might prefer to provide both view values and ask the bx system to find a suitable hidden state consistent with both at once. This can be accommodated, by providing a third initialisation function:

$$initBoth :: \alpha \to \beta \to \tau\ (Maybe\ \sigma)$$

However, *initBoth* and $init_L$, $init_R$ are not interdefinable: *initBoth* requires both initial values, so is no help in defining a function that has access only to one; and conversely, given both initial values, there are in general two different ways to initialise from one of them (and two more to initialise from one and then set with the other). Furthermore, it is not clear how to define *initBoth* for the composition of two bx equipped with *initBoth*. ◇

5.2 Basic Constructions and Combinators

It is obviously desirable – and essential in the design of any future bx programming language – to be able to build up bx from components using combinators that preserve interesting properties, and therefore avoid having to prove well-behavedness from scratch for each bx. Symmetric lenses [13] admit several standard constructions, involving constants, duality, pairing, sum types, and lists. We show that these constructions can be generalised to *StateTBX*, and establish that they preserve well-behavedness. For most combinators, the initialisation operations are straightforward; in the interests of brevity, they and obvious duals are omitted in what follows.

Definition 5.3 (duality). Trivially, we can dualise any bx:

$$
\begin{aligned}
&\mathit{dual} :: \mathit{StateTBX}\ \tau\ \sigma\ \alpha\ \beta \rightarrow \mathit{StateTBX}\ \tau\ \sigma\ \beta\ \alpha \\
&\mathit{dual}\ bx = \mathit{BX}\ bx.\mathit{get}_R\ bx.\mathit{set}_R\ bx.\mathit{get}_L\ bx.\mathit{set}_L
\end{aligned}
$$

This simply exchanges the left and right operations; it preserves transparency and overwritability of the underlying bx. ♢

Definition 5.4 (constant and pair combinators). *StateTBX* also admits constant, pairing and projection operations:

$$
\begin{aligned}
&\mathit{constBX} :: \mathit{Monad}\ \tau \Rightarrow \alpha \rightarrow \mathit{StateTBX}\ \tau\ \alpha\ ()\ \alpha \\
&\mathit{fstBX} \quad :: \mathit{Monad}\ \tau \Rightarrow \mathit{StateTBX}\ \tau\ (\alpha, \beta)\ (\alpha, \beta)\ \alpha \\
&\mathit{sndBX} \quad :: \mathit{Monad}\ \tau \Rightarrow \mathit{StateTBX}\ \tau\ (\alpha, \beta)\ (\alpha, \beta)\ \beta
\end{aligned}
$$

These straightforwardly generalise to bx the corresponding operations for symmetric lenses. If they are to be initialisable, *fstBX* and *sndBX* also have to take a parameter for the initial value of the opposite side:

$$
\begin{aligned}
&\mathit{fstIBX} \ :: \mathit{Monad}\ \tau \Rightarrow \beta \rightarrow \mathit{InitStateTBX}\ \tau\ (\alpha, \beta)\ (\alpha, \beta)\ \alpha \\
&\mathit{sndIBX} :: \mathit{Monad}\ \tau \Rightarrow \alpha \rightarrow \mathit{InitStateTBX}\ \tau\ (\alpha, \beta)\ (\alpha, \beta)\ \beta
\end{aligned}
$$

Pairing is defined as follows:

$$
\begin{aligned}
&\mathit{pairBX} :: \mathit{Monad}\ \tau \Rightarrow \mathit{StateTBX}\ \tau\ \sigma_1\ \alpha_1\ \beta_1 \rightarrow \mathit{StateTBX}\ \tau\ \sigma_2\ \alpha_2\ \beta_2 \rightarrow \\
&\qquad\qquad \mathit{StateTBX}\ \tau\ (\sigma_1, \sigma_2)\ (\alpha_1, \alpha_2)\ (\beta_1, \beta_2) \\
&\mathit{pairBX}\ bx_1\ bx_2 = \mathit{BX}\ \mathit{gl}\ \mathit{sl}\ \mathit{gr}\ \mathit{sr}\ \mathbf{where} \\
&\quad \mathit{gl} = \mathbf{do}\ \{ a_1 \leftarrow \mathit{left}\ (bx_1.\mathit{get}_L); a_2 \leftarrow \mathit{right}\ (bx_2.\mathit{get}_L); \mathit{return}\ (a_1, a_2) \} \\
&\quad \mathit{sl}\ (a_1, a_2) = \mathbf{do}\ \{ \mathit{left}\ (bx_1.\mathit{set}_L\ a_1); \mathit{right}\ (bx_2.\mathit{set}_L\ a_2) \} \\
&\quad \mathit{gr} \qquad\quad = \ldots \quad \text{-- dual} \\
&\quad \mathit{sr} \qquad\quad = \ldots \quad \text{-- dual}
\end{aligned}
$$

♢

Other operations based on isomorphisms, such as associativity of pairs, can be lifted to *StateTBX*s without problems. Well-behavedness is immediate for *constBX*, *fstBX*, *sndBX* and for any other bx that can be obtained from an asymmetric or symmetric lens. For the *pairBX* combinator we need to verify preservation of transparency:

Proposition 5.5. If bx_1 and bx_2 are transparent (and hence well-behaved), then so is *pairBX* bx_1 bx_2. ♢

Remark 5.6. The pair combinator does not necessarily preserve overwritability. For this to be the case, we need to be able to commute the *set* operations of the component bx, including any effects in T. Moreover, the pairing combinator is not in general uniquely determined for non-commutative T, because the effects of bx_1 and bx_2 can be applied in different orders. ♢

Definition 5.7 (sum combinators). Similarly, we can define combinators analogous to the 'retentive sum' symmetric lenses and injection operations [13]. The injection operations relate an α and either the same α or some unrelated β; the old α value of the left side is retained when the right side is a β.

$$
\begin{array}{l}
\mathit{inlBX} :: \mathit{Monad}\ \tau \Rightarrow \alpha \to \mathit{StateTBX}\ \tau\ (\alpha, \mathit{Maybe}\ \beta)\ \alpha\ (\mathit{Either}\ \alpha\ \beta) \\
\mathit{inrBX} :: \mathit{Monad}\ \tau \Rightarrow \beta \to \mathit{StateTBX}\ \tau\ (\beta, \mathit{Maybe}\ \alpha)\ \beta\ (\mathit{Either}\ \alpha\ \beta)
\end{array}
$$

The *sumBX* combinator combines two underlying bx and allows switching between them; the state of both (including that of the bx that is not currently in focus) is retained.

$$
\begin{array}{l}
\mathit{sumBX} :: \mathit{Monad}\ \tau \Rightarrow \mathit{StateTBX}\ \tau\ \sigma_1\ \alpha_1\ \beta_1 \to \mathit{StateTBX}\ \tau\ \sigma_2\ \alpha_2\ \beta_2 \to \\
\qquad\qquad \mathit{StateTBX}\ \tau\ (\mathit{Bool}, \sigma_1, \sigma_2)\ (\mathit{Either}\ \alpha_1\ \alpha_2)\ (\mathit{Either}\ \beta_1\ \beta_2) \\
\mathit{sumBX}\ bx_1\ bx_2 = \mathit{BX}\ gl\ sl\ gr\ sr\ \mathbf{where} \\
\quad gl = \mathbf{do}\ \{(b, s_1, s_2) \leftarrow \mathit{get}; \\
\qquad\qquad \mathbf{if}\ b\ \mathbf{then}\ \mathbf{do}\ \{(a_1, \_) \leftarrow \mathit{lift}\ (bx_1.\mathit{get}_L\ s_1); \mathit{return}\ (\mathit{Left}\ a_1)\} \\
\qquad\qquad\qquad \mathbf{else}\ \ \mathbf{do}\ \{(a_2, \_) \leftarrow \mathit{lift}\ (bx_2.\mathit{get}_L\ s_2); \mathit{return}\ (\mathit{Right}\ a_2)\}\} \\
\quad sl\ (\mathit{Left}\ a_1)\ \ = \mathbf{do}\ \{(b, s_1, s_2) \leftarrow \mathit{get}; \\
\qquad\qquad\qquad ((), s_1') \leftarrow \mathit{lift}\ ((bx_1.\mathit{set}_L\ a_1)\ s_1); \\
\qquad\qquad\qquad \mathit{set}\ (\mathit{True}, s_1', s_2)\} \\
\quad sl\ (\mathit{Right}\ a_2) = \mathbf{do}\ \{(b, s_1, s_2) \leftarrow \mathit{get}; \\
\qquad\qquad\qquad ((), s_2') \leftarrow \mathit{lift}\ ((bx_2.\mathit{set}_L\ a_2)\ s_2); \\
\qquad\qquad\qquad \mathit{set}\ (\mathit{False}, s_1, s_2')\} \\
\quad gr = ...\quad \text{-- dual} \\
\quad sr = ...\quad \text{-- dual}
\end{array}
$$

◇

Proposition 5.8. If bx_1 and bx_2 are transparent, then so is $\mathit{sumBX}\ bx_1\ bx_2$. ◇

Finally, we turn to building a bx that operates on lists from one that operates on elements. The symmetric lens list combinators [13] implicitly regard the length of the list as data that is shared between the two views. The forgetful list combinator forgets all data beyond the current length. The retentive version maintains list elements beyond the current length, so that they can be restored if the list is lengthened again. We demonstrate the (more interesting) retentive version, making the shared list length explicit. Several other variants are possible.

Definition 5.9 (retentive list combinator). This combinator relies on the initialisation functions to deal with the case where the new values are inserted into the list, because in this case we need the capability to create new values on the other side (and new states linking them).

$$
\begin{array}{l}
\mathit{listIBX} :: \mathit{Monad}\ \tau \Rightarrow \\
\qquad\qquad \mathit{InitStateTBX}\ \tau\ \sigma\ \alpha\ \beta \to \mathit{InitStateTBX}\ \tau\ (\mathit{Int}, [\sigma])\ [\alpha]\ [\beta] \\
\mathit{listIBX}\ bx = \mathit{InitStateTBX}\ gl\ sl\ il\ gr\ sr\ ir\ \mathbf{where} \\
\quad gl \quad = \mathbf{do}\ \{(n, cs) \leftarrow \mathit{get}; \mathit{mapM}\ (\mathit{lift} \cdot \mathit{eval}\ bx.\mathit{get}_L)\ (\mathit{take}\ n\ cs)\}
\end{array}
$$

$$
\begin{aligned}
&sl\ as = \mathbf{do}\ \{(\_, cs) \leftarrow get;\\
&\qquad cs' \leftarrow lift\ (sets\ (exec \cdot bx.set_L)\ bx.init_L\ as\ cs);\\
&\qquad set\ (length\ as, cs')\}\\
&il\ as = \mathbf{do}\ \{cs \leftarrow mapM\ (bx.init_L)\ as; return\ (length\ as, cs)\}\\
&gr \quad = \ldots \quad \text{-- dual}\\
&sr\ bs = \ldots \quad \text{-- dual}\\
&ir\ bs = \ldots \quad \text{-- dual}
\end{aligned}
$$

Here, the standard Haskell function *mapM* sequences a list of computations, and *sets* sequentially updates a list of states from a list of views, retaining any leftover states if the view list is shorter:

$$
\begin{aligned}
&sets :: Monad\ \tau \Rightarrow (\alpha \rightarrow \gamma \rightarrow \tau\ \gamma) \rightarrow (\alpha \rightarrow \tau\ \gamma) \rightarrow [\alpha] \rightarrow [\gamma] \rightarrow \tau\ [\gamma]\\
&sets\ set\ init\ []\quad cs \quad = return\ cs\\
&sets\ set\ init\ (x : xs)\ (c : cs) = \mathbf{do}\ \{c' \leftarrow set\ x\ c; cs' \leftarrow sets\ set\ init\ xs\ cs;\\
&\qquad return\ (c' : cs')\}\\
&sets\ set\ init\ xs \quad [] \quad = mapM\ init\ xs
\end{aligned}
$$

◇

Proposition 5.10. If *bx* is transparent, then so is *listIBX bx*. ◇

5.3 Effectful bx

We now consider examples of bx that make nontrivial use of monadic effects. The careful consideration we paid earlier to the requirements for composability give rise to some interesting and non-obvious constraints on the definitions, which we highlight as we go.

For accessibility, we use specific monads in the examples in order to state and prove properties; for generality, the accompanying code abstracts from specific monads using Haskell type class constraints instead. In the interests of brevity, we omit dual cases and initialisation functions, but these are defined in the online code supplement.

Example 5.11 (environment). The *Reader* or environment monad is useful for modelling global parameters. Some classes of bidirectional transformations are naturally parametrised; for example, Voigtländer *et al.*'s approach [36] uses a *bias* parameter to determine how to merge changes back into lists.

Suppose we have a family of bx indexed by some parameter γ, over a monad *Reader* γ. Then we can define

$$
\begin{aligned}
&switch :: (\gamma \rightarrow StateTBX\ (Reader\ \gamma)\ \sigma\ \alpha\ \beta) \rightarrow StateTBX\ (Reader\ \gamma)\ \sigma\ \alpha\ \beta\\
&switch\ f = BX\ gl\ sl\ gr\ sr\ \mathbf{where}\\
&\quad gl \quad = \mathbf{do}\ \{c \leftarrow lift\ ask; (f\ c).get_L\}\\
&\quad sl\ a = \mathbf{do}\ \{c \leftarrow lift\ ask; (f\ c).set_L\ a\}\\
&\quad gr \quad = \ldots \quad \text{-- dual}\\
&\quad sr \quad = \ldots \quad \text{-- dual}
\end{aligned}
$$

where the standard *ask* :: *Reader* γ operation reads the γ value. ◇

Proposition 5.12. If $f\ c :: \mathit{StateTBX}\ (\mathit{Reader}\ C)\ S\ A\ B$ is transparent for any $c :: C$, then $\mathit{switch}\ f$ is a well-behaved, but not necessarily transparent, $\mathit{StateTBX}\ (\mathit{Reader}\ C)\ S\ A\ B$. ◇

Remark 5.13. Note that $\mathit{switch}\ f$ is well-behaved but not necessarily transparent. This is because the *get* operations read not only from the designated state of the *StateTBX* but also from the *Reader* environment, and so they are not $(\mathit{Reader}\ C)$-pure. This turns out not to be a big problem in this case, because $\mathit{Reader}\ C$ is a commutative monad. But suppose that the underlying monad were not $\mathit{Reader}\ C$ but a non-commutative monad such as $\mathit{State}\ C$, maintaining some flag that may be changed by the *set* operations; in this scenario, it is not difficult to construct a counterexample to the identity laws for composition. Such counterexamples are why we have largely restricted attention in this paper to transparent bx. (Besides, one what argue that it is never necessary for the get operations to depend on the context; any such dependencies could be handled entirely by the set operations.) ◇

Example 5.14 (exceptions). We turn next to the possibility of failure. Conventionally, the functions defining a bx are required to be total, but often it is not possible to constrain the source and view types enough to make this literally true; for example, consider a bx relating two *Rational* views whose consistency relation is $\{(x, {}^1\!/_x) \mid x \neq 0\}$. A principled approach to failure is to use the *Maybe* (exception) monad, so that an attempt to divide by zero yields *Nothing*.

$$
\begin{array}{l}
\mathit{invBX} :: \mathit{StateTBX}\ \mathit{Maybe}\ \mathit{Rational}\ \mathit{Rational}\ \mathit{Rational} \\
\mathit{invBX} = \mathit{BX}\ \mathit{get}\ \mathit{set}_L\ (\mathit{gets}\ (\lambda a.\, {}^1\!/_a))\ \mathit{set}_R\ \mathbf{where} \\
\quad \mathit{set}_L\ a = \mathbf{do}\ \{\mathit{lift}\ (\mathit{guard}\ (a \not\equiv 0)); \mathit{set}\ a\} \\
\quad \mathit{set}_R\ b = \mathbf{do}\ \{\mathit{lift}\ (\mathit{guard}\ (b \not\equiv 0)); \mathit{set}\ ({}^1\!/_b)\}
\end{array}
$$

where $\mathit{guard}\ b = \mathbf{do}\ \{\mathbf{if}\ b\ \mathbf{then}\ \mathit{Just}\ ()\ \mathbf{else}\ \mathit{Nothing}\}$ is a standard operation in the *Maybe* monad. As another example, suppose we know that A is in the *Read* and *Show* type classes, so each A value can be printed to and possibly read from a string. We can define:

$$
\begin{array}{l}
\mathit{readSomeBX} :: (\mathit{Read}\ \alpha, \mathit{Show}\ \alpha) \Rightarrow \mathit{StateTBX}\ \mathit{Maybe}\ (\alpha, \mathit{String})\ \alpha\ \mathit{String} \\
\mathit{readSomeBX} = \mathit{BX}\ (\mathit{gets}\ \mathit{fst})\ \mathit{set}_L\ (\mathit{gets}\ \mathit{snd})\ \mathit{set}_R\ \mathbf{where} \\
\quad \mathit{set}_L\ a' = \mathit{set}\ (a', \mathit{show}\ a') \\
\quad \mathit{set}_R\ b' = \mathbf{do}\ \{(\_, b) \leftarrow \mathit{get}; \\
\qquad\qquad\qquad \mathbf{if}\ b == b'\ \mathbf{then}\ \mathit{return}\ ()\ \mathbf{else}\ \mathbf{case}\ \mathit{reads}\ b'\ \mathbf{of} \\
\qquad\qquad\qquad\quad ((a', \texttt{""}) : \_) \rightarrow \mathit{set}\ (a', b') \\
\qquad\qquad\qquad\quad \_ \qquad\qquad\quad\ \rightarrow \mathit{lift}\ \mathit{Nothing}\}
\end{array}
$$

(The function *reads* returns a list of possible parses with remaining text.) Note that the get operations are *Maybe*-pure: if there is a *Read* error, it is raised instead by the set operations.

The same approach can be generalised to any monad T having a polymorphic error value $err :: \forall \alpha . T\ \alpha$ and any pair of partial inverse functions $f :: A \rightarrow Maybe\ B$ and $g :: B \rightarrow Maybe\ A$ (*i.e.*, $f\ a = Just\ b$ if and only if $g\ b = Just\ a$, for all a, b):

$$
\begin{array}{l}
partialBX :: Monad\ \tau \Rightarrow (\forall \alpha . \tau\ \alpha) \rightarrow (\alpha \rightarrow Maybe\ \beta) \rightarrow (\beta \rightarrow Maybe\ \alpha) \rightarrow \\
\qquad\qquad\qquad StateTBX\ \tau\ (\alpha, \beta)\ \alpha\ \beta \\
partialBX\ err\ f\ g = BX\ (gets\ fst)\ set_L\ (gets\ snd)\ set_R\ \mathbf{where} \\
\quad set_L\ a' = \mathbf{case}\ f\ a'\ \mathbf{of}\ Just\ b' \ \ \rightarrow set\ (a', b') \\
\qquad\qquad\qquad\qquad\qquad\quad Nothing \rightarrow lift\ err \\
\quad set_R\ b' = \mathbf{case}\ g\ b'\ \mathbf{of}\ Just\ a' \ \ \rightarrow set\ (a', b') \\
\qquad\qquad\qquad\qquad\qquad\quad Nothing \rightarrow lift\ err
\end{array}
$$

Then we could define *invBX* and a stricter variation of *readSomeBX* (one that will *read* only a string that it *shows*—rejecting alternative renderings, whitespace, and so on) as instances of *partialBX*. ◊

Proposition 5.15. Let $f :: A \rightarrow Maybe\ B$ and $g :: B \rightarrow Maybe\ A$ be partial inverses and let *err* be a zero element for T. Then $partialBX\ err\ f\ g :: StateTBX\ T\ S\ A\ B$ is well-behaved, where $S = \{(a, b) \mid f\ a = Just\ b\}$. ◊

Example 5.16 (nondeterminism—Scenario 1.1 revisited). For simplicity's sake, we model nondeterminism via the list monad: a 'nondeterministic function' from A to B is represented as a pure function of type $A \rightarrow [B]$. The following bx is parametrised on a predicate *ok* that checks consistency of two states, a fix-up function *bs* that returns the B values consistent with a given A, and symmetrically a fix-up function *as*.

$$
\begin{array}{l}
nondetBX :: (\alpha \rightarrow \beta \rightarrow Bool) \rightarrow (\alpha \rightarrow [\beta]) \rightarrow (\beta \rightarrow [\alpha]) \rightarrow \\
\qquad\qquad\qquad StateTBX\ [\,]\ (\alpha, \beta)\ \alpha\ \beta \\
nondetBX\ ok\ bs\ as = BX\ (gets\ fst)\ set_L\ (gets\ snd)\ set_R\ \mathbf{where} \\
\quad set_L\ a' = \mathbf{do}\ \{(a, b) \leftarrow get; \\
\qquad\qquad\qquad\quad \mathbf{if}\ ok\ a'\ b\ \mathbf{then}\ set\ (a', b)\ \mathbf{else} \\
\qquad\qquad\qquad\qquad \mathbf{do}\ \{b' \leftarrow lift\ (bs\ a'); set\ (a', b')\}\} \\
\quad set_R\ b' = \mathbf{do}\ \{(a, b) \leftarrow get; \\
\qquad\qquad\qquad\quad \mathbf{if}\ ok\ a\ b'\ \mathbf{then}\ set\ (a, b')\ \mathbf{else} \\
\qquad\qquad\qquad\qquad \mathbf{do}\ \{a' \leftarrow lift\ (as\ b'); set\ (a', b')\}\}
\end{array}
$$

◊

Proposition 5.17. Given *ok*, $S = \{(a, b) \mid ok\ a\ b\}$, and *as* and *bs* satisfying

$$
\begin{array}{l}
a \in as\ b \Rightarrow ok\ a\ b \\
b \in bs\ a \Rightarrow ok\ a\ b
\end{array}
$$

then $nondetBX\ ok\ bs\ as :: StateTBX\ [\,]\ S\ A\ B$ is well-behaved (indeed, it is clearly transparent). It is not necessary for the two conditions to be equivalences. ◊

Remark 5.18. Note that, in addition to choice, the list monad also allows for failure: the fix-up functions can return the empty list. From a semantic point of view, nondeterminism is usually modelled using the monad of finite nonempty sets. If we had used the nonempty set monad instead of lists, then failure would not be possible. ◇

Example 5.19 (signalling). We can define a bx that sends a signal every time either side changes:

$$\begin{array}{l}
\mathit{signalBX} :: (\mathit{Eq}\ \alpha, \mathit{Eq}\ \beta, \mathit{Monad}\ \tau) \Rightarrow (\alpha \to \tau\ ()) \to (\beta \to \tau\ ()) \to \\
\qquad\qquad\quad \mathit{StateTBX}\ \tau\ \sigma\ \alpha\ \beta \to \mathit{StateTBX}\ \tau\ \sigma\ \alpha\ \beta \\
\mathit{signalBX}\ \mathit{sigA}\ \mathit{sigB}\ \mathit{bx} = \mathit{BX}\ (\mathit{bx}.\mathit{get}_L)\ \mathit{sl}\ (\mathit{bx}.\mathit{get}_R)\ \mathit{sr}\ \mathbf{where} \\
\quad \mathit{sl}\ a' = \mathbf{do}\ \{ a \leftarrow \mathit{bx}.\mathit{get}_L; \mathit{bx}.\mathit{set}_L\ a'; \\
\qquad\qquad\qquad \mathit{lift}\ (\mathbf{if}\ a \not\equiv a'\ \mathbf{then}\ \mathit{sigA}\ a'\ \mathbf{else}\ \mathit{return}\ ())\} \\
\quad \mathit{sr}\ b' = \mathbf{do}\ \{ b \leftarrow \mathit{bx}.\mathit{get}_R; \mathit{bx}.\mathit{set}_R\ b'; \\
\qquad\qquad\qquad \mathit{lift}\ (\mathbf{if}\ b \not\equiv b'\ \mathbf{then}\ \mathit{sigB}\ b'\ \mathbf{else}\ \mathit{return}\ ())\}
\end{array}$$

Note that *sl* checks to see whether the new value a' equals the old value a, and does nothing if so; only if they are different does it perform $\mathit{sigA}\ a'$. If the bx is to be well-behaved, then no action can be performed in the case that $a == a'$.

For example, instantiating the underlying monad to *IO* we have:

$$\begin{array}{l}
\mathit{alertBX} :: (\mathit{Eq}\ \alpha, \mathit{Eq}\ \beta) \Rightarrow \mathit{StateTBX}\ \mathit{IO}\ \sigma\ \alpha\ \beta \to \mathit{StateTBX}\ \mathit{IO}\ \sigma\ \alpha\ \beta \\
\mathit{alertBX} = \mathit{signalBX}\ (\lambda_.\ \mathit{putStrLn}\ \texttt{"Left"})\ (\lambda_.\ \mathit{putStrLn}\ \texttt{"Right"})
\end{array}$$

which prints a message whenever one side changes. This is well-behaved; the *set* operations are side-effecting, but the side-effects only occur when the state is changed. It is not overwritable, because multiple changes may lead to different signals from a single change.

As another example, we can define a logging bx as follows:

$$\begin{array}{l}
\mathit{logBX} :: (\mathit{Eq}\ \alpha, \mathit{Eq}\ \beta) \Rightarrow \mathit{StateTBX}\ (\mathit{Writer}\ [\mathit{Either}\ \alpha\ \beta])\ \sigma\ \alpha\ \beta \to \\
\qquad\qquad\qquad\qquad\qquad\quad \mathit{StateTBX}\ (\mathit{Writer}\ [\mathit{Either}\ \alpha\ \beta])\ \sigma\ \alpha\ \beta \\
\mathit{logBX} = \mathit{signalBX}\ (\lambda a.\ \mathit{tell}\ [\mathit{Left}\ a])\ (\lambda b.\ \mathit{tell}\ [\mathit{Right}\ b])
\end{array}$$

where $\mathit{tell} :: \sigma \to \mathit{Writer}\ \sigma\ ()$ is a standard operation in the *Writer* monad that writes a value to the output. This bx logs a list of all of the views as they are changed. Wrapping a component of a chain of composed bx with *log* can provide insight into how changes at the ends of the chain propagate through that component. If memory use is a concern, then we could limit the length of the list to record only the most recent updates – lists of bounded length also form a monoid. ◇

Proposition 5.20. If A and B are types equipped with a well-behaved notion of equality (in the sense that $(a == b) = \mathit{True}$ if and only if $a = b$), and $\mathit{bx} :: \mathit{StateTBX}\ T\ S\ A\ B$ is well-behaved, then $\mathit{signalBX}\ \mathit{sigA}\ \mathit{sigB}\ \mathit{bx} :: \mathit{StateTBX}\ T\ S\ A\ B$ is well-behaved. Moreover, *signalBX* preserves transparency. ◇

Example 5.21 (interaction—Scenario 1.2 revisited). For this example, we need to record both the current state (an A and a B) and the learned collection of consistency restorations. The latter is represented as two lists; the first list contains a tuple $((a', b), b')$ for each invocation of $set_L\ a'$ on a state $(\_, b)$ resulting in an updated state (a', b'); the second is symmetric, for $set_R\ b'$ invocations. The types A and B must each support equality, so that we can check for previously asked questions. We abstract from the base monad; we parametrise the bx on two monadic functions, each somehow determining a consistent match for one state.

```
dynamicBX :: (Eq α, Eq β, Monad τ) ⇒
             (α → β → τ β) → (α → β → τ α) →
             StateTBX τ ((α, β), [((α, β), β)], [((α, β), α)]) α β
dynamicBX f g = BX (gets (fst · fst3)) setL (gets (snd · fst3)) setR where
  setL a' = do {((a, b), fs, bs) ← get;
                if a == a' then return () else
                  case lookup (a', b) fs of
                    Just b'  → set ((a', b'), fs, bs)
                    Nothing → do {b' ← lift (f a' b);
                                  set ((a', b'), ((a', b), b') : fs, bs)}}
  setR b' = ...   -- dual
```

where $fst3\ (a, b, c) = a$. For example, the bx below finds matching states by asking the user, writing to and reading from the terminal.

```
dynamicIOBX :: (Eq α, Eq β, Show α, Show β, Read α, Read β) ⇒
               StateTBX IO ((α, β), [((α, β), β)], [((α, β), α)]) α β
dynamicIOBX = dynamicBX matchIO (flip matchIO)
matchIO :: (Show α, Show β, Read β) ⇒ α → β → IO β
matchIO a b = do {putStrLn ("Setting " ++ show a);
                  putStr ("Replacement for " ++ show b ++ "?");
                  s ← getLine; return (read s)}
```

An alternative way to find matching states, for a finite state space, would be to search an enumeration $[minBound..maxBound]$ of the possible values, checking against a fixed oracle p:

```
dynamicSearchBX ::
  (Eq α, Eq β, Enum α, Bounded α, Enum β, Bounded β) ⇒
  (α → β → Bool) →
  StateTBX Maybe ((α, β), [((α, β), β)], [((α, β), α)]) α β
dynamicSearchBX p = dynamicBX (search p) (flip (search (flip p)))
search :: (Enum β, Bounded β) ⇒ (α → β → Bool) → α → β → Maybe β
search p a _ = find (p a) [minBound..maxBound]
```
◇

Proposition 5.22. For any f, g, the bx $dynamicBX\ f\ g$ is well-behaved (it is clearly transparent). ◇

6 Related Work

Bidirectional programming. This has a large literature; work on view update flourished in the early 1980s, and the term 'lens' was coined in 2005 [9]. The GRACE report [6] surveys work since. We mention here only the closest related work.

Pacheco *et al.* [27] present 'putback-style' asymmetric lenses; *i.e.* their laws and combinators focus only on the 'put' functions, of type $Maybe\ s \to v \to m\ s$, for some monad m. This allows for effects, and they include a combinator *effect* that applies a monad morphism to a lens. Their laws assume that the monad m admits a membership operation $(\in) :: a \to m\ a \to Bool$. For monads such as *List* or *Maybe* that support such an operation, their laws are similar to ours, but their approach does not appear to work for other important monads such as *IO* or *State*. In Diviánsky's monadic lens proposal [8], the *get* function is monadic, so in principle it too can have side-effects; as we have seen, this possibility significantly complicates composition.

Johnson and Rosebrugh [16] analyse symmetric lenses in a general setting of categories with finite products, showing that they correspond to pairs of (asymmetric) lenses with a common source. Our composition for *StateTBX*s uses a similar idea; however, their construction does not apply directly to monadic lenses, because the Kleisli category of a monad does not necessarily have finite products. They also identify a different notion of equivalence of symmetric lenses.

Elsewhere, we have considered a coalgebraic approach to bx [2]. Relating such an approach to the one presented here, and investigating their associated equivalences, is an interesting future direction of research.

Macedo *et al.* [23] observe that most bx research deals with just two models, but many tools and specifications, such as QVT-R [26], allow relating multiple models. Our notion of bx generalises straightforwardly to such multidirectional transformations, provided we only update one source value at a time.

Monads and Algebraic Effects. The vast literature on combining and reasoning about monads [17,21,22,25] stems from Moggi's work [24]; we have shown that bidirectionality can be viewed as another kind of computational effect, so results about monads can be applied to bidirectional computation.

A promising area to investigate is the *algebraic* treatment of effects [28], particularly recent work on combining effects using operations such as sum and tensor [15] and *handlers* of algebraic effects [3,19,29]. It appears straightforward to view entangled state as generated by operations and equations analogous to the bx laws. What is less clear is whether operations such as composition can be defined in terms of effect handlers: so far, the theory underlying handlers [29] does not support 'tensor-like' combinations of computations. We therefore leave this investigation for future work.

The relationship between lenses and state monad morphisms is intriguing, and hints of it appear in previous work on *compositional references* by Kagawa [18]. The fact that lenses determine state monad morphisms (Definition 4.1) appears to be folklore; Shkaravska [30] stated this result in a talk, and it is implicit in the design of the Haskell `Data.Lens` library [20], but we are not aware of any previous published proof.

7 Conclusions and Further Work

We have presented a semantic framework for effectful bidirectional transformations (bx). Our framework encompasses symmetric lenses, which (as is well-known) in turn encompass other approaches to bx such as asymmetric lenses [10] and relational bx [31]; we have also given examples of other monadic effects. This is an advance on the state of the art of bidirectional transformations: ours is the first formalism to reconcile the stateful behavior of bx with other effects such as nondeterminism, I/O or exceptions with due attention paid to the corresponding laws. We have defined composition for effectful bx and shown that composition is associative and satisfies identity laws, up to a suitable notion of equivalence based on monad isomorphisms. We have also demonstrated some combinators suitable for grounding the design of future bx languages based on our approach.

In future we plan to investigate equivalence, and the relationship with the work of Johnson and Rosebrugh [16], further. The equivalence we present here is finer than theirs, and also finer than the equivalence for symmetric lenses presented by Hofmann *et al.* [13]. Early investigations, guided by an alternative coalgebraic presentation [2] of our framework, suggest that the situation for bx may be similar to that for processes given as labelled transition systems: it is possible to give many different equivalences which are 'right' according to different criteria. We think the one we have given here is the finest reasonable, equating just enough bx to make composition work. Another interesting area for exploration is formalisation of our (on-paper) proofs.

Our framework provides a foundation for future languages, libraries, or tools for effectful bx, and there are several natural next steps in this direction. In this paper we explored only the case where the get and set operations read or write complete states, but our framework allows for generalisation beyond the category Set and hence, perhaps, into delta-based bx [7], edit lenses [14] and ordered updates [12], in which the operations record state changes rather than complete states. Another natural next step is to explore different *witness structures* encapsulating the dependencies between views, in order to formulate candidate principles of Least Change (informally, that "a bx should not change more than it has to in order to restore consistency") that are more practical and flexible than those that can be stated in terms of views alone.

Acknowledgements. Preliminary work on this topic was presented orally at the BIRS workshop 13w5115 in December 2013; a four-page abstract [4] of some of the ideas in this paper appeared at the Athens BX Workshop in March 2014; and a short presentation on an alternative coalgebraic approach [2] was made at CMCS 2014. We thank the organisers of and participants at those meetings and the anonymous reviewers for their helpful comments. The work was supported by the UK EPSRC-funded project *A Theory of Least Change for Bidirectional Transformations* [34] (EP/K020218/1, EP/K020919/1).

References

1. Abou-Saleh, F., Cheney, J., Gibbons, J., McKinna, J., Stevens, P.: Notions of bidirectional computation and entangled state monads. Technical report, TLCBX project (2015), extended version with proofs, available from http://arxiv.org/abs/1505.02579
2. Abou-Saleh, F., McKinna, J.: A coalgebraic approach to bidirectional transformations (2014). Short presentation at CMCS
3. Brady, E.: Programming and reasoning with algebraic effects and dependent types. In: ICFP, pp. 133–144. ACM (2013)
4. Cheney, J., McKinna, J., Stevens, P., Gibbons, J., Abou-Saleh, F.: Entangled state monads (abstract). In: Terwilliger and Hidaka [33]
5. Cicchetti, A., Di Ruscio, D., Eramo, R., Pierantonio, A.: JTL: a bidirectional and change propagating transformation language. In: Malloy, B., Staab, S., van den Brand, M. (eds.) SLE 2010. LNCS, vol. 6563, pp. 183–202. Springer, Heidelberg (2011)
6. Czarnecki, K., Foster, J.N., Hu, Z., Lämmel, R., Schürr, A., Terwilliger, J.F.: Bidirectional transformations: a cross-discipline perspective. In: Paige, R.F. (ed.) ICMT 2009. LNCS, vol. 5563, pp. 260–283. Springer, Heidelberg (2009)
7. Diskin, Z., Xiong, Y., Czarnecki, K.: From state- to delta-based bidirectional model transformations: the asymmetric case. JOT **10**(6), 1–25 (2011)
8. Diviánszky, P.: LGtk API correction, April 2013. http://people.inf.elte.hu/divip/LGtk/CorrectedAPI.html
9. Foster, J.N., Greenwald, M.B., Moore, J.T., Pierce, B.C., Schmitt, A.: Combinators for bidirectional tree transformations: A linguistic approach to the view update problem. In: POPL, pp. 233–246. ACM (2005)
10. Foster, J.N., Greenwald, M.B., Moore, J.T., Pierce, B.C., Schmitt, A.: Combinators for bidirectional tree transformations: A linguistic approach to the view-update problem. ACM TOPLAS **29**(3), 17 (2007). Extended version of [9]
11. Gibbons, J., Hinze, R.: Just do it: Simple monadic equational reasoning. In: ICFP, pp. 2–14. ACM (2011)
12. Hegner, S.J.: An order-based theory of updates for closed database views. Ann. Math. Art. Int. **40**, 63–125 (2004)
13. Hofmann, M., Pierce, B.C., Wagner, D.: Symmetric lenses. In: POPL, pp. 371–384. ACM (2011)
14. Hofmann, M., Pierce, B.C., Wagner, D.: Edit lenses. In: POPL, pp. 495–508. ACM (2012)
15. Hyland, M., Plotkin, G.D., Power, J.: Combining effects: Sum and tensor. TCS **357**(1–3), 70–99 (2006)
16. Johnson, M., Rosebrugh, R.: Spans of lenses. In: Terwilliger and Hidaka [33]
17. Jones, M.P., Duponcheel, L.: Composing monads. Technical report RR-1004, DCS, Yale (1993)
18. Kagawa, K.: Compositional references for stateful functional programming. In: ICFP, pp. 217–226 (1997)
19. Kammar, O., Lindley, S., Oury, N.: Handlers in action. In: ICFP, pp. 145–158. ACM (2013)
20. Kmett, E.: `lens-4.0.4` library. http://hackage.haskell.org/package/lens
21. Liang, S., Hudak, P., Jones, M.P.: Monad transformers and modular interpreters. In: POPL, pp. 333–343 (1995)

22. Lüth, C., Ghani, N.: Composing monads using coproducts. In: ICFP, pp. 133–144. ACM (2002)
23. Macedo, N., Cunha, A., Pacheco, H.: Toward a framework for multidirectional model transformations. In: Terwilliger and Hidaka [33]
24. Moggi, E.: Notions of computation and monads. Inf. Comp. **93**(1), 55–92 (1991)
25. Mossakowski, T., Schröder, L., Goncharov, S.: A generic complete dynamic logic for reasoning about purity and effects. FAC **22**(3–4), 363–384 (2010)
26. OMG: MOF 2.0 Query/View/Transformation specification (QVT), version 1.1, January 2011. http://www.omg.org/spec/QVT/1.1/
27. Pacheco, H., Hu, Z., Fischer, S.: Monadic combinators for "putback" style bidirectional programming. In: PEPM, pp. 39–50. ACM (2014). http://doi.acm.org/10.1145/2543728.2543737
28. Plotkin, G., Power, J.: Notions of computation determine monads. In: Nielsen, M., Engberg, U. (eds.) FOSSACS 2002. LNCS, vol. 2303, pp. 342–356. Springer, Heidelberg (2002)
29. Plotkin, G.D., Pretnar, M.: Handling algebraic effects. LMCS **9**(4), 1–36 (2013)
30. Shkaravska, O.: Side-effect monad, its equational theory and applications (2005), seminar slides available at: http://www.ioc.ee/~tarmo/tsem05/shkaravska1512-slides.pdf
31. Stevens, P.: Bidirectional model transformations in QVT: Semantic issues and open questions. SoSyM **9**(1), 7–20 (2010)
32. Stevens, P., McKinna, J., Cheney, J.: 'Composers' example (2014). http://bx-community.wikidot.com/examples:composers
33. Terwilliger, J., Hidaka, S. (eds.): BX Workshop (2014). http://ceur-ws.org/Vol-1133/#bx
34. TLCBX Project: A theory of least change for bidirectional transformations (2013–2016). http://www.cs.ox.ac.uk/projects/tlcbx/, http://groups.inf.ed.ac.uk/bx/
35. Varró, D.: Model transformation by example. In: Wang, J., Whittle, J., Harel, D., Reggio, G. (eds.) MoDELS 2006. LNCS, vol. 4199, pp. 410–424. Springer, Heidelberg (2006)
36. Voigtländer, J., Hu, Z., Matsuda, K., Wang, M.: Enhancing semantic bidirectionalization via shape bidirectionalizer plug-ins. JFP **23**(5), 515–551 (2013)

A Clear Picture of Lens Laws
Functional Pearl

Sebastian Fischer[1](✉), Zhenjiang Hu[2], and Hugo Pacheco[3]

[1] Christian-Albrechts-Universität, Kiel, Germany
sebf@informatik.uni-kiel.de
[2] National Institute of Informatics, Tokyo, Japan
hu@nii.ac.jp
[3] Cornell University, Ithaca, NY, USA
hpacheco@cs.cornell.edu

Abstract. A lens is an optical device which refracts light. Properly adjusted, it can be used to project sharp images of objects onto a screen—a principle underlying photography as well as human vision. Striving for clarity, we shift our focus to lenses as abstractions for bidirectional programming. By means of standard mathematical terminology as well as intuitive properties of bidirectional programs, we observe different ways to characterize lenses and show exactly how their laws interact. Like proper adjustment of optical lenses is essential for taking clear pictures, proper organization of lens laws is essential for forming a clear picture of different lens classes. Incidentally, the process of understanding bidirectional lenses clearly is quite similar to the process of taking a good picture.

By showing that it is exactly the backward computation which defines lenses of a certain standard class, we provide an unusual perspective, as contemporary research tends to focus on the forward computation.

1 Scene Selection

Select the scene to determine the topic of what will be seen in your picture.

A lens [3] is a program that can be run forwards and backwards. The forward computation extracts a *view* from a *source* and the backward computation updates a given source according to a given view producing a new source. Using Haskell notation [1], a lens consists of a view extraction function $get :: Source \to View$ and a source update function $put :: Source \to View \to Source$.

Not every combination of such functions is reasonable. Different algebraic laws between *get* and *put* have been proposed to define a hierarchy of lens classes. For the sake of clarity, we present all laws in terms of total functions. Some programmers find it easier to define partial functions for certain problems, and our results can be adjusted to this setting.

The smallest among the lens classes we investigate is that of *bijective lenses* where *get* and $put\ s$ are bijective inverse functions of each other for all sources s.

R. Hinze and J. Voigtländer (Eds.): MPC 2015, LNCS 9129, pp. 215–223, 2015.
DOI:10.1007/978-3-319-19797-5_10

$$\forall\ s, s' \quad put\ s\ (get\ s') = s' \qquad \textsf{StrongGetPut}$$
$$\forall\ s, v \quad get\ (put\ s\ v) = v \qquad \textsf{PutGet}$$

Bijective lenses require a one-to-one correspondence between the *Source* and *View* types which does not fit the asymmetric types of *get* and *put*. In fact, the source argument of *put* is not needed for bijective lenses because the *put* function builds the new source using only the view and discarding the input source.

In the broader class of *very well-behaved lenses*, the *get* function is allowed to discard information. Very well-behaved lenses satisfy the above PutGet law as well as the two following laws.

$$\forall\ s \quad put\ s\ (get\ s) = s \qquad \textsf{GetPut}$$
$$\forall\ s, v, v' \quad put\ (put\ s\ v')\ v = put\ s\ v \qquad \textsf{PutPut}$$

The GetPut law is a more permissive variant of the StrongGetPut law because it passes the same source values to *get* and *put*, allowing (and requiring) the *put* function to reconstruct information discarded by *get*. The PutPut law requires that source updates overwrite the effect of previous source updates.

That law is dropped in an even broader class of *well-behaved lenses* which satisfy only the PutGet and GetPut laws, requiring that source updates reflect changes in given views but do not perform any changes if views do not change.

The presented lens classes are closed under lens composition [3], i.e., for any two lenses in a class there is a composed lens (where the *get* function is the composition of the underlying *get* functions) that is itself in that class.

Example 1. As an example lens consider the following pair of functions.

$$\begin{aligned} &getFirst\ (x, \_) &&= x \\ &putFirst\ (\_, y)\ v &&= (v, y) \end{aligned}$$

Applying *putFirst* is analogous to changing the width of a picture without changing its height. We can verify the PutGet and GetPut laws as follows.

$$\begin{aligned} &\quad getFirst\ (putFirst\ (x, y)\ v) \\ &= \quad \{\text{ by definition of } putFirst \ \} \\ &\quad getFirst\ (v, y) \\ &= \quad \{\text{ by definition of } getFirst \ \} \\ &\quad v \end{aligned}$$

$$\begin{aligned} &\quad putFirst\ (x, y)\ (getFirst\ (x, y)) \\ &= \quad \{\text{ by definition of } getFirst \ \} \\ &\quad putFirst\ (x, y)\ x \\ &= \quad \{\text{ by definition of } putFirst \ \} \\ &\quad (x, y) \end{aligned}$$

Consequently, *getFirst* and *putFirst* form a well-behaved lens between an arbitrary type of pairs and the type of their first components. Specialized to the type $(\mathbb{R}^+, \mathbb{R}^+) \to \mathbb{R}^+$, the function *getFirst* also forms a well-behaved lens with a different *put* function:

$$putScaled\ (x, y)\ v = (v, v * y\ /\ x)$$

The function *putScaled* maintains the ratio of the components of the pair, which is analogous to resizing a picture without distorting it.

Both *put* functions satisfy the PutPut law because the result of an update does not depend on previous updates.

Example 1 shows an inherent ambiguity in the definition of (very) well-behaved lenses based on *get* alone: the same *get* function can be combined with different *put* functions to form a valid lens. One cannot completely specify the behaviour of all lenses by only writing forward computations.

We select our scene to show certain laws of asymmetric bidirectional transformations and soon cast new light on the presented laws using old searchlights.

2 Camera Setup

Set up the camera to meet technical requirements for capturing your picture.

For taking our picture, we remind ourselves of mathematical properties that we later use to describe implications of the lens laws.

Surjectivity is a property of functions which requires that every element in the codomain of a function is the image of some element in its domain. More formally, a function $f :: A \to B$ is surjective if and only if the following property holds.

$$\forall\ b :: B\ \exists\ a :: A \quad f\ a = b$$

Surjectivity of $f :: A \to B$, therefore, requires that A is *at least as big as* B.

Conversely, injectivity is a property of functions which requires different arguments to be mapped to different results. More formally, a function $f :: A \to B$ is injective if and only if the following property holds.

$$\forall\ a, a' :: A \quad f\ a = f\ a' \Rightarrow a = a'$$

Injectivity of $f :: A \to B$, therefore, requires that B is *at least as big as* A.

The technical requirements for taking our picture are standard properties of functions. We hope that describing lens laws based on established mathematical terminology helps to gain simple intuitions about lenses in different classes.

3 Perspective

Choose a perspective exposing elements of your picture you want to highlight.

The *put* functions are essential elements in our picture of lenses. We now highlight certain necessary conditions which must be satisfied for *put* functions in all lens classes discussed in this paper.

The GetPut law requires *put* to satisfy the following law.

$$\forall\ s\ \exists\ v \quad put\ s\ v = s \qquad \textsf{SourceStability}$$

Proof. To verify that GetPut *implies* SourceStability*, let s be an arbitrary source value and $v = get\ s$ in the following calculation.*

$$\begin{aligned} & put\ s\ v \\ = & \quad \{ \text{ by definition of } v \ \} \\ & put\ s\ (get\ s) \\ = & \quad \{ \textsf{ GetPut } \} \\ & s \end{aligned}$$

SourceStability requires that every source is stable under a certain update. It implies that *put* is surjective in the following sense (equivalent to surjectivity of *uncurry put*.)

$$\forall\ s\ \exists\ s', v \quad put\ s'\ v = s \qquad \textsf{PutSurjectivity}$$

SourceStability is stronger than PutSurjectivity. For example, the *putShift* function defined below, which places the view in the first component of the source and moves the old first component to the second, satisfies the latter property but not the former, because there is no v such that $putShift\ (1,2)\ v = (1,2)$.

$$putShift\ (x, \_)\ v = (v, x)$$

Consequently, *putShift* cannot be part of a well-behaved lens.

PutGet also implies a necessary condition on well-behaved *put* functions.

$$\forall\ s, s', v, v' \quad put\ s\ v = put\ s'\ v' \Rightarrow v = v' \qquad \textsf{ViewDetermination}$$

Proof. We can use the PutGet *law to conclude* ViewDetermination *as follows.*

$$\begin{aligned} & put\ s\ v = put\ s'\ v' \\ \Rightarrow & \quad \{ \text{ by applying } get \text{ to each side } \} \\ & get\ (put\ s\ v) = get\ (put\ s'\ v') \\ \Rightarrow & \quad \{ \textsf{ PutGet } \} \\ & v = v' \end{aligned}$$

ViewDetermination requires that a view used to update a source can be determined from the result of an update, regardless of the original source. It implies the following injectivity property of *put* functions.

$$\forall\ s \quad put\ s \text{ is injective} \qquad \text{PutInjectivity}$$

ViewDetermination is stronger than PutInjectivity. For example, the *putSum* function defined below, which adds the source to the view to produce an updated source, satisfies the latter property but not the former, because $putSum\ 2\ 3 = putSum\ 1\ 4$ and $3 \neq 4$.

$$putSum\ m\ n = m + n$$

Consequently, *putSum* cannot be part of a well-behaved lens.

PutInjectivity confirms our previous intuition about the asymmetry of the *Source* and *View* types of a well-behaved lens: the *Source* type needs to be at least as big as the *View* type. From our examples, we already concluded that the opposite is not necessary: *get* may discard information that *put* is able to reconstruct based on a given source.

We finally observe yet another necessary condition on *put* functions that is implied by the combination of the PutGet and GetPut laws.

$$\forall\ s, v \quad put\ (put\ s\ v)\ v = put\ s\ v \qquad \text{PutTwice}$$

Proof.

$$\begin{aligned} & put\ (put\ s\ v)\ v \\ = & \quad \{\ \textit{PutGet}\ \} \\ & put\ (put\ s\ v)\ (get\ (put\ s\ v)) \\ = & \quad \{\ \textit{GetPut}\ \} \\ & put\ s\ v \end{aligned}$$

PutTwice is weaker than PutPut, because it only requires source updates to be independent of previous updates that were using the same view. For example, the *putChanges* function defined below places the given view in the first component of the source and counts the number of changes in the second, which is anologous to counting how often a picture (in the first component) has been resized. It satisfies PutTwice but not PutPut.

$$putChanges\ (x, c)\ v = (v, \mathbf{if}\ x == v\ \mathbf{then}\ c\ \mathbf{else}\ c + 1)$$

So, although *putChanges* forms a well-behaved lens with *getFirst* it cannot be part of a very well-behaved one.

The following proposition summarizes the properties of *put* functions that we observed in this section.

Proposition 1. *Every put function of a well-behaved lens satisfies SourceStability, ViewDetermination, PutSurjectivity, PutInjectivity, and PutTwice.*

We choose a perspective that highlights the *put* function by masking the *get* function in implications of the standard formulation of lens laws. Our focus will be justified by the central role of *put* which we are about to reveal.

4 Composition

Compose picture elements to convey your message, applying appropriate technique.

We now align our camera to put the described properties into a position that clarifies their role in our picture. We do not concern ourselves with the composition of lenses (which is orthogonal to our focus on *put* functions) but with the composition of lens laws to characterize well-behaved lenses using the necessary conditions on *put* functions presented in Sect. 3.

The following theorem shows that the *put* function in a well-behaved lens determines the corresponding *get* function and that every *put* function which satisfies the properties identified in Sect. 3 is part of a well-behaved lens.

Theorem 1. *Let put* :: *Source* → *View* → *Source be a function that satisfies* SourceStability *and* ViewDetermination. *Then there is a unique get function that forms a well-behaved lens with put.*

Proof. **Uniqueness** *Let* get_1, get_2 :: *Source* → *View be functions that form a well-behaved lens with put. Then the following equation holds for all sources s.*

$$\begin{aligned} & get_1\ s \\ = & \quad \{\ by\ \mathsf{GetPut}\ law\ for\ get_2\ \} \\ & get_1\ (put\ s\ (get_2\ s)) \\ = & \quad \{\ by\ \mathsf{PutGet}\ law\ for\ get_1\ \} \\ & get_2\ s \end{aligned}$$

Existence *Let s be a source value. Due to* SourceStability, *there is a view v such that* $put\ s\ v = s$. *Because of* ViewDetermination *this view v is unique because for a view w with* $put\ s\ w = s$ *we have* $put\ s\ v = put\ s\ w \Rightarrow v = w$. *Hence, setting* $get\ s = v$ *for the v such that* $put\ s\ v = s$ *defines get s uniquely.*

To conclude well-behavedness, we observe the GetPut *and* PutGet *laws for this definition of get. For all sources s, the following equation, i.e., the* GetPut *law, follows from the definition of get based on* SourceStability.

$$\begin{aligned} & put\ s\ (get\ s) \\ = & \quad \{\ by\ definition\ of\ get\ \} \\ & put\ s\ v \quad \{\ with\ v\ such\ that\ put\ s\ v = s\ \} \\ = & \quad \{\ put\ s\ v = s\ \} \\ & s \end{aligned}$$

For all sources s and views v, the following equation, i.e., the PutGet *law, follows from the definition of get and* ViewDetermination.

$$\begin{array}{ll} & get\ (put\ s\ v) \\ = & \quad \{\ by\ definition\ of\ get\ \} \\ & v' \quad \{\ with\ v'\ such\ that\ put\ (put\ s\ v)\ v' = put\ s\ v\ \} \\ = & \quad \{\ \mathsf{ViewDetermination}\ \} \\ & v \end{array}$$

Choosing a perspective that highlights properties of *put* functions, we observed that some of them can be composed to characterize well-behaved lenses. We used equational reasoning as our techique to show that certain properties of *put* functions uniquely determine *get* functions in well-behaved lenses. We neither delve into the question (raised by the non-constructive way of defining *get*) whether there is always a *computable get* function nor do we reflect on how a corresponding *get* function could be derived from *put*.

5 Post-editing

Apply changes during post-editing to refine your picture.

In order to increase the contrast in our picture, we investigate the internal structure of the properties of *put* functions presented in Sect. 3. The following theorem characterizes the conditions of Theorem 1 to provide an alternative way to specify the sufficient and necessary conditions on *put* functions in well-behaved lenses based on standard mathematical terminology.

Theorem 2. *For put :: Source → View → Source, the following conjunctions are equivalent.*

1. SourceStability ∧ ViewDetermination
2. PutSurjectivity ∧ PutInjectivity ∧ PutTwice

Proof. 1 ⇒ 2 *Because of Theorem 1, put is part of a well-behaved lens. Hence, it satisfies all conditions summarized in Proposition 1.* 2 ⇒ 1 *First, we conclude* SourceStability *from* PutSurjectivity *and* PutTwice.

$$\begin{array}{ll} & s \\ = & \quad \{\ by\ \mathsf{PutSurjectivity},\ choosing\ s'\ and\ v\ with\ put\ s'\ v = s\ \} \\ & put\ s'\ v \\ = & \quad \{\ \mathsf{PutTwice}\ \} \\ & put\ (put\ s'\ v)\ v \\ = & \quad \{\ put\ s'\ v = s\ \} \\ & put\ s\ v \end{array}$$

Now, we conclude ViewDetermination *from* PutInjectivity *and* PutTwice.

$$
\begin{aligned}
& put\ s\ v = put\ s'\ v' \\
\Rightarrow\ & \{\ \textit{by applying}\ \mathsf{PutTwice}\ \textit{on each side}\ \} \\
& put\ (put\ s\ v)\ v = put\ (put\ s'\ v')\ v' \\
\Rightarrow\ & \{\ put\ s\ v = put\ s'\ v'\ \} \\
& put\ (put\ s\ v)\ v = put\ (put\ s\ v)\ v' \\
\Rightarrow\ & \{\ \textit{by injectivity of put}\ (put\ s\ v)\ \} \\
& v = v'
\end{aligned}
$$

Theorem 2 provides an alternative characterization of *put* functions that are part of well-behaved lenses. It shows how the PutTwice law closes the gap between the combination of SourceStability with ViewDetermination and the weaker combination of PutSurjectivity with PutInjectivity.

By incorporating PutPut into each of the conjunctions describing well-behaved *put* functions, we can also characterize very well-behaved lenses based on *put*. The difference that makes well-behaved lenses *very* well-behaved is by definition already expressed using only the *put* function.

After post editing, our picture shows how simple mathematical properties of functions play together to give rise to established laws for asymmetric bidirectional transformations.

6 Perception

Focus on essentials to let others perceive the message of your picture.

Our picture provides a novel perspective on bidirectional lenses. While there may be in general more than one *put* function that can be combined with a given *get* function to form a (very) well-behaved lens, our view on the lens laws emphasizes that *put* functions determine corresponding *get* functions uniquely in all lens classes discussed in this paper. While the idea of *put* determining *get* may be folklore in the circle of lens specialists, we are the first to put it in focus. Our contribution, therefore, is to a lesser extent based on directly applicable technical novelty than it aims to clarify formal properties of lenses in simple terms to gain insights into an already established theory.

Theorem 1 provides a concise characterization of well-behaved lenses only based on the *put* function. Theorem 2 provides an alternative, more elegant, *put*-based characterization in terms of standard mathematical terminology. As far as we know, there has not been a peer-reviewed publication of a *put*-based lens characterization.

Foster gives a similar characterization in his PhD thesis [2] based on properties we present in Sect. 3. He does not use one of the conjunctions characterized in Theorem 2 but a redundant and, therefore, less revealing combination. We believe that the clarification of the role of PutTwice in our Theorem 2 is novel.

The contemporary focus on *get* functions is incomplete, because a *put*-based characterization shows that it is *exactly* the *put* function that programmers need to define in order to specify (very) well-behaved lenses completely. Similar to how lens combinators allow to express compositions of bidirectional programs in a single definition, defining *put* and deriving *get* would allow to specify primitive lenses without giving a separate implementation for each direction.

Implementations of lens combinators in Haskell such as the `lens`[1] or the `fclabels`[2] packages usually require very well-behaved implementations of both directions of primitive lenses. Deriving the *get* function automatically from *put* functions would be beneficial especially for supporting more complex synchronization strategies, because programmers would have to maintain only one function (not two corresponding ones) for each strategy. Our observations show that a *put*-based approach supports even less restrictive lens combinators that only require well-behavedness.

Acknowledgements. We appreciate the help of Janis Voigtländer, Jeremy Gibbons, Alcino Cunha, Nikita Danilenko, Insa Stucke, and José Nuno Oliveira who have looked at drafts of our picture.

References

1. Bird, R.S.: Introduction to Functional Programming Using Haskell. Prentice-Hall (1998). http://www.cs.ox.ac.uk/publications/books/functional/
2. Foster, J.: Bidirectional programming languages. Ph.D. thesis. University of Pennsylvania, December 2009
3. Foster, J.N., Greenwald, M.B., Moore, J.T., Pierce, B.C., Schmitt, A.: Combinators for bidirectional tree transformations: a linguistic approach to the view-update problem. ACM Trans. Program. Lang. Syst. **29**(3), 17 (2007)

[1] http://hackage.haskell.org/package/lens.

[2] http://hackage.haskell.org/package/fclabels.

Regular Varieties of Automata and Coequations

J. Salamanca[1(✉)], A. Ballester-Bolinches[2], M.M. Bonsangue[1,3], E. Cosme-Llópez[2], and J.J.M.M. Rutten[1,4]

[1] CWI Amsterdam, Amsterdam, The Netherlands
salamanc@cwi.nl
[2] Universitat de València, Valencia, Spain
{Adolfo.Ballester,Enric.Cosme}@uv.es
[3] LIACS - Leiden University, Leiden, The Netherlands
marcello@liacs.nl
[4] Radboud University Nijmegen, Nijmegen, The Netherlands
jjmmrutten@gmail.com

Abstract. In this paper we use a duality result between equations and coequations for automata, proved by Ballester-Bolinches, Cosme-Llópez, and Rutten to characterize nonempty classes of deterministic automata that are closed under products, subautomata, homomorphic images, and sums. One characterization is as classes of automata defined by regular equations and the second one is as classes of automata satisfying sets of coequations called varieties of languages. We show how our results are related to Birkhoff's theorem for regular varieties.

1 Introduction

Initial algebras provide minimal canonical models for inductive data types, recursive definition principles of functions and induction as a corresponding proof principle [7]. Over the last decade, coalgebras have emerged as a mathematical structure suitable for capturing infinite data structures and infinite computations [12]. Final coalgebras are the categorical dual of initial algebras. They represent infinite data or behavior defined by observations rather than constructors, and come equipped with corecursive definitions of functions and coinduction as a dual proof principle of induction [1].

More generally, algebraic theories of free algebras are specified by equations. Dually, cofree coalgebras generalize final coalgebras, where coequations are used instead of equations. Intuitively, coequations can be thought of as behaviours, or pattern specifications [12], that a coalgebra is supposed to exhibit or adhere to. By Birkhoff's celebrated theorem [5], a class of algebras is equationally defined if and only if it is a variety, i.e., closed under homomorphic images, subalgebras and products. Dually, on the coalgebraic side [3,8,12], generally less is known about the notions of coequations and covarieties.

In the present paper, we study deterministic automata both from an algebraic perspective and a coalgebraic one. From the algebraic perspective, deterministic automata are algebras with unary operations. In this context, an equation is just

R. Hinze and J. Voigtländer (Eds.): MPC 2015, LNCS 9129, pp. 224–237, 2015.
DOI:10.1007/978-3-319-19797-5_11

a pair of words, and it holds in an automaton if for every initial state, the states reached from that state by both words are the same. Coalgebraically, an automaton is a deterministic transition system with final states (the observations). A coequation is then a set of languages, and an automaton satisfies a coequation if for every possible observation (colouring the states as either final or not) the language accepted by the automaton is within the specified coequation.

In [4] a subset of the authors have established a new duality result between equations and coequations. Building on their work, we show that classes of deterministic automata closed under products, subautomata, homomorphic images, and sums are definable both by congruences and by varieties of languages. The first characterization, by congruences, is algebraic. In this case, congruences are equational theories of *regular equations* which give rise to regular varieties. An equation $e_1 = e_2$ is regular if the sets of variables occurring in e_1 and e_2 are the same. It is worth mentioning that another characterization of regular varieties was given by Taylor [13]; we will show how that characterization relates to the one we will present here.

The second characterization is a coalgebraic one. Here coequations are used to define classes of automata. Coequations are given by sets of languages, and a central concept in our characterization is that of variety of languages: sets of languages that are both (complete atomic) Boolean algebras and closed under right and left derivatives (details will follow). The coalgebraic characterization will look less familiar than the algebraic one. It is interesting since it will turn out to be equivalent to definitions by so-called regular equations, thus yielding a novel restriction of Birkhoff's theorem.

As a consequence, classes of automata closed under products, subautomata, homomorphic images, and sums can be defined by both equations and coequations. In fact, the first three closure properties characterize an algebraic variety (cf. Birkhoff theorem [5]), whereas the last three closure properties define a coalgebraic covariety [3,8,12]. Our result fits into the recent line of work which uses Stone-like duality as a tool for proving the correspondence between local varieties of regular languages and local pseudovarieties of monoids [2,9]. The main difference is that we do not impose any restriction on the state space of the automata and on the size of the input alphabet.

2 Preliminaries

In this section we introduce the notation and main concepts we will use in the paper. (See [4] for more details). Given two sets X and Y we define

$$Y^X = \{f \mid f : X \to Y\}$$

For a function $f \in Y^X$, we define the *kernel* and the *image* of f by

$$\ker(f) = \{(x_1, x_2) \in X \times X \mid f(x_1) = f(x_2)\}$$

and

$$\mathrm{Im}(f) = \{f(x) \mid x \in X\}.$$

For $Y_0 \subseteq Y$, the set $f^{-1}(Y_0) \subseteq X$ is defined as

$$f^{-1}(Y_0) = \{x \in X \mid f(x) \in Y_0\}.$$

We define the set $2 = \{0, 1\}$ and, for any set X and $B \subseteq X$, we define the function $\chi_B : X \to 2$ by

$$\chi_B(x) = \begin{cases} 1 & \text{if} x \in B, \\ 0 & \text{if} x \notin B. \end{cases}$$

If $B \subseteq X$ then $\chi_B \in 2^X$, and for any $f \in 2^X$ we get the subset $f^{-1}(\{1\})$ of X. The previous correspondence between subsets of X and elements in 2^X is bijective, so elements in 2^X and subsets of X will often be identified.

For any family of sets $\{X_i\}_{i \in I}$, we define their disjoint union by

$$\sum_{i \in I} X_i = \bigcup_{i \in I} \{i\} \times X_i$$

For any set A we denote by A^* the free monoid with generators A, its identity element will be denoted by ϵ. Elements in 2^{A^*} are called *languages* over A, which can also be seen as subsets of A^*. Given a language $L \in 2^{A^*}$ and $w \in L$, we define the *left derivative* ${}_wL$ of L with respect to w and the *right derivative* L_w of L with respect to w as the elements ${}_wL, L_w \in 2^{A^*}$ such that for every $u \in A^*$

$${}_wL(u) = L(uw) \text{ and } L_w(u) = L(wu).$$

Let A be a (not necessarily finite) alphabet. A *deterministic automaton* on A is a pair (X, α) where $\alpha : X \times A \to X$ is a function. Let (X, α) and (Y, β) be deterministic automata on A. We say that (X, α) is a *subautomaton* of (Y, β) if $X \subseteq Y$ and for every $x \in X$ and $a \in A$, $\alpha(x, a) = \beta(x, a) \in X$. A function $h : X \to Y$ is a *homomorphism* from (X, α) to (Y, β) if for every $x \in X$ and $a \in A$, $h(\alpha(x, a)) = \beta(h(x), a)$. We say that (Y, β) is a *homomorphic image* of (X, α) if there exists a surjective homomorphism $h : X \to Y$.

The *product* of a family $\mathcal{X} = \{(X_i, \alpha_i)\}_{i \in I}$ of deterministic automata is the deterministic automaton $\prod_{i \in I}(X_i, \alpha_i) = (\prod_{i \in I} X_i, \bar{\alpha})$ where

$$\bar{\alpha}(f, a)(i) = \alpha_i(f(i), a),$$

Furthermore, the *sum* of the family $\mathcal{X}$ is defined as the deterministic automaton $\sum_{i \in I}(X_i, \alpha_i) = (\sum_{i \in I} X_i, \hat{\alpha})$ where

$$\hat{\alpha}((i, x)(a)) = (i, \alpha_i(x, a)).$$

Given an automaton (X, α) we can add an initial state $x_0 \in X$ or a colouring $c : X \to 2$ to get a *pointed automaton* (X, x_0, α) or a *coloured automaton* (X, c, α), respectively. For a deterministic automaton (X, α), $x \in X$, and $u \in A^*$, we define $u(x) \in X$ inductively as follows

$$u(x) = \begin{cases} x & \text{if} u = \epsilon, \\ \alpha(w(x), a) & \text{if} u = wa, \end{cases}$$

thus $u(x)$ is the state we reach from x by processing the word u.

By using the correspondence

$$\alpha : X \times A \to X \;\Leftrightarrow\; \alpha' : X \to X^A$$

given by $\alpha(x,a) = \alpha'(x)(a)$, we have that pointed automata are F-algebras for the endofunctor F on Set given by $F(X) = 1 + (A \times X)$. Dually, coloured automata are G-coalgebras for the endofunctor G on Set given by $G(X) = 2 \times X^A$.

The initial F–algebra is the pointed automaton (A^*, ϵ, τ), where the states are strings over A, the empty string ϵ is the initial state, and the transition function τ is concatenation, that is $\tau(w,a) = wa$, for all $w \in A^*$ and $a \in A$. For any pointed automaton (X, x_0, α), the unique F–algebra morphism

$$r_{x_0} : (A^*, \epsilon, \tau) \to (X, x_0, \alpha)$$

is given by $r_{x_0}(w) = w(x_0)$. The F–algebra morphism r_{x_0} is called the *reachability* map, and it maps every word w to the state $w(x_0)$ which is the state we reach from x_0 by processing the word w.

Dually, the final G–coalgebra is the coloured automaton $(2^{A^*}, \hat{\epsilon}, \hat{\tau})$, where states are languages over A, accepting states are only the languages that contain the empty word ϵ, i.e. $\hat{\epsilon}(L) = L(\epsilon)$, and the transition function $\hat{\tau} : 2^{A^*} \to (2^{A^*})^A$ is the right derivative operation, that is $\hat{\tau}(L)(a) = L_a$, for all $L \in 2^{A^*}$ and $a \in A$. For any coloured automaton (X, c, α), the unique G–coalgebra morphism

$$o_c : (X, c, \alpha) \to (2^{A^*}, \hat{\epsilon}, \hat{\tau})$$

is given by $o_c(x) = \lambda w.c(w(x)) \in 2^{A^*}$. The G–coalgebra morphism o_c is called the *observability* map, and it maps every state x to the language $o_c(x)$ accepted from the state x according to the colouring c.

Example 1. Consider the deterministic automaton (X, α) on $A = \{a, b\}$ given by:

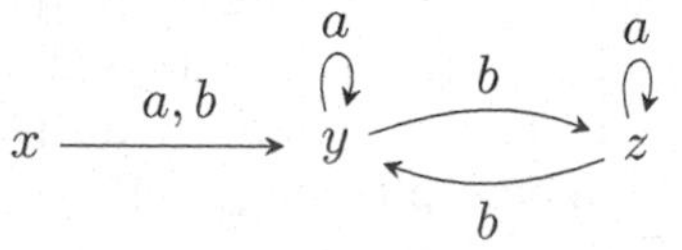

Then we have:

(i) The image of the reachability map r_x, with x as initial state, on the word $aabb$ is $r_x(aabb) = y$. Similar calculations, for possible different initial states, are the following:

$$r_x(ba^5ba) = z, \quad r_y(ba^5ba) = y, \quad r_y(b^{11}) = z, \quad r_z(aba) = y, \quad r_x(\epsilon) = x.$$

(ii) The image of the observability map o_c for the colouring $c = \{x, z\}$ on the state x (i.e. the language accepted by the automaton if the initial state is x and the set of accepting states is $\{x, z\}$) is

$$o_c(x) = \epsilon \cup (a \cup b)a^*b(a^*ba^*b)^*a^*. \qquad \square$$

3 Equations and Coequations

In this section we summarize some concepts and facts from [4] that will be used in the following sections.

Let (X, α) be a deterministic automaton on A. An *equation* is a pair $(u, v) \in A^* \times A^*$, sometimes also denoted by $u = v$. Given $(u, v) \in A^* \times A^*$, we define $(X, \alpha) \models_{\mathsf{e}} (u, v)$ – and say: (X, α) satisfies the equation (u, v) – as follows:

$$(X, \alpha) \models_{\mathsf{e}} (u, v) \Leftrightarrow \forall x \in X \;\; u(x) = v(x) \Leftrightarrow \forall x \in X \;\; (u, v) \in \ker(r_x),$$

and for any set of equations $E \subseteq A^* \times A^*$ we write $(X, \alpha) \models_{\mathsf{e}} E$ if $(X, \alpha) \models_{\mathsf{e}} (u, v)$ for every $(u, v) \in E$. Basically, an equation (u, v) is satisfied by an automaton if the states reached by u and v from any initial state $x \in X$, are the same.

An equivalence relation C on A^* is a *congruence* on A^* if for any $t, u, v, w \in A^*$, $(t, v) \in C$ and $(u, w) \in C$ imply $(tu, vw) \in C$. If C is a congruence on A^*, the *congruence quotient* A^*/C has a pointed automaton structure $A^*/C = (A^*/C, [\epsilon], \alpha)$ with transition function given by $\alpha([w], a) = [wa]$, which is well defined since C is a congruence.

A set of *coequations* is a subset $D \subseteq 2^{A^*}$. We define $(X, \alpha) \models_{\mathsf{c}} D$ – and say: (X, α) satisfies the set of coequations D – as follows:

$$(X, \alpha) \models_{\mathsf{c}} D \Leftrightarrow \forall c \in 2^X, x \in X \;\; o_c(x) \in D \Leftrightarrow \forall c \in 2^X \;\; \mathrm{Im}(o_c) \subseteq D.$$

In other words, an automaton satisfies a coequation D if for every colouring, the language accepted by the automaton, starting from any state, belongs to D. Note that, categorically, coequations are dual to equations [12].

Next we show how to construct the maximum set of equations and the minimum set of coequations satisfied by an automaton (X, α). To get the maximum set of equations of (X, α), we define the pointed deterministic automaton $\mathbf{free}(X, \alpha)$ as follows:

1. Define the pointed deterministic automaton $\prod(X, \alpha) = (\prod_{x \in X} X, \Delta, \hat{\alpha})$ where $\hat{\alpha}$ is the product of α $|X|$ times, that is $\hat{\alpha}(\theta, a)(x) = \alpha(\theta(x), a)$, and $\Delta \in \prod_{x \in X} X$ is given by $\Delta(x) = x$. Then, by initiality of $A^* = (A^*, \epsilon, \tau)$, we get a unique F-algebra homomorphism $r_\Delta : A^* \to \prod(X, \alpha)$.
2. Define $\mathbf{free}(X, \alpha)$ and $\mathbf{Eq}(X, \alpha)$ as

$$\mathbf{free}(X, \alpha) := A^*/\ker(r_\Delta) \text{ and } \mathbf{Eq}(X, \alpha) := \ker(r_\Delta)$$

Notice that $\mathbf{free}(X, \alpha)$ has the structure of a pointed automaton.

Example 2. Let $A = \{a, b\}$ and consider the following automaton (X, α) on A:

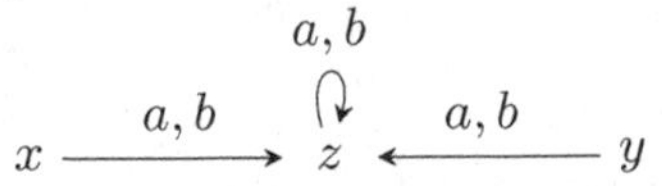

Then by definition $\mathbf{free}(X, \alpha) = A^*/\ker(r_\Delta) \cong \mathrm{Im}(r_\Delta)$. So in order to construct $\mathbf{free}(X, \alpha)$ we only need to construct the reachable part of $\prod(X, \alpha)$ from the state $\Delta = (x, y, z)$, which gives us the following automaton $\mathrm{Im}(r_\Delta)$:

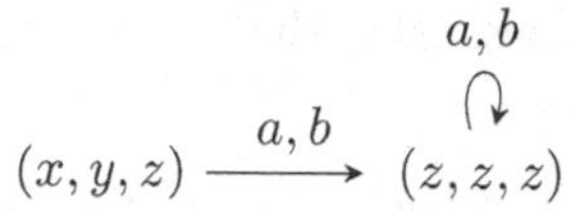

In this case $\mathbf{free}(X, \alpha) = A^*/\ker(r_\Delta)$, where $\ker(r_\Delta)$ is the equivalence relation that corresponds to the partition $\{\{\epsilon\}(a \cup b)^+\}$ of A^*. Hence, $\mathbf{free}(X, \alpha)$ is isomorphic to the automaton (†) in which $(x, y, z) \mapsto [\epsilon]$ and $(z, z, z) \mapsto [a]$. □

By construction, we have the following theorem.

Theorem 3. *(Proposition 6, [4])* $\mathbf{Eq}(X, \alpha)$ *is the maximum set of equations satisfied by* (X, α).

Note that the above equations are just identities for algebras with unary operations in which both the left and right terms use the same variable, that is, identities of the form $p(x) \approx q(y)$ where $p, q \in A^*$, and x and y are variables with $x = y$ (see [7, Definition 11.1]). A similar result as the above theorem can be obtained for any identity $p(x) \approx q(y)$. In order to do that one should consider the functor $F'(X) = A \times X$, where A is a fixed alphabet. Clearly, the free F'-algebra on S generators is the initial algebra for the functor $F'_S(X) := S + F'(X) = S + (A \times X)$. Furthermore, as every identity uses at most two variables it is enough to consider the free F'-algebra on 2 generators in order to express the left and the right term of every identity.

We show next how to construct the minimum set of coequations satisfied by (X, α). In this case, we construct the coloured deterministic automaton $\mathbf{cofree}(X, \alpha) \subseteq 2^{A^*}$, by taking the following steps:

1. Define the coloured automaton $\sum(X, \alpha) = (\sum_{c \in 2^X} X, \Phi, \tilde{\alpha})$ where $\tilde{\alpha}$ and Φ are given by $\tilde{\alpha}(c, x)(a) = (c, a(x))$ and $\Phi(c, x) = c(x)$. Then, by finality of $2^{A^*} = (2^{A^*}, \hat{\epsilon}, \hat{\tau})$, we get a unique G-coalgebra homomorphism $o_\Phi : \sum(X, \alpha) \to 2^{A^*}$.
2. Define $\mathbf{cofree}(X, \alpha)$ and $\mathbf{coEq}(X, \alpha)$ as

$$\mathbf{cofree}(X, \alpha) = \mathbf{coEq}(X, \alpha) := \mathrm{Im}(o_\Phi).$$

Similarly as in the case of equations we have the following theorem.

Theorem 4. *(Proposition 6, [4])* $\mathbf{coEq}(X, \alpha)$ *is the minimum set of coequations satisfied by* (X, α).

Example 5. Let $A = \{a, b\}$ and consider the following automaton (X, α) on A:

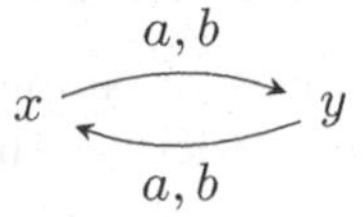

By definition of $\mathbf{cofree}(X, \alpha)$, we have that

$$\mathbf{cofree}(X, \alpha) = \{o_c(z) \mid c \in 2^X, z \in X = \{x, y\}\} = \{\emptyset, L_{\mathrm{odd}}, L_{\mathrm{even}}, A^*\}$$

where L_{odd} and L_{even} are the sets of words in A^* with an odd and even number of symbols, respectively. □

Next we define varieties of languages, that is, the kind of coequations that we will use in the following section.

Definition 6. *A variety of languages is a set $V \subseteq 2^{A^*}$ such that:*

(i) V is a complete atomic Boolean subalgebra of $2^{A^*} = (2^{A^*}, (\ )', \cup, \cap, \emptyset, A^*)$.
(ii) V is closed under left and right derivatives: if $L \in V$ then ${}_wL \in V$ and $L_w \in V$, for all $w \in A^*$.

This notion is related to but different from that of *local variety of regular languages*, as defined in [9]: according to our definition above, a variety may contain languages that are non-regular; moreover, a variety has the structure of a *complete atomic* Boolean algebra rather than just a Boolean algebra.

Our main result, in the next section, will be that regular varieties of automata (i.e., defined by regular equations) are characterized by varieties of languages.

There is a correspondence between congruences of A^* and varieties of languages that can be stated as follows.

Theorem 7. *([4]) Let C be a congruence on A^* and let $V \subseteq 2^{A^*}$ be a variety of languages. Then*

(i) $\mathbf{cofree}(A^*/C)$ *is a variety of languages.*
(ii) $\mathbf{Eq}(V)$ *is a congruence on A^*.*
(iii) $\mathbf{free} \circ \mathbf{cofree}(A^*/C) = A^*/C$.
(iv) $\mathbf{cofree} \circ \mathbf{free}(V) = V$.

Notice that every variety of languages V has a coloured automaton structure $V = (V, \hat{\epsilon}, \hat{\tau})$ because it is closed under the right derivatives. (In (ii), we write $\mathbf{Eq}(V)$ rather than $\mathbf{Eq}(V, \hat{\tau})$.) Thus $\mathbf{cofree}(A^*/C)$ has the structure of a coloured automaton, and so expressions (iii) and (iv) of the theorem above are well defined.

Additionally, we have that for a congruence C on A^*, $\mathbf{cofree}(A^*/C)$ is the complete Boolean subalgebra of 2^{A^*} whose set of atoms is A^*/C. Conversely, given a variety of languages V, $\mathbf{free}(V)$ is the congruence quotient whose associated congruence corresponds to the partition given by the set of atoms of V.

As an application of the previous facts we have the following:

Example 8. Given a family of languages $\mathcal{L} \subseteq 2^{A^*}$, we can construct an automaton (X, α) representing that family in the following sense:

For every $L \in \mathcal{L}$ there exists $x \in X$ and $c \in 2^X$ such that $o_c(x) = L$.

For any given family, we can construct an automaton such that it has the minimum number of states and moreover satisfies the following stronger property:

There exists $x \in X$ such that for every $L \in \mathcal{L}$ there exists $c \in 2^X$ such that $o_c(x) = L$.

The construction is a follows: let $V(\mathcal{L})$ be the least variety of languages containing $\mathcal{L}$, which always exists. Then the automaton $\mathbf{free}(V(\mathcal{L})) = A^*/\mathbf{Eq}(V(\mathcal{L}))$ has the desired property. In fact, by [4, Lemma 13], there exists, for every $L \in V(\mathcal{L})$, a colouring $c_L : A^*/\mathbf{Eq}(V(\mathcal{L})) \to 2$ such that $o_{c_L}([\epsilon]) = L$. □

The previous example gives us a way to construct a single program (automaton) for a specific set of behaviours (set of languages) in an efficient way (minimum number of states) with the property that the initial configuration (initial state) of the program is the same for every desired behaviour. Here is a small illustration of this fact.

Example 9. Let $A = \{a, b\}$ and consider the following family of languages on A^*

$$\mathcal{L} = \{(a \cup b)^+, L_{\text{odd}}, L_{\text{even}}\}$$

We would like to construct a pointed automaton (X, x_0, α) with the property that for every $L \in \mathcal{L}$ there exists $c_L \in 2^X$ such that $o_{c_L}(x_0) = L$. According to the previous example, we only need to construct the least variety of languages V containing $\mathcal{L}$. In this case, V is the variety of languages (with 8 elements) whose atoms are

$$A_1 = \{\epsilon\}, A_2 = (a \cup b)[(a \cup b)(a \cup b)]^*, \text{ and } A_3 = (a \cup b)(a \cup b)[(a \cup b)(a \cup b)]^*$$

Clearly $\mathcal{L} \subseteq V$ since $(a \cup b)^+ = A_2 \cup A_3$, $L_{\text{odd}} = A_2$, and $L_{\text{even}} = A_1 \cup A_3$. Then the pointed automaton we are looking for is **free**(V) which is given by

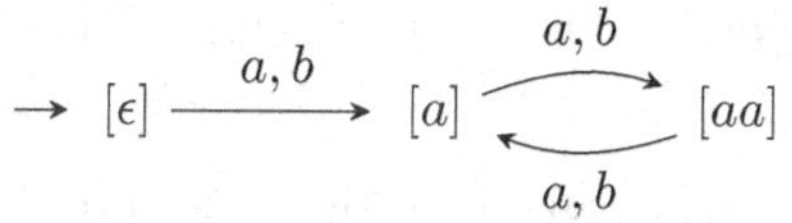

Clearly, for the colourings $c_1 = \{[a], [aa]\}$, $c_2 = \{[a]\}$, and $c_3 = \{[\epsilon], [aa]\}$ we have that

$$o_{c_1}([\epsilon]) = (a \cup b)^+,\ o_{c_2}([\epsilon]) = L_{\text{odd}}, \text{and } o_{c_3}([\epsilon]) = L_{\text{even}}.$$

□

4 Characterization of Regular Varieties of Automata

In this section we show that classes of deterministic automata that are closed under subautomata, products, homomorphic images, and sums are the same as classes of regular varieties of automata (defined below). Furthermore we will give a characterization in terms of coequations.

For an alphabet A let τ_A be the type $\tau_A = \{f_a\}_{a \in A}$ where each f_a is a unary operation symbol. Clearly, an algebra of type τ_A is a deterministic automaton over A since we have the correspondence

$$(X, \alpha : X \times A \to X) \iff (X, \{f_a : X \to X\}_{a \in A})$$

where $f_a(x) = \alpha(x, a)$, for all $a \in A$ and $x \in X$.

Every term $f_{a_n} f_{a_{n-1}} \cdots f_{a_1}(x)$ of type τ_A will be written as $u(x)$ where $u = a_1 \cdots a_{n-1} a_n$. For any $u, v \in A^*$ there are two possible *identities*

$$u(x) \approx v(x) \text{ and } u(x) \approx v(y)$$

which correspond to the formulas

$$\forall x[u(x) = v(x)] \text{ and } \forall x, y[u(x) = v(y)].$$

Identities of the form $u(x) \approx v(x)$ are called *regular identities.* They were first introduced by Płonka [11] and can be identified with pairs $(u, v) \in A^* \times A^*$ as in the previous section.

For any set of identities E of type τ_A we define the class $M_{\mathsf{e}}(E)$ of automata satisfying E by

$$M_{\mathsf{e}}(E) = \{(X, \alpha) \mid (X, \alpha) \models_{\mathsf{e}} E\},$$

where E can include identities of the form $u(x) \approx v(y)$. We will write $E \subseteq A^* \times A^*$ if all the identities in E are of the form $u(x) \approx v(x)$, that is if they are regular identities.

Clearly, for any set E of identities, $M_{\mathsf{e}}(E)$ is a *variety*, that is, a class of automata that is closed under products, subautomata, and homomorphic images. (Note that now, we are talking of a variety of *automata*, as opposed to our earlier notion of variety of *languages*.) By Birkhoff's theorem [5], any variety V of automata on A is of the form $V = M_{\mathsf{e}}(E)$ for some set E of identities. Classes of automata of the form $M_{\mathsf{e}}(R)$ where R is a set of regular identities are called *regular varieties of automata.* The next example shows a variety of automata that is not regular.

Example 10. Let $A = \{a, b\}$, and consider the variety V_1 generated by the automaton (X, α) on A given by

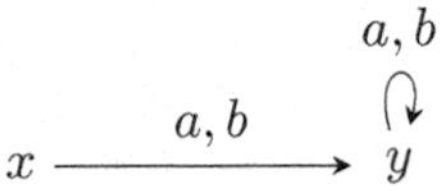

that is V_1 is the least variety containing (X, α). Then, an automaton $(Y, \sigma) \in V_1$ if and only if there exists $s \in Y$ such that for every $z \in Y$, $\sigma(z, a) = \sigma(z, b) = s$, that is, an automaton is in V_1 if and only if there is no difference between a and b transitions, and there is a state ('*sink*') that is reachable from any state by inputting the letter a (or, equivalently, that is reachable from any state by inputting the letter b).

Let E be a set of identities such that $V_1 = M_{\mathsf{e}}(E)$. If $E \subseteq A^* \times A^*$ then $(X, \alpha) + (X, \alpha) \in M_{\mathsf{e}}(E)$ but $(X + X, \hat{\alpha}) = (X, \alpha) + (X, \alpha) \notin V_1$ (as $\hat{\alpha}((0, x), a) = (0, y) \neq (1, y) = \hat{\alpha}((1, x), a)$ meaning that there is no sink, a contradiction). Observe that a set of defining identities for V_1 is $E = \{a(x) \approx b(y)\}$. □

It's worth mentioning that a Birkhoff-like theorem for regular varieties was formulated by Taylor in [13, p. 4]. Applied to the present situation, it says that a class K of automata on A is a regular variety if and only if K is closed under products, subalgebras, homomorphic images, and $\mathbf{2}_\emptyset \in K$, where $\mathbf{2}_\emptyset$ is the algebra

$$\mathbf{2}_\emptyset = (\{0, 1\}, \{f_a : 2 \to 2\}_{a \in A})$$

where f_a is the identity function on the set 2, for every $a \in A$. If $A = \{a, b\}$ then the algebra (automaton) $\mathbf{2}_\emptyset$ is given by

This automaton has, in general, the property that an identity holds in $\mathbf{2}_\emptyset$ if and only if it is regular [10, Lemma 2.1].

Disconnected automata cannot satisfy equations of the form $u(x) \approx v(y)$, which property is the key fact to obtain the following theorem.

Theorem 11. *Let K be a nonempty class of automata on A. The following are equivalent:*

(i) K is closed under products, subalgebras, homomorphic images, and sums.
(ii) $K = M_{\mathsf{e}}(C)$ for some congruence $C \subseteq A^* \times A^*$ on A^*. That is, K is a regular variety.

Proof. (*i*) $\Rightarrow$ (*ii*) By Birkhoff's theorem, $K = M_{\mathsf{e}}(E)$ for some set of equations. Now, E cannot contain identities of the form $u(x) \approx v(y)$ because K is closed under sums. Clearly, $M_{\mathsf{e}}(E) = M_{\mathsf{e}}(\langle E \rangle)$ where $\langle E \rangle$ denotes the least congruence containing E.

(*ii*) $\Rightarrow$ (*i*) Identities of the form $u(x) \approx v(x)$ are preserved under products, subalgebras, homomorphic images, and sums. □

Combining the previous theorem with the characterization of regular varieties given by Taylor, we have that a variety of automata is closed under sums if and only if it contains $\mathbf{2}_\emptyset$. Which can be proved directly by noticing that $\mathbf{2}_\emptyset$ is the sum of the trivial (one element) algebra and, conversely, that the sum of a family $\{(X_i, \alpha_i)\}_{i \in I}$ can be obtained as a homomorphic image of the algebra $\prod_{i \in I}(X_i, \alpha_i) \times \prod_{i \in I} \mathbf{2}_\emptyset$. In fact, let $\phi : I \to \prod_{i \in I} 2$ be an injective function and $i_0 \in I$ a fixed element, then the function $h : \prod_{i \in I} X_i \times \prod_{i \in I} 2 \to \sum_{i \in I} X_i$ defined by

$$h(f, p) = \begin{cases} (i_0, f(i_0)) & \text{if} \, p \notin \mathrm{Im}(\phi), \\ (i, f(i)) & \text{if} \, p \in \mathrm{Im}(\phi) \text{ and } \phi(i) = p, \end{cases}$$

is a surjective homomorphism onto $\sum_{i \in I}(X_i, \alpha_i)$. Notice that h is well-defined since ϕ is injective.

Similarly to the equational case, for a set of coequations $D \subseteq 2^{A^*}$, we define the class $M_{\mathsf{c}}(D)$ of automata satisfying the set of coequations D by

$$M_{\mathsf{c}}(D) = \{(X, \alpha) \mid (X, \alpha) \models_{\mathsf{c}} D\}.$$

Lemma 12. *Let C be a congruence on A^*, then*

$$M_{\mathsf{c}}(\mathbf{cofree}(A^*/C)) = M_{\mathsf{e}}(C)$$

Proof. $M_{\mathsf{c}}(\mathbf{cofree}(A^*/C)) \subseteq M_{\mathsf{e}}(C)$: Let (X, α) be an automaton such that $(X, \alpha) \models_{\mathsf{c}} \mathbf{cofree}(A^*/C)$. We have to show that $(X, \alpha) \models_{\mathsf{e}} C$. Fix an equation

$(u, v) \in C$ and assume by contradiction that there exists $x \in X$ such that $u(x) \neq v(x)$. Consider the colouring $\delta_{u(x)} : X \to 2$ given by

$$\delta_{u(x)}(z) = \begin{cases} 1 & \text{if } z = u(x), \\ 0 & \text{if } z \neq u(x). \end{cases}$$

Then $o_{\delta_{u(x)}}(x) \in \mathbf{cofree}(A^*/C)$ since $(X, \alpha) \models_{\mathsf{c}} \mathbf{cofree}(A^*/C)$. Clearly we have that

$$(\star) \qquad o_{\delta_{u(x)}}(x)(u) = 1 \neq 0 = o_{\delta_{u(x)}}(x)(v).$$

By applying Lemma 13 from [4] we get a colouring $c : A^*/C \to 2$ such that $o_c([\epsilon]) = o_{\delta_{u(x)}}(x)$, and as $[u] = [v]$ we get that $o_c([\epsilon])(u) = o_c([\epsilon])(v)$, which contradicts $(\star)$.

$M_{\mathsf{e}}(C) \subseteq M_{\mathsf{c}}(\mathbf{cofree}(A^*/C))$: Let (X, α) be an automaton such that $(X, \alpha) \models_{\mathsf{e}} C$. We have to show that $(X, \alpha) \models_{\mathsf{c}} \mathbf{cofree}(A^*/C)$. Fix a colouring $c : X \to 2$ and $x \in X$. Define the colouring $\tilde{c} : A^*/C \to 2$ as $\tilde{c}([w]) := c(w(x))$ which is well-defined since $(X, \alpha) \models_{\mathsf{e}} C$. One easily shows that $o_c(x) = o_{\tilde{c}}([\epsilon])$ which is an element of $\mathbf{cofree}(A^*/C)$. □

By using Theorem 7 and the previous lemma we obtain a dual version of Theorem 11.

Theorem 13. *Let K be a nonempty class of automata on A. The following are equivalent:*

(i) K is closed under products, subalgebras, homomorphic images, and sums.
(ii) $K = M_{\mathsf{c}}(V)$ for some variety of languages V.

Proof. (i) ⇒ (ii) By Theorem 11, $K = M_{\mathsf{e}}(C)$ for some congruence $C \subseteq A^* \times A^*$ on A^*. Put $V = \mathbf{cofree}(A^*/C)$, then by the previous lemma $M_{\mathsf{e}}(C) = M_{\mathsf{c}}(V)$ where V is a variety of languages by Theorem 7.

(ii) ⇒ (i) Assume that $K = M_{\mathsf{c}}(V)$ for some variety of languages V, then one easily shows that K is closed under subalgebras, homomorphic images and sums. By Theorem 7 $V = \mathbf{cofree}(A^*/C)$ for the congruence $C = \mathbf{Eq}(V)$, then by Lemma 12 $M_{\mathsf{c}}(V) = M_{\mathsf{e}}(C)$ which implies that K is closed under products. □

It is worth mentioning that the property that the class $K = M_{\mathsf{c}}(V)$ is closed under products can be proved directly from the fact that V is a variety of languages as follows: Consider a family $\{(X_i, \alpha_i) \mid i \in I\} \subseteq M_{\mathsf{c}}(V)$ and let $X = (\prod_{i \in I} X_i, \alpha)$ be the product of that family. Fix a colouring $c : \prod_{i \in I} X_i \to 2$ and $x \in \prod_{i \in I} X_i$, we want to show that $o_c(x) \in V$, which follows the fact that V is a complete Boolean algebra and from the equality

$$o_c(x) = \bigvee_{y \in c^{-1}(1)} \left(\bigwedge_{i \in I} o_{\delta_{y(i)}}(x(i)) \right)$$

where $\delta_{y(i)} : X_i \to 2$. In fact,

$$\begin{aligned} w \in o_c(x) &\Leftrightarrow \exists y \in c^{-1}(1) \;\; w(x) = y \\ &\Leftrightarrow \exists y \in c^{-1}(1) \forall i \in I \;\; w(x(i)) = y(i) \\ &\Leftrightarrow \exists y \in c^{-1}(1) \forall i \in I \;\; w \in o_{\delta_{y(i)}}(x(i)) \\ &\Leftrightarrow w \in \bigvee_{y \in c^{-1}(1)} \left(\bigwedge_{i \in I} o_{\delta_{y(i)}}(x(i)) \right). \end{aligned}$$

From Lemma 12 we obtain the following corollary.

Corollary 14. *For any automaton* (X, α) *and any congruence* C *we have that*

$$(X, \alpha) \models_{\mathsf{e}} C \quad \Leftrightarrow \quad (X, \alpha) \models_{\mathsf{c}} \mathbf{coEq}(A^*/C).$$

Similarly, using the fact that $\mathbf{cofree} \circ \mathbf{free}(V) = V$ for every variety of languages, we get the following corollary.

Corollary 15. *For any automaton* (X, α) *and any variety of languages* V *we have that*

$$(X, \alpha) \models_{\mathsf{c}} V \quad \Leftrightarrow \quad (X, \alpha) \models_{\mathsf{e}} \mathbf{Eq}(V).$$

Example 16. Let $A = \{a, b\}$, and consider the regular variety V_2 generated by the automaton (X, α) on A given by

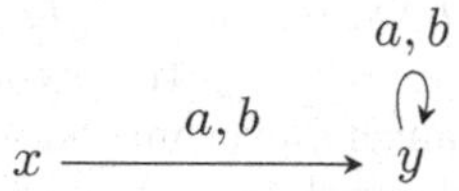

which by Theorems 11 and 13 can be described in three diferent ways, namely:

(i) As the closure under products, subautomata, homomorphic images, and sums of the set $\{(X, \alpha)\}$, which in this case implies that an automaton $(Y, \beta) \in V_2$ if and only if (Y, β) is the sum of elements in V_1 (see Example 10).
(ii) $V_2 = M_{\mathsf{e}}(C)$ where C is the congruence generated by $\{a = b, aa = a\}$.
(iii) $V_2 = M_{\mathsf{c}}(V)$ where V is the variety of languages $V = \{\emptyset, \{\epsilon\}, A^+, A^*\}$ where $A^+ = A^* \setminus \{\epsilon\}$. □

We can summarize the results of this section in one theorem as follows.

Theorem 17. *Let* K *be a nonempty class of automata on* A*. The following are equivalent:*

(i) K *is a regular variety, that is* $K = M_{\mathsf{e}}(R)$ *where* R *is a set of regular identities, which can be taken to be a congruence on* A^**.*
(ii) K *is closed under products, subalgebras, homomorphic images, and sums.*
(iii) $K = M_{\mathsf{c}}(V)$ *for some variety of languages* $V \subseteq 2^{A^*}$*.*

5 Conclusion

Algebras and coalgebras are in general different structures. Deterministic automata have the advantage that they can be defined both as algebras and coalgebras. This not only gives us the advantage of using all the machinery available in those areas but also gives us the possibility to understand and connect unrelated areas and, in some cases, create new results. The results of the present paper are an example of that. (Other examples can be found, for instance, in [4,6]).

Homomorphic images and substructures are characterizing properties common to both varieties and covarieties, but varieties are closed under products and tipically not under sums [5], while, dually, covarieties are closed under sums and, in general, not under products [8]. The fact that deterministic automata can be seen as both algebras and coalgebras allowed us to define classes of automata closed under *all* those four constructions. In the present paper, such classes were characterized both equationally and coequationally, and surprisingly, they turned out to be the same as regular varieties of automata, which were studied and characterized by Taylor [13].

As future work we intend to investigate similar results for other structures that can be viewed both as algebras and coalgebras at the same time, such as weighted automata and tree automata.

Acknowledgements. The research of Julian Salamanca is funded by NWO project 612.001.210. The research of Adolfo Ballester-Bolinches has been supported by the grant 11271085 from the *National Natural Science Foundation of China.* The research of Enric Cosme-Llópez has been supported by the predoctoral grant AP2010-2764 from the *Ministerio de Educación* (Spanish Government) and by an internship from *CWI.* The research of Adolfo Ballester-Bolinches, Enric Cosme-Llópez, and Jan Rutten has been supported by the grant MTM2014-54707-C3-1-P from the *Ministerio de Economía y Competitividad* (Spanish Government).

References

1. Abel, A., Pientka, B., Thibodeau, D., Setzer, A.: Copatterns: programming infinite structures by observations. In: Giacobazzi, R., Cousot, R. (eds.) Proceedings of the 40th Annual ACM Symposium on Principles of Programming Languages (POPL '13), pp. 27–38, ACM (2013)
2. Adámek, J., Milius, S., Myers, R.S.R., Urbat, H.: Generalized Eilenberg theorem I: local varieties of languages. In: Muscholl, A. (ed.) FOSSACS 2014 (ETAPS). LNCS, vol. 8412, pp. 366–380. Springer, Heidelberg (2014)
3. Awodey, S., Hughes, J.: Modal operators and the formal dual of Birkhoff's completeness theorem. Math. Struct. Comput. Sci. **13**(2), 233–258 (2003)
4. Ballester-Bolinches, A., Cosme-Llópez, E., Rutten, J.J.M.M.: The dual equivalence of equations and coequations for automata. CWI Technical Report report FM-1403, pp. 1–30 (2014)
5. Birkhoff, G.: On the structure of abstract algebras. Proc Camb. Philos. Soc. **31**, 433–454 (1935)

6. Bonchi, F., Bonsangue, M., Hansen, H., Panangaden, P., Rutten, J., Silva, A.: Algebra-coalgebra duality in Brzozowski's minimization algorithm. ACM Trans. Comput. Logic **15**(1), 1–27 (2014)
7. Burris, S.N., Sankappanavar, H.P.: A Course in Universal Algebra, Graduate Texts in Mathematics. Springer, New York (1981)
8. Clouston, R., Goldblatt, R.: Covarieties of coalgebras: comonads and coequations. In: Van Hung, D., Wirsing, M. (eds.) ICTAC 2005. LNCS, vol. 3722, pp. 288–302. Springer, Heidelberg (2005)
9. Gehrke, M., Grigorieff, S., Pin, J.É.: Duality and equational theory of regular languages. In: Aceto, L., Damgård, I., Goldberg, L.A., Halldórsson, M.M., Ingólfsdóttir, A., Walukiewicz, I. (eds.) ICALP 2008, Part II. LNCS, vol. 5126, pp. 246–257. Springer, Heidelberg (2008)
10. Graczyńska, E.: Birkhoff's theorems for regular varieties. Bull. Sect. Logic, Univ. Łódź **26**(4), 210–219 (1997)
11. Płonka, J.: On a method of construction of abstract algebras. Fund. Math. **61**, 183–189 (1967)
12. Rutten, J.J.M.M.: Universal coalgebra: a theory of systems. Theor. Comput. Sci. **249**(1), 3–80 (2000)
13. Taylor, W.: Equational logic. Houston J. Math. **5**, 1–83 (1979)

Column-Wise Extendible Vector Expressions and the Relational Computation of Sets of Sets

Rudolf Berghammer(✉)

Institut Für Informatik, Universität Kiel, 24098 Kiel, Germany
rub@informatik.uni-kiel.de

Abstract. We present a technique for the relational computation of sets of sets. It is based on specific vector expressions, which form the syntactical counterparts of B. Kehden's vector predicates. Compared with the technique that directly solves a posed problem by the development of a vector expression of type $2^X \leftrightarrow \mathbf{1}$ from a formal logical problem description, we reduce the solution to the development of inclusions between vector expressions of type $X \leftrightarrow \mathbf{1}$. Frequently, this is a lot simpler. The transition from the inclusions to the desired vector expression of type $2^X \leftrightarrow \mathbf{1}$ is then immediately possible by means of a general result. We apply the technique to some examples from different areas and show how the solutions behave with regard to running time if implemented and evaluated by the Kiel RELVIEW tool.

1 Introduction

For the solution of problems it is sometimes necessary to determine a large set of sets; the computation of Banks winners in social choice theory (see [1]), of the concept lattice in formal concept analysis (see [13]) and of the permanent of a Boolean matrix via the set of all perfect matchings of a specific bipartite graph (see [15]) are prominent examples. It is known that reduced ordered binary decision diagrams (ROBDDs) are a very efficient data structure for the representation of sets and relations. This is also proved by numerous applications of RELVIEW, a ROBDD-based specific purpose computer algebra system for the manipulation and visualization of relations and relational programming (see [4]). The use of ROBDDs often leads to an amazing computational power, in particular, if a hard problem is to solve and its solution is done by the computation of a very large set of 'interesting objects' and a subsequent search through it.

Computation of a very large set usually means that of a subset $\mathcal{A}$ of a powerset 2^X. A technique to solve such a task by a combination of logic, relation algebra and RELVIEW, which we successfully apply since many years and in case of rather different problem areas (like graph-theory [2,4–6,11], simple games and social choice theory [10,12], orders and lattices [2,5,8,9], Petri nets [2,3], Boolean logic [7]), is as follows: We start with a formal logical description of a set $Y \in 2^X$ to belong to $\mathcal{A}$. Then we transform this formula step-by-step into the form $\mathfrak{v}_Y$, where $\mathfrak{v}$ is a relation-algebraic expression that evaluates to a vector of type $2^X \leftrightarrow \mathbf{1}$. Hence, this vector models the set $\mathcal{A}$. During the transformation

R. Hinze and J. Voigtländer (Eds.): MPC 2015, LNCS 9129, pp. 238–256, 2015.
DOI:10.1007/978-3-319-19797-5_12

process we are targeted on the use of standard basic relations and operations of relation algebra only, for instance those introduced in [19]. This enables a straightforward translation of $\mathfrak{v}$ into RELVIEW-code and the tool can be used to compute $\mathcal{A}$ and to visualize it, e.g., via the columns of a Boolean matrix.

The relation-algebraic modeling of sets by vectors has a long tradition. But the main area of application seems not to be the computation of sets and sets of sets. Instead vectors are mainly used to specify properties of subsets of a universe X in a succinct way, that is well-suited for algebraic reasoning. See e.g., [18,19]. Also in such applications the first step consists in the translation of the property in question into the language of relation algebra. Usually the results are inclusions $\mathfrak{w} \subseteq \mathfrak{u}$, with relation-algebraic expressions that now evaluate to vectors of type $X \leftrightarrow \mathbf{1}$, and the developments are rather simple when compared with those of the aforementioned technique.

During our investigations we have noticed that in many cases there is a close connection between the inclusion $\mathfrak{w} \subseteq \mathfrak{u}$ specifying a certain property of a set and the vector expression $\mathfrak{v}$ modeling the set of all sets with this property. Based on this, we have developed a general technique to compute the latter from the former in case that both sides $\mathfrak{w}$ and $\mathfrak{u}$ of the inclusion are of a specific syntactic form that, in Boolean matrix terminology, allows to apply them column-wisely. Our specific vector expressions can be seen as syntactical counterparts of B. Kehden's vector predicates, introduced in [17]. In [17] vector predicates are motivated by the relational treatment of evolutionary algorithms in [16] and their use to model sets of search points with a specific property. In contrast with [17], we use our specific vector expressions for the relational computation of sets of sets to avoid the often intricate developments of the aforementioned vector expression $\mathfrak{v}$. Nevertheless, they also can be applied in the sense of [17].

2 Relation-Algebraic Preliminaries

In the following we recall those facts about heterogeneous relation algebra that are used in the remainder of the paper (for more details, see e.g., [18,19]). In doing so, we use $R : X \leftrightarrow Y$ to denote that R is a (binary) relation with source X and target Y, i.e., a subset of the direct product $X \times Y$, call $X \leftrightarrow Y$ the *type* of R and write $[X \leftrightarrow Y]$ for the set of all relations of type $X \leftrightarrow Y$, i.e., for the powerset $2^{X \times Y}$. If X and Y are finite sets of size m and n, respectively, we may consider $R : X \leftrightarrow Y$ as a Boolean $m \times n$ matrix. Such an interpretation is well suited for many purposes and also used by RELVIEW as one of its possibilities to depict relations. Therefore, we frequently will use matrix terminology and matrix notation. Specifically, we talk about the rows, columns and 0- or 1-entries of a relation and will also write $R_{x,y}$ instead of $(x, y) \in R$ or $x\,R\,y$.

In its original form (as generalization of A. Tarski's homogeneous relation algebra [21] to relations on more than one set) heterogeneous relation algebra has three basic relations. The *identity relation* $\mathsf{I} : X \leftrightarrow X$ satisfies for all $x, y \in X$ that $\mathsf{I}_{x,y}$ iff $x = y$, for the *universal relation* $\mathsf{L} : X \leftrightarrow Y$ it holds $\mathsf{L}_{x,y}$ for all $x \in X$ and $y \in Y$ and for the *empty relation* $\mathsf{O} : X \leftrightarrow Y$ it holds $\mathsf{O}_{x,y}$ for no $x \in X$ and

$y \in Y$. To improve readability, in relation-algebraic expressions we will overload the symbols for the basic relations, i.e., avoid the binding of types to them. If necessary, then we will state their types in the surrounding text.

Relations are sets. As a consequence, we can form the *complement* $\overline{R} : X \leftrightarrow Y$, the *union* $R \cup S : X \leftrightarrow Y$ and the *intersection* $R \cap S : X \leftrightarrow Y$ of $R, S : X \leftrightarrow Y$ and to state the *inclusion* $R \subseteq S$ and *equality* $R = S$. Using subscripts to express relationships, we have for all $x \in X$ and $y \in Y$ that $\overline{R}_{x,y}$ iff $\neg R_{x,y}$, that $(R \cup S)_{x,y}$ iff $R_{x,y}$ or $S_{x,y}$ and that $(R \cap S)_{x,y}$ iff $R_{x,y}$ and $S_{x,y}$. Furthermore, $R \subseteq S$ holds iff for all $x \in X$ and $y \in Y$ from $R_{x,y}$ it follows $S_{x,y}$, and $R = S$ holds iff for all $x \in X$ and $y \in Y$ the relationships $R_{x,y}$ and $S_{x,y}$ are equivalent. A lot of the expressive power of relation algebra is due to the *composition* $R; S : X \leftrightarrow Z$ of $R : X \leftrightarrow Y$ and $S : Y \leftrightarrow Z$ and the *transposition* $R^{\mathsf{T}} : Y \leftrightarrow X$ of R. Their definitions are that for all $x \in X$ and $z \in Z$ we have $(R; S)_{x,z}$ iff there exists $y \in Y$ such that $R_{x,y}$ and $S_{y,z}$, and that for all $x \in X$ and $y \in Y$ we have $R^{\mathsf{T}}_{x,y}$ iff $R_{y,x}$. As a derived operation we will use the *symmetric difference* of $R, S : X \leftrightarrow Y$, defined by $R \oplus S := (R \cap \overline{S}) \cup (S \cap \overline{R})$. We assume that complementation and transposition bind stronger than composition and composition binds stronger than union, intersection and symmetric difference.

Relation algebra as just introduced can express exactly those formulae of first-order predicate logic which contain at most two free variables and all in all at most three variables. The expressive power of full first-order logic is obtained if projection relations or equivalent notions (like fork) are assumed to exist. See [22]. In the present paper we denote for a direct product $X \times Y$ by $\pi : X \times Y \leftrightarrow X$ and $\rho : X \times Y \leftrightarrow Y$ the corresponding *projection relations*. Furthermore, we always assume a pair u from a direct product to be of the form $u = (u_1, u_2)$. This allows to specify the meaning of π and ρ by $\pi_{u,x}$ iff $u_1 = x$ and $\rho_{u,y}$ iff $u_2 = y$, for all $u \in X \times Y$, $x \in X$ and $y \in Y$. We also will use the relation-level equivalents of the set-theoretic symbol '$\in$' as basic relations. These are the *membership relations* $\mathsf{M} : X \leftrightarrow 2^X$ and defined by $\mathsf{M}_{x,Y}$ iff $x \in Y$, for all $x \in X$ and $Y \in 2^X$. As last extension we will use *size-comparison relations* $\mathsf{S} : 2^X \leftrightarrow 2^X$, which are specified by $\mathsf{S}_{Y,Z}$ iff $|Y| \leq |Z|$, for all $Y, Z \in 2^X$. If projection relations and membership relations are assumed as additional basic relations, then by this kind of heterogeneous relation algebra the expressive power of monadic second-order predicate logic is obtained. By the further addition of size-comparison relations this expressive power is once more strictly increased.

3 Modeling Via Vectors and Column-Wise Enumerations

There are several possibilities to model sets via relations; see e.g., [18,19]. In the present paper we will model sets using *vectors*. These are relations v such that $v = v; \mathsf{L}$. Since in our applications the targets of vectors will always be irrelevant, we assume a specific singleton set $\mathbf{1} := \{\bot\}$ as common target. Such vectors correspond to Boolean column vectors. Therefore, we prefer lower case letters for vectors, as in linear algebra, and omit almost always the second subscript, again as in linear algebra. That is, in case of $v : X \leftrightarrow \mathbf{1}$ and $x \in X$ we write v_x

instead of $v_{x,\perp}$. Then we get for all $v, w : X \leftrightarrow \mathbf{1}$ and $x, y \in X$ that $v_x \wedge w_y$ iff $(v; w^{\mathsf{T}})_{x,y}$. Specific vectors are the universal vectors $\mathsf{L} : X \leftrightarrow \mathbf{1}$ and the empty vectors $\mathsf{O} : X \leftrightarrow \mathbf{1}$. There are precisely two vectors of type $\mathbf{1} \leftrightarrow \mathbf{1}$. They may be used to represent truth-values, where $\mathsf{L} : \mathbf{1} \leftrightarrow \mathbf{1}$ stands for 'true' and $\mathsf{O} : \mathbf{1} \leftrightarrow \mathbf{1}$ stands for 'false'. With this interpretation the relation-algebraic operations $\cup$, $\cap$ and $\overline{}$ correspond to the Boolean connectives $\vee$, $\wedge$ and $\neg$.

Let $v : X \leftrightarrow \mathbf{1}$ be a vector. Then we say that v *models* (or *is a model of*) the subset Y of X if for all $x \in X$ we have $x \in Y$ iff v_x. This means that precisely those entries of v are 1, which correspond to an element of Y. It can easily be verified that the mapping $\mathit{set} : [X \leftrightarrow \mathbf{1}] \to 2^X$, defined by $\mathit{set}(v) = \{x \in X \mid v_x\}$, is a Boolean lattice isomorphism between the Boolean lattices $([X \leftrightarrow \mathbf{1}], \cup, \cap, \overline{})$ and $(2^X, \cup, \cap, \overline{})$, with $\mathit{set}^{-1}(Y) = Y \times \{\perp\}$, for all $Y \in 2^X$. If $v : X \times Y \leftrightarrow \mathbf{1}$ is a vector with a direct product as source, then it models a subset of $X \times Y$, i.e., a relation of type $X \leftrightarrow Y$. In such a case instead of $\mathit{set}(v)$ the notation $\mathit{rel}(v)$ is used, leading to v_u iff $\mathit{rel}(v)_{u_1,u_2}$, for all $u \in X \times Y$. There exists a relation-algebraic specification of the mapping $\mathit{rel} : [X \times Y \leftrightarrow \mathbf{1}] \to [X \leftrightarrow Y]$: With $\pi : X \times Y \leftrightarrow X$ and $\rho : X \times Y \leftrightarrow Y$ as the projection relations of $X \times Y$ we have $\mathit{rel}(v) = \pi^{\mathsf{T}}; (v; \mathsf{L} \cap \rho)$, for all $v : X \times Y \leftrightarrow \mathbf{1}$.

Let again $v : X \leftrightarrow \mathbf{1}$ be a vector. Then $\mathit{inj}(v) : \mathit{set}(v) \leftrightarrow X$ denotes the *embedding relation* of $\mathit{set}(v)$ into X. This means that for all $y \in \mathit{set}(v)$ and $x \in X$ we have $\mathit{inj}(v)_{y,x}$ iff $y = x$. Thus, $\mathit{inj}(v)$ is nothing else than the relation-level equivalent of the identity mapping from the set that is modeled by v to the source of v. As a Boolean matrix $\mathit{inj}(v)$ is obtained from $\mathsf{I} : X \leftrightarrow X$ by a removal of precisely those rows which do not correspond to an element of $\mathit{set}(v)$.

Given $\mathsf{M} : X \leftrightarrow 2^X$, each column models, if considered as a vector of type $X \leftrightarrow \mathbf{1}$, exactly one set of 2^X. We therefore say that M *column-wisely enumerates* the powerset 2^X. A combination of M with embedding relations allows such a column-wise enumeration via relations of type $X \leftrightarrow \mathcal{A}$ also for subsets $\mathcal{A}$ of 2^X. Namely if the vector $v : 2^X \leftrightarrow \mathbf{1}$ models $\mathcal{A}$ in the sense defined above, i.e., $\mathit{set}(v) = \mathcal{A}$, and we define a relation $S : X \leftrightarrow \mathcal{A}$ by $S := \mathsf{M}; \mathit{inj}(v)^{\mathsf{T}}$, then we get for all $x \in X$ and $Y \in \mathcal{A}$ that $S_{x,Y}$ iff $x \in Y$. This means that S results from M by a removal of precisely those columns which do not correspond to an element of $\mathcal{A}$. With regard to the development of relation-algebraic solutions of problems, embedding relations are frequently used for type adaptions. E.g., if $R : X \leftrightarrow Y$ is given and $Z \subseteq X$ is modeled by $v : X \leftrightarrow \mathbf{1}$, then $\mathit{inj}(v); R : Z \leftrightarrow Y$ restricts the source of R to Z. The latter means that $R_{x,y}$ iff $(\mathit{inj}(v); R)_{x,y}$, for all $x \in Z$ and $y \in Y$. In the context of RelView column-wise enumerations are frequently applied for the visualization of results if sets of sets are to be computed.

4 A Motivating Example

In the following we want to prepare our approach and the general result of the next section by an example. To this end, we assume a directed graph G to be given and consider the task of computing the set of all absorbant subsets of its vertex set X, where a set of vertices $Y \in 2^X$ is *absorbant* if from each vertex outside of Y there starts an edge that leads to a vertex inside of Y. For solving this problem relation-algebraically, we suppose $R : X \leftrightarrow X$ to be the relation of G, that is, for all vertices $x, y \in X$ we suppose that $R_{x,y}$ iff there is an edge from x to y in G. Our goal is to compute a vector of type $2^X \leftrightarrow \mathbf{1}$ that models the subset $\mathcal{A}_R$ of 2^X consisting of all absorbant subsets of X.

A 'direct' approach to a solution, which has been applied in case of many problems from rather different areas (as already mentioned in the introduction, cf. again the references given there), is as follows: Given an arbitrary set $Y \in 2^X$, we start with a formal logical description of Y to be an absorbant set of G. Then we transform this formula step-by-step into the form $\mathfrak{v}_Y$, where $\mathfrak{v}$ is a relation-algebraic expression of type $2^X \leftrightarrow \mathbf{1}$ (that is, a *vector expression*) built over the relation R and the vocabulary (i.e., the basic relations and operations) introduced in Sect. 2. Because of the definition of the equality of relations, finally, $\mathfrak{v}$ specifies the desired vector. Below there is the calculation of the relationship $\mathfrak{v}_Y$, where the variables x and y of all quantifications range over the set X:

$$
\begin{array}{rlr}
Y \in \mathcal{A}_R & \iff \forall x : x \notin Y \to \exists y : y \in Y \land R_{x,y} & \text{logical description} \\
& \iff \forall x : \neg \mathsf{M}_{x,Y} \to \exists y : \mathsf{M}_{y,Y} \land R_{x,y} & \text{def. } \mathsf{M} \\
& \iff \forall x : \neg \mathsf{M}_{x,Y} \to (R;\mathsf{M})_{x,Y} & \text{def. composition} \\
& \iff \neg\exists x : \neg(\mathsf{M}_{x,Y} \lor (R;\mathsf{M})_{x,Y}) & \\
& \iff \neg\exists x : \neg(\mathsf{M} \cup R;\mathsf{M})_{x,Y} & \text{def. union} \\
& \iff \neg\exists x : \mathsf{L}_{\perp,x} \land \overline{\mathsf{M} \cup R;\mathsf{M}}_{x,Y} & \text{def. complement and } \mathsf{L} \\
& \iff \neg(\mathsf{L};\overline{\mathsf{M} \cup R;\mathsf{M}})_{\perp,Y} & \text{def. composition} \\
& \iff \overline{\mathsf{L};\overline{\mathsf{M} \cup R;\mathsf{M}}}_{\perp,Y} & \text{def. complement} \\
& \iff \overline{\mathsf{L};\overline{\mathsf{M} \cup R;\mathsf{M}}}^{\mathsf{T}}_{Y,\perp} & \text{def. transposition} \\
& \iff \overline{\mathsf{L};\overline{\mathsf{M} \cup R;\mathsf{M}}}^{\mathsf{T}}_{Y} & \text{abbreviatory notation}
\end{array}
$$

This shows $\mathcal{A}_R = \mathit{set}(\mathfrak{v})$, with $\mathfrak{v} := \overline{\mathsf{L};\overline{\mathsf{M} \cup R;\mathsf{M}}}^{\mathsf{T}}$.

Now, we use an 'indirect' approach. Instead of a vector expression of type $2^X \leftrightarrow \mathbf{1}$ we calculate, given a vector $s : X \leftrightarrow \mathbf{1}$, a relation-algebraic formula that holds iff s models an absorbant subset of X. The corresponding calculation (where the variables x and y of all quantifications range again over X) is a lot simpler than the above one and the result is known (see e.g., [18]):

$$
\begin{array}{rlr}
\mathit{set}(s) \in \mathcal{A}_R & \iff \forall x : x \notin \mathit{set}(s) \to \exists y : y \in \mathit{set}(s) \land R_{x,y} & \text{logical description} \\
& \iff \forall x : \neg s_x \to \exists y : s_y \land R_{x,y} & \text{def. } \mathit{set}(s) \\
& \iff \forall x : \neg s_x \to (R;s)_x & \text{def. composition} \\
& \iff \forall x : \overline{s}_x \to (R;s)_x & \text{def. complement} \\
& \iff \overline{s} \subseteq R;s & \text{def. inclusion}
\end{array}
$$

It is known that with the help of the Tarski rule of [18] (saying that $S \neq \mathsf{O}$ implies $\mathsf{L};S;\mathsf{L} = \mathsf{L}$, for all relations S and all universal relations of appropriate type) each Boolean combination of relation-algebraic inclusions can be transformed into an equivalent relation-algebraic equation with $\mathsf{L} : \mathbf{1} \leftrightarrow \mathbf{1}$ as one of its sides.

In our example the equation is calculated as follows:

$$\overline{s} \subseteq R;s \iff \overline{s \cup R;s} = \mathsf{O} \iff \mathsf{L};\overline{s \cup R;s} = \mathsf{O} \iff \overline{\mathsf{L};\overline{s \cup R;s}} = \mathsf{L}$$

Only the direction "$\Leftarrow$" of the second step needs an explanation. The proof uses contraposition and assumes $\overline{s \cup R;s} \neq \mathsf{O}$. The Tarski rule then yields $\mathsf{L};\overline{s \cup R;s};\mathsf{L} = \mathsf{L}$, i.e., $\mathsf{L};\overline{s \cup R;s};\mathsf{L} \neq \mathsf{O}$, and, since $\overline{s \cup R;s}$ is a vector, we get $\mathsf{L};\overline{s \cup R;s} \neq \mathsf{O}$ as desired.

The above computations show that the vector expression $\mathfrak{b} := \overline{\mathsf{L};\overline{s \cup R;s}}$ of type $\mathbf{1} \leftrightarrow \mathbf{1}$ evaluates to L (i.e., 'true') if s models an absorbant subset of X and to O (i.e., 'false') if this is not the case. A specific feature of $\overline{s}$ and $R;s$ and, hence, also of $\mathfrak{b}$ is that all are built from s using complement, union and left-composition with another relation only. Consequently, if we replace s in $\mathfrak{b}$ with $\mathsf{M} : X \leftrightarrow 2^X$, then we get a relation $\overline{\mathsf{L};\overline{\mathsf{M} \cup R;\mathsf{M}}} : \mathbf{1} \leftrightarrow 2^X$ such that its Y-column is $\mathsf{L} : \mathbf{1} \leftrightarrow \mathbf{1}$ iff the Y-column of M models an absorbant set, for all $Y \in 2^X$. This shows again $\mathcal{A}_R = set(\overline{\mathsf{L};\overline{\mathsf{M} \cup R;\mathsf{M}}}^{\mathsf{T}})$.

5 Specific Vector Expressions and a General Result

Via an example, in the last section we have demonstrated how to obtain from an inclusion $\mathfrak{w} \subseteq \mathfrak{u}$ that specifies an arbitrary vector $s : X \leftrightarrow \mathbf{1}$ as a model of a set from a subset $\mathcal{A}$ of 2^X a vector expression $\mathfrak{b}$ of type $\mathbf{1} \leftrightarrow \mathbf{1}$ (a truth-value) that immediately yields a vector expression $\mathfrak{v}$ of type $2^X \leftrightarrow \mathbf{1}$ such that $\mathfrak{v}$ models $\mathcal{A}$ and is equal to the result of the direct approach. To get $\mathfrak{v}$, we only had to replace all occurrences of s in $\mathfrak{b}$ with $\mathsf{M} : X \leftrightarrow 2^X$ and then to transpose the result. Decisive for the correctness of the step from $\mathfrak{b}$ to $\mathfrak{v}$ was the syntactic structure of both sides of the original inclusion. They are built from s using complement and left-composition only. In this section we prove that the approach works in general and also when other vectors, unions and intersections are allowed as means for constructing vector expressions. To distinguish our syntactic vector expressions of type $Y \leftrightarrow \mathbf{1}$ from B. Kehden's vector predicates, which are mappings from vectors to $[\mathbf{1} \leftrightarrow \mathbf{1}]$ in the usual mathematical sense and correspond to the above $\mathfrak{b}$, we call them column-wise extendible vector expressions.

Definition 1. *Given $s : X \leftrightarrow \mathbf{1}$, the set* $\mathsf{VE}(s)$ *of typed* column-wise extendible vector expressions *over s is inductively defined as follows:*

(a) We have $s \in \mathsf{VE}(s)$ and its type is $X \leftrightarrow \mathbf{1}$.
(b) If $v : Y \leftrightarrow \mathbf{1}$ is different from s (as a vector), then $v \in \mathsf{VE}(s)$ and its type is $Y \leftrightarrow \mathbf{1}$.
(c) If $\mathfrak{v} \in \mathsf{VE}(s)$ is of type $Y \leftrightarrow \mathbf{1}$, then $\overline{\mathfrak{v}} \in \mathsf{VE}(s)$ and its type is $Y \leftrightarrow \mathbf{1}$.
(d) If $\mathfrak{v}, \mathfrak{w} \in \mathsf{VE}(s)$ are of type $Y \leftrightarrow \mathbf{1}$, then $\mathfrak{v} \cup \mathfrak{w} \in \mathsf{VE}(s)$ and $\mathfrak{v} \cap \mathfrak{w} \in \mathsf{VE}(s)$ and their types are $Y \leftrightarrow \mathbf{1}$.
(e) If $\mathfrak{v} \in \mathsf{VE}(s)$ is of type $Y \leftrightarrow \mathbf{1}$ and $\mathfrak{R}$ is a relation-algebraic expression of type $Z \leftrightarrow Y$ in which s does not occur, then $\mathfrak{R};\mathfrak{v} \in \mathsf{VE}(s)$ and its type is $Z \leftrightarrow \mathbf{1}$.

In a column-wise extendible vector expression over s the vector s can be seen as a variable in the logical sense. Using this interpretation, next we define how to replace s by a relation of appropriate type.

Definition 2. *Given $s : X \leftrightarrow \mathbf{1}$, $\mathfrak{v} \in \mathsf{VE}(s)$ and $R : X \leftrightarrow Z$, we define $\mathfrak{v}[R/s]$ to be the relation-algebraic expression obtained by* replacing *each occurrence of s in $\mathfrak{v}$ with R. Using induction on the structure of $\mathfrak{v}$, formally, we define*

(a) $s[R/s] = R$,
(b) $v[R/s] = v; \mathsf{L}$, *with* $\mathsf{L} : \mathbf{1} \leftrightarrow Z$,
(c) $\overline{\mathfrak{v}}[R/s] = \overline{\mathfrak{v}[R/s]}$,
(d) $(\mathfrak{v} \cup \mathfrak{w})[R/s] = \mathfrak{v}[R/s] \cup \mathfrak{w}[R/s]$ *and* $(\mathfrak{v} \cap \mathfrak{w})[R/s] = \mathfrak{v}[R/s] \cap \mathfrak{w}[R/s]$,
(e) $(\mathfrak{R}; \mathfrak{v})[R/s] = \mathfrak{R}; (\mathfrak{v}[R/s])$,

for all vectors v different from s, vector expressions $\mathfrak{v}, \mathfrak{w} \in \mathsf{VE}(s)$ and relation-algebraic expressions $\mathfrak{R}$ of appropriate type in which s does not occur.

The next two results state basic properties of the replacement operation. Lemma 1 concerns the type of the result and Lemma 2 shows that the column-wise extendible vector expressions are closed under replacements.

Lemma 1. *For all $s : X \leftrightarrow \mathbf{1}$, $\mathfrak{v} \in \mathsf{VE}(s)$ of type $Y \leftrightarrow \mathbf{1}$ and $R : X \leftrightarrow Z$ the type of $\mathfrak{v}[R/s]$ is $Y \leftrightarrow Z$.*

Proof. We use induction on the structure of $\mathfrak{v}$.

The first case of the induction base is that $\mathfrak{v}$ equals s. Then we have $Y = X$. Using the definition of replacement we obtain $s[R/s] = R$ and, hence, the type of $s[R/s]$ is $X \leftrightarrow Z$, i.e., $Y \leftrightarrow Z$. The second case is that $\mathfrak{v}$ equals a vector $v : Y \leftrightarrow \mathbf{1}$ not being s. Here the definition of replacement yields $v[R/s] = v; \mathsf{L}$, with $\mathsf{L} : \mathbf{1} \leftrightarrow Z$, and, thus, $Y \leftrightarrow Z$ as its type using the typing rule of composition.

The induction step follows from the typing rules of complement, union, intersection and composition. We only consider the case that $\mathfrak{v}$ is a complement $\overline{\mathfrak{w}}$, with $\mathfrak{w} \in \mathsf{VE}(s)$ having type $Y \leftrightarrow \mathbf{1}$. Here we get $\overline{\mathfrak{w}}[R/s] = \overline{\mathfrak{w}[R/s]}$ due to the definition of replacement. By the induction hypothesis $\mathfrak{w}[R/s]$ has type $Y \leftrightarrow Z$. This yields $Y \leftrightarrow Z$ as type of $\overline{\mathfrak{w}[R/s]}$ and, hence, also as type of $\overline{\mathfrak{w}}[R/s]$. □

Lemma 2. *For all $s : X \leftrightarrow \mathbf{1}$ and $\mathfrak{v}, \mathfrak{w} \in \mathsf{VE}(s)$, with $Y \leftrightarrow \mathbf{1}$ as type of $\mathfrak{v}$ and $X \leftrightarrow \mathbf{1}$ as type of $\mathfrak{w}$, we have $\mathfrak{v}[\mathfrak{w}/s] \in \mathsf{VE}(s)$.*

Proof. From the type of $\mathfrak{w}$ and Lemma 1 we get $Y \leftrightarrow \mathbf{1}$ as type of $\mathfrak{v}[\mathfrak{w}/s]$. We use again induction on the structure of $\mathfrak{v}$ to prove the claim.

For the induction base, let first $\mathfrak{v}$ be s and assume $\mathfrak{w} \in \mathsf{VE}(s)$. By the definition of replacement then we have $s[\mathfrak{w}/s] = \mathfrak{w}$ and, hence, $s[\mathfrak{w}/s] \in \mathsf{VE}(s)$. Next, let $\mathfrak{v}$ be a vector v different from s and assume $\mathfrak{w} \in \mathsf{VE}(s)$. Here the definition of replacement shows $v[\mathfrak{w}/s] = v; \mathsf{L}$, with $\mathsf{L} : \mathbf{1} \leftrightarrow \mathbf{1}$ because of the target $\mathbf{1}$ of $\mathfrak{w}$, which in turn yields $v[\mathfrak{w}/s] = v$ so that $v[\mathfrak{w}/s] \in \mathsf{VE}(s)$.

The first case of the induction step is that $\mathfrak{v}$ is a complement $\overline{\mathfrak{u}}$, with $\mathfrak{u} \in \mathsf{VE}(s)$. Assume $\mathfrak{w} \in \mathsf{VE}(s)$. Then the definition of replacement shows $\overline{\mathfrak{u}}[\mathfrak{w}/s] = \overline{\mathfrak{u}[\mathfrak{w}/s]}$.

The induction hypothesis yields $\mathfrak{u}[\mathfrak{w}/s] \in \mathsf{VE}(s)$ and this implies $\overline{\mathfrak{u}[\mathfrak{w}/s]} \in \mathsf{VE}(s)$, that is $\overline{\mathfrak{u}}[\mathfrak{w}/s] \in \mathsf{VE}(s)$. In the same way the remaining cases of $\mathfrak{v}$ being of the form $\mathfrak{u} \cup \mathfrak{x}$, $\mathfrak{u} \cap \mathfrak{x}$, $\mathfrak{R}; \mathfrak{u}$, with $\mathfrak{u}, \mathfrak{x} \in \mathsf{VE}(s)$ and $\mathfrak{R}$ being a relation-algebraic expression of appropriate type in which s does not occur, can be treated. □

The next lemma states the decisive property of column-wise extendible vector expressions, which is also the reason for this name. Informally, it states the following: Assume a column-wise extendible vector expression $\mathfrak{v}$ over $s : X \leftrightarrow \mathbf{1}$ to be given and $Y \leftrightarrow \mathbf{1}$ to be its type. Furthermore, let $\Phi : [X \leftrightarrow \mathbf{1}] \to [Y \leftrightarrow \mathbf{1}]$ be the mapping induced by s via $\Phi(s) = \mathfrak{v}$. If all columns of $\mathsf{M} : X \leftrightarrow 2^X$ are interpreted as vectors of type $X \leftrightarrow \mathbf{1}$ and those of $\mathfrak{v}[\mathsf{M}/s] : Y \leftrightarrow 2^X$ as vectors of type $Y \leftrightarrow \mathbf{1}$, then Φ maps the Z-column of M to the Z-column of $\mathfrak{v}[\mathsf{M}/s]$, for all $Z \in 2^X$. I.e., column-wise extendible vector expressions allow to extend Φ to a mapping that computes, by means of $\mathfrak{v}[\mathsf{M}/s]$, the vectors $\Phi(v)$ for all columns v of M in parallel.

Lemma 3. *For all $s : X \leftrightarrow \mathbf{1}$, $\mathfrak{v} \in \mathsf{VE}(s)$ of type $Y \leftrightarrow \mathbf{1}$, $x \in Y$ and $\mathsf{M} : X \leftrightarrow 2^X$ we have $\mathfrak{v}_x$ iff $\mathfrak{v}[\mathsf{M}/s]_{x,set(s)}$.*

Proof. Lemma 1 implies $Y \leftrightarrow 2^X$ as type of $\mathfrak{v}[\mathsf{M}/s]$. We use again induction on the structure of $\mathfrak{v}$ to prove the claim.

For the induction base, let first $\mathfrak{v}$ be s. Then this implies $X = Y$ and we get for all $x \in Y$ the result as follows:

$$\begin{aligned} s_x &\iff x \in set(s) && x \in X \text{anddef. } set(s) \\ &\iff \mathsf{M}_{x,set(s)} && \text{def. } \mathsf{M} \\ &\iff s[\mathsf{M}/s]_{x,set(s)} && \text{def. replacement} \end{aligned}$$

Secondly, we assume that $\mathfrak{v}$ is a vector $v : Y \leftrightarrow \mathbf{1}$ different from s. In this case we get for all $x \in Y$ the following equivalence, where the variable y of the existential quantification ranges over the set $\mathbf{1}$:

$$\begin{aligned} v_x &\iff v_{x,\perp} \wedge \mathsf{L}_{\perp,set(s)} && \text{abbreviatory notation, def. } \mathsf{L} \\ &\iff \exists y : v_{x,y} \wedge \mathsf{L}_{y,set(s)} \\ &\iff (v;\mathsf{L})_{x,set(s)} && \text{def. composition} \\ &\iff v[\mathsf{M}/s]_{x,set(s)} && \text{def. replacement} \end{aligned}$$

The first case of the induction step is again that $\mathfrak{v}$ is a complement $\overline{\mathfrak{w}}$, with $\mathfrak{w} \in \mathsf{VE}(s)$. Here we get for all $x \in Y$ the result as follows:

$$\begin{aligned} \overline{\mathfrak{w}}_x &\iff \neg\mathfrak{w}_x && \text{def. complement} \\ &\iff \neg\mathfrak{w}[\mathsf{M}/s]_{x,set(s)} && \text{induction hypothesis} \\ &\iff \overline{\mathfrak{w}[\mathsf{M}/s]}_{x,set(s)} && \text{def. complement} \\ &\iff \overline{\mathfrak{w}}[\mathsf{M}/s]_{x,set(s)} && \text{def. replacement} \end{aligned}$$

Next, we assume $\mathfrak{v}$ to be a union $\mathfrak{w} \cup \mathfrak{u}$ such that $\mathfrak{w}, \mathfrak{u} \in \mathsf{VE}(s)$. In this case the following calculation shows for all $x \in Y$ the claim:

$$\begin{aligned} (\mathfrak{w} \cup \mathfrak{u})_x &\iff \mathfrak{w}_x \vee \mathfrak{u}_x && \text{def. union} \\ &\iff \mathfrak{w}[\mathsf{M}/s]_{x,set(s)} \vee \mathfrak{u}[\mathsf{M}/s]_{x,set(s)} && \text{induction hypothesis} \\ &\iff (\mathfrak{w}[\mathsf{M}/s] \cup \mathfrak{u}[\mathsf{M}/s])_{x,set(s)} && \text{def. union} \\ &\iff (\mathfrak{w} \cup \mathfrak{u})[\mathsf{M}/s]_{x,set(s)} && \text{def. replacement} \end{aligned}$$

In the same way the case of $\mathfrak{v}$ being an intersection $\mathfrak{w} \cap \mathfrak{u}$ can be treated. The final case is that $\mathfrak{v}$ equals a composition $\mathfrak{R};\mathfrak{w}$, with $\mathfrak{w} \in \mathsf{VE}(s)$ and a relation-algebraic expression $\mathfrak{R}$ of type $Z \leftrightarrow Y$ in which s does not occur. Then for all $x \in Y$ a proof of the claim is as follows, where the variable y of the existential quantifications ranges over the set Y:

$$\begin{aligned}
(\mathfrak{R};\mathfrak{w})_x &\iff \exists y : \mathfrak{R}_{x,y} \wedge \mathfrak{w}_y && \text{def. composition}\\
&\iff \exists y : \mathfrak{R}_{x,y} \wedge \mathfrak{w}[\mathsf{M}/s]_{y,set(s)} && \text{induction hypothesis}\\
&\iff (\mathfrak{R};(\mathfrak{w}[\mathsf{M}/s]))_{x,set(s)} && \text{def. composition}\\
&\iff (\mathfrak{R};\mathfrak{w})[\mathsf{M}/s]_{x,set(s)} && \text{def. replacement}
\end{aligned}$$ □

There is also a purely algebraic version of this result. It says that for all *injective* ($R;R^\mathsf{T} \subseteq \mathsf{I}$) and *surjective* ($\mathsf{L} \subseteq \mathsf{L};R$) relations R (especially for all points $p : 2^X \leftrightarrow \mathbf{1}$ in the sense of [18]) it holds $\mathfrak{v}[\mathsf{M};R/s] = \mathfrak{v}[\mathsf{M}/s];R$. The proof by induction uses $(Q \cap S);R = Q;R \cap S;R$ and $\overline{Q;R} = \overline{Q};R$. But we won't go into detail, since this fact does not help us to reach our original goal of proving that the indirect approach of the example of Sect. 4 works in general, provided both sides of the inclusion specifying a property of sets/vectors are column-wise extendible vector expressions. Here is a proof of our goal using Lemma 3.

Theorem 1. *Assume $\mathcal{A}$ to be specified as $\mathcal{A} = \{set(s) \mid s : X \leftrightarrow \mathbf{1} \wedge \mathfrak{w} \subseteq \mathfrak{u}\}$, with $\mathfrak{w}, \mathfrak{u} \in \mathsf{VE}(s)$ being of type $Y \leftrightarrow \mathbf{1}$. Using $\mathsf{L} : \mathbf{1} \leftrightarrow Y$ and $\mathsf{M} : X \leftrightarrow 2^X$, then*

$$\mathcal{A} = set(\mathfrak{v}), \ \text{where } \mathfrak{v} := \overline{\mathsf{L};(\mathfrak{w}[\mathsf{M}/s] \cap \overline{\mathfrak{u}[\mathsf{M}/s]})}^{\mathsf{T}}.$$

Proof. Lemma 1 and the typing rules of the relational operations yield $2^X \leftrightarrow \mathbf{1}$ as type of $\mathfrak{v}$. Now, let an arbitrary set $Z \in 2^X$ be given. Since $set : [X \leftrightarrow \mathbf{1}] \to 2^X$ is surjective, there exists $s : X \leftrightarrow \mathbf{1}$ with $set(s) = Z$. Using s, we can calculate as given below, where the variable x of all quantifications ranges over the set Y:

$$\begin{aligned}
Z \in \mathcal{A} &\iff set(s) \in \mathcal{A} && \text{as } Z = set(s)\\
&\iff \mathfrak{w} \subseteq \mathfrak{u} && \text{specification of } \mathcal{A}\\
&\iff \forall x : \mathfrak{w}_x \to \mathfrak{u}_x && \text{def. inclusion}\\
&\iff \forall x : \mathfrak{w}[\mathsf{M}/s]_{x,set(s)} \to \mathfrak{u}[\mathsf{M}/s]_{x,set(s)} && \text{Lemma 5.3}\\
&\iff \forall x : \mathfrak{w}[\mathsf{M}/s]_{x,Z} \to \mathfrak{u}[\mathsf{M}/s]_{x,Z} && \text{as } Z = set(s)\\
&\iff \neg\exists x : \mathfrak{w}[\mathsf{M}/s]_{x,Z} \wedge \neg\mathfrak{u}[\mathsf{M}/s]_{x,Z} && \\
&\iff \neg\exists x : \mathfrak{w}[\mathsf{M}/s]_{x,Z} \wedge \overline{\mathfrak{u}[\mathsf{M}/s]}_{x,Z} && \text{def. complement}\\
&\iff \neg\exists x : \mathsf{L}_{\perp,x} \wedge (\mathfrak{w}[\mathsf{M}/s] \cap \overline{\mathfrak{u}[\mathsf{M}/s]})_{x,Z} && \text{def. intersection and } \mathsf{L}\\
&\iff \neg(\mathsf{L};(\mathfrak{w}[\mathsf{M}/s] \cap \overline{\mathfrak{u}[\mathsf{M}/s]}))_{\perp,Z} && \text{def. composition}\\
&\iff \overline{\mathsf{L};(\mathfrak{w}[\mathsf{M}/s] \cap \overline{\mathfrak{u}[\mathsf{M}/s]})}_{\perp,Z} && \text{def. complement}\\
&\iff \overline{\mathsf{L};(\mathfrak{w}[\mathsf{M}/s] \cap \overline{\mathfrak{u}[\mathsf{M}/s]})}^{\mathsf{T}}_{Z,\perp} && \text{def. transposition}\\
&\iff \overline{\mathsf{L};(\mathfrak{w}[\mathsf{M}/s] \cap \overline{\mathfrak{u}[\mathsf{M}/s]})}^{\mathsf{T}}_{Z} && \text{abbreviatory notation}\\
&\iff Z \in set(\overline{\mathsf{L};(\mathfrak{w}[\mathsf{M}/s] \cap \overline{\mathfrak{u}[\mathsf{M}/s]})}^{\mathsf{T}}) && \text{def. } set(\cdots)
\end{aligned}$$

Now, the definition of $\mathfrak{v}$ shows the desired result $\mathcal{A} = set(\mathfrak{v})$. □

Applied to the inclusion $\overline{s} \subseteq R;s$ of Sect. 4, Theorem 1 yields $\overline{\mathsf{L};(\overline{\mathsf{M}} \cap \overline{R;\mathsf{M}})}^{\mathsf{T}}$ as model of the set $\mathcal{A}_R$ of all absorbant sets. By a de Morgan law, however, this vector is equal to the vector we have obtained in Sect. 4.

Using the $\cap$-distributivity of the mapping *set*, Theorem 1 can be generalized as follows: If the set $\mathcal{A}$ is specified as $\mathcal{A} = \{set(s) \mid s : X \leftrightarrow \mathbf{1} \wedge \bigwedge_{i=1}^{k} \mathfrak{w}_i \subseteq \mathfrak{u}_i\}$, with $\mathfrak{w}_i, \mathfrak{u}_i \in \mathsf{VE}(s)$ for all $1 \in \{1 \ldots, k\}$, then we get

$$\mathcal{A} = set(\mathfrak{v}),\ where\ \mathfrak{v} := \bigcap_{i=1}^{k} \overline{\mathsf{L};(\mathfrak{w}_i[\mathsf{M}/s] \cap \overline{\mathfrak{u}_i[\mathsf{M}/s]})}^{\mathsf{T}}.$$

For the specific case $k = 2$, $\mathfrak{w}_1 = \mathfrak{u}_2$ and $\mathfrak{w}_2 = \mathfrak{u}_1 = s$ this fact, a de Morgan law and the definitions of replacement and symmetric difference imply that from $\mathcal{A} = \{set(s) \mid s : X \leftrightarrow \mathbf{1} \wedge \mathfrak{w} = s\}$, with $\mathfrak{w} \in \mathsf{VE}(s)$, it follows

$$\mathcal{A} = set(\mathfrak{v}),\ where\ \mathfrak{v} := \overline{\mathsf{L};(\mathfrak{w}[\mathsf{M}/s] \oplus \mathsf{M})}^{\mathsf{T}}.$$

Hence, $\mathcal{A}$ consists precisely of those sets $set(s)$, for which the vector s is a fixed point of the mapping $\Phi : [X \leftrightarrow \mathbf{1}] \to [X \leftrightarrow \mathbf{1}]$, defined by $\Phi(s) = \mathfrak{w}$. Fixed point specifications of vectors appear in many applications; see e.g., [18,19]. We will provide examples in the next section.

6 Some Further Applications

By means of Theorem 1 many of the relation-algebraic specifications of specific sets of vertices of a graph G given in [18,19] immediately lead to vector models of the set of all these sets. Below we present some examples. In doing so, we assume again $R : X \leftrightarrow X$ to be the relation of the graph G in the sense of Sect. 4 and $\mathsf{M} : X \leftrightarrow 2^X$ to be a membership relation.

A set $Y \in 2^X$ is *stable* if no two vertices of Y are connected by an edge and Y is a *kernel* if it is absorbant and stable. For all $s : X \leftrightarrow \mathbf{1}$ this means that $set(s)$ is stable iff $R;s \subseteq \overline{s}$ and $set(s)$ is a kernel iff $\overline{R;s} = s$. Since all sides of these relation-algebraic formulae are from $\mathsf{VE}(s)$, Theorem 1 yields the vector

$$\mathfrak{stab} := \overline{\mathsf{L};(R;\mathsf{M} \cap \mathsf{M})}^{\mathsf{T}} : 2^X \leftrightarrow \mathbf{1}$$

as model of the set of all stable sets of G and, in its extended form for fixed points specifications $\mathfrak{w} = s$, the vector

$$\mathfrak{ker} := \overline{\mathsf{L};(\overline{R;\mathsf{M}} \oplus \mathsf{M})}^{\mathsf{T}} : 2^X \leftrightarrow \mathbf{1}$$

as model of the set of all kernels of G. For the last two examples we assume the graph G to be undirected, relation-algebraically described by R to be *symmetric* ($R \subseteq R^{\mathsf{T}}$) and *irreflexive* ($R \subseteq \overline{\mathsf{I}}$). Then a set $Y \in 2^X$ is a *vertex cover* if it contains at least one endpoint of each edge of G. In terms of relation algebra this means that $s : X \leftrightarrow \mathbf{1}$ models a vertex cover iff $R;\overline{s} \subseteq s$. Both sides of this inclusion are from $\mathsf{VE}(s)$ and by Theorem 1 the vector

$$\mathfrak{cov} := \overline{\mathsf{L};(R;\overline{\mathsf{M}} \cap \overline{\mathsf{M}})}^{\mathsf{T}} : 2^X \leftrightarrow \mathbf{1}$$

models the set of all vertex covers of G. Finally, a set $Y \in 2^X$ is a *clique* of G if each pair of different vertices of Y is connected via an edge. This means that the set complement $\overline{Y} := X \setminus Y$ is a vertex cover of the complement graph of G, or, relation-algebraically, that $s : X \leftrightarrow \mathbf{1}$ models a clique iff $\overline{R \cup \mathsf{I}};s \subseteq \overline{s}$. Since $\overline{R \cup \mathsf{I}};s$ and $\overline{s}$ are from $\mathsf{VE}(s)$, again Theorem 1 is applicable and a vector that models the set of all cliques of G is:

$$\mathfrak{clique} := \overline{\mathsf{L};(\overline{R \cup \mathsf{I}};\mathsf{M} \cap \mathsf{M})}^{\mathsf{T}} : 2^X \leftrightarrow \mathbf{1}$$

In most applications not all of the particular sets of vertices we have introduced so far are of interest, but only extremal ones. For instance, in case of absorbant sets and vertex covers one is mainly interested in the minimum ones and in case of stable sets and cliques one is mainly interested in the maximum ones. Having computed the above vectors, a selection of extremal sets by relation-algebraic means is rather simple. Given a *reflexive* ($\mathsf{I} \subseteq Q$) and *transitive* ($Q;Q \subseteq Q$) relation $Q : Z \leftrightarrow Z$, that is, a pre-ordered set (Z, Q), and $v : Z \leftrightarrow \mathbf{1}$, the column-wise extendible vector expression $min_Q(v) := v \cap \overline{\overline{R};v}$ models the set of all least elements of $set(v)$ w.r.t. Q; see [18]. Using it in combination with the size-comparison relation $\mathsf{S} : 2^X \leftrightarrow 2^X$ of Sect. 2, for instance, the vector $min_{\mathsf{S}}(\mathfrak{cov}) : 2^X \leftrightarrow \mathbf{1}$ models the set of all minimum vertex covers of G and the vector $min_{\mathsf{S}^{\mathsf{T}}}(\mathfrak{clique}) : 2^X \leftrightarrow \mathbf{1}$ models the set of all maximum cliques of G.

Now, we apply our technique to a problem from order theory. Assume (X, R) to be an ordered set, that is, $R : X \leftrightarrow X$ to be reflexive, transitive and *antisymmetric* ($R \cap R^{\mathsf{T}} \subseteq \mathsf{I}$). A set $Y \in 2^X$ is a *Dedekind cut* if it coincides with the lower bounds of its upper bounds. How to specify lower and upper bounds relation-algebraically is known. Given $s : X \leftrightarrow \mathbf{1}$, the column-wise extendible vector expression $lbds_R(s) := \overline{\overline{R};s}$ models the set of all lower bounds of $set(s)$ and $ubds_R(s) := lbds_{R^{\mathsf{T}}}(s)$ does the same for the set of all upper bounds. So, s models a Dedekind cut iff $lbds_R(ubds_R(s)) = s$. Due to Lemma 2 both sides of this fixed point specification are from $\mathsf{VE}(s)$ and due to the extension of Theorem 1 to fixed point specifications the vector (where $\mathsf{M} : X \leftrightarrow 2^X$)

$$\mathfrak{cut} := \overline{\mathsf{L};(lbds_R(ubds_R(\mathsf{M})) \oplus \mathsf{M})}^{\mathsf{T}} : 2^X \leftrightarrow \mathbf{1}$$

models the set $\mathcal{C}_R$ of all Dedekind cuts of (X, R). Using the technique described in Sect. 3, the relation $S := \mathsf{M};inj(\mathfrak{cut})^{\mathsf{T}} : X \leftrightarrow \mathcal{C}_R$ column-wisely enumerates the set $\mathcal{C}_R$ and the relation $\overline{S^{\mathsf{T}};\overline{S}} : \mathcal{C}_R \leftrightarrow \mathcal{C}_R$ relation-algebraically specifies the inclusion on $\mathcal{C}_R$, that is, the order relation of the cut completion $(\mathcal{C}_R, \subseteq)$ of (X, R). In [2] the direct approach is used to develop a vector model of the set

$\mathcal{C}_R$. The corresponding relation-algebraic expression contains a sub-expression of type $2^X \leftrightarrow 2^X$. In the above specification the 'largest' type of an intermediate result is $X \leftrightarrow 2^X$ and this makes, as also RelView experiments have shown, the new solution superior to that of [2].

In the examples we have treated so far, the types of $\mathfrak{w}, \mathfrak{u} \in \mathsf{VE}(s)$ in the inclusions $\mathfrak{w} \subseteq \mathfrak{u}$ always equal the type of s. Next, we present an example with different types. We consider the static structure of a *Petri net* N and model it by two relations $R : P \leftrightarrow T$, for the edges from the places to the transitions, and $S : T \leftrightarrow P$, for the edges from the transitions to the places. As shown in [2,3], many of the static properties of a Petri net can be tested using relation algebra. In the following we demonstrate how to compute *deadlocks*, which are sets of places $Y \in 2^P$ such that each predecessor of Y is also a successor of Y. The identification of deadlocks (and traps, the dual notion) might help identify certain structural and dynamic properties of the system modeled by the Petri net prior to implementing it.

Assume $R : P \leftrightarrow T$ and $S : T \leftrightarrow P$ as introduced above and let $s : P \leftrightarrow \mathbf{1}$ be given. Then $S; s : T \leftrightarrow \mathbf{1}$ models the set of all predecessors of $set(s)$ and $R^\mathsf{T}; s : T \leftrightarrow \mathbf{1}$ models the set of all successors of $set(s)$. As a consequence, s models a deadlock of N iff $S; s \subseteq R^\mathsf{T}; s$. Since $S; s$ and $R^\mathsf{T}; s$ are from $\mathsf{VE}(s)$, Theorem 1 is applicable and (with $\mathsf{M} : P \leftrightarrow 2^P$) we get the following vector model of the set of all deadlocks of N:

$$\mathfrak{dead} := \overline{\mathsf{L}; (S; \mathsf{M} \cap \overline{R^\mathsf{T}; \mathsf{M}})}^\mathsf{T} : 2^P \leftrightarrow \mathbf{1}$$

In practical applications, usually minimum non-empty deadlocks are of interest. A little reflection shows that the vector $(\mathsf{L}; \mathsf{M})^\mathsf{T} \cap \mathfrak{dead}$ models the set of all non-empty deadlocks of N and, hence, $min_\mathsf{S}((\mathsf{L}; \mathsf{M})^\mathsf{T} \cap \mathfrak{dead})$ models the net's set of all minimum non-empty deadlocks.

In the last series of examples we treat the computation of specific relations, that is, subsets of a set $2^{X \times Y}$ or, in our notation, of a set $[X \leftrightarrow Y]$. In doing so, we suppose $\pi : X \times Y \leftrightarrow X$ and $\rho : X \times Y \leftrightarrow Y$ as first and second projection relation, respectively, and $\mathsf{M} : X \times Y \leftrightarrow [X \leftrightarrow Y]$ as membership relation.

We start with *univalence* and assume $s : X \times Y \leftrightarrow \mathbf{1}$ to model a relation of type $X \leftrightarrow Y$. The next derivation is the first step towards an inclusion over $\mathsf{VE}(s)$ that specifies the univalence of $rel(s)$, i.e., of the relation modeled by s. Here the variables u, v of the universal quantifications range over the set $X \times Y$:

$$\begin{array}{lr}
rel(s)\text{univalent} & \\
\iff \forall u, v : rel(s)_{u_1,u_2} \wedge rel(s)_{v_1,v_2} \wedge u_1 = v_1 \rightarrow u_2 = v_2 & \text{logical description} \\
\iff \forall u, v : s_u \wedge s_v \wedge u_1 = v_1 \rightarrow u_2 = v_2 & \text{def. } rel(s) \\
\iff \forall u, v : (s; s^\mathsf{T})_{u,v} \rightarrow \neg(u_1 = v_1) \vee u_2 = v_2 & \\
\iff \forall u, v : (s; s^\mathsf{T})_{u,v} \rightarrow \overline{\pi; \pi^\mathsf{T}}_{u,v} \vee (\rho; \rho^\mathsf{T})_{u,v} & \text{def. } \pi \text{and } \rho \\
\iff \forall u, v : (s; s^\mathsf{T})_{u,v} \rightarrow (\overline{\pi; \pi^\mathsf{T}} \cup \rho; \rho^\mathsf{T})_{u,v} & \text{def. union} \\
\iff s; s^\mathsf{T} \subseteq \overline{\pi; \pi^\mathsf{T}} \cup \rho; \rho^\mathsf{T} & \text{def. inclusion}
\end{array}$$

The left-hand side $s; s^\mathsf{T}$ of the inclusion is not from $\mathsf{VE}(s)$ and, as a consequence, Theorem 1 is not applicable. But with the help of a Schröder equivalence (stating that $Q;R^{\mathsf{T}} \subseteq S$ iff $\overline{S};R \subseteq \overline{Q}$, for all relations Q, R and S of appropriate type; see [18]) and two de Morgan laws we get an equivalent inclusion with both sides from $\mathsf{VE}(s)$ as follows:

$$s;s^{\mathsf{T}} \subseteq \overline{\pi;\pi^{\mathsf{T}} \cup \rho;\rho^{\mathsf{T}}} \iff \overline{\overline{\pi;\pi^{\mathsf{T}} \cup \rho;\rho^{\mathsf{T}}}};s \subseteq \overline{s} \iff (\pi;\pi^{\mathsf{T}} \cap \overline{\rho;\rho^{\mathsf{T}}});s \subseteq \overline{s}$$

Now, Theorem 1 is applicable. It yields the following vector as model of the set of all univalent relations of type $X \leftrightarrow Y$:

$$\mathfrak{univ} := \overline{\mathsf{L};((\pi;\pi^{\mathsf{T}} \cap \overline{\rho;\rho^{\mathsf{T}}});\mathsf{M} \cap \overline{\mathsf{M}})}^{\mathsf{T}} : [X \leftrightarrow Y] \leftrightarrow \mathbf{1},$$

To get the same for the *total relations*, we assume again $s : X \times Y \leftrightarrow \mathbf{1}$ to model a relation of type $X \leftrightarrow Y$ and calculate as follows, where the variable x of the universal quantifications ranges over the set X and the variable u of the existential quantifications ranges over the set $X \times Y$:

$$\begin{aligned}
rel(s)\text{total} &\iff \forall x : \exists u : rel(s)_{u_1,u_2} \wedge x = u_1 && \text{logical description}\\
&\iff \forall x : \exists u : s_u \wedge x = u_1 && \text{def. } rel(s)\\
&\iff \forall x : \exists u : s_u \wedge \pi_{u,x} && \text{def. } \pi\\
&\iff \forall x : \exists u : \pi^{\mathsf{T}}_{x,u} \wedge s_u && \text{def. transposition}\\
&\iff \forall x : (\pi^{\mathsf{T}};s)_x && \text{def. composition}\\
&\iff \mathsf{L} \subseteq \pi^{\mathsf{T}};s && \text{def. } \mathsf{L} \text{ and inclusion}
\end{aligned}$$

Again Theorem 1 can be applied and, using $\mathsf{L} \cap \overline{\pi^{\mathsf{T}};\mathsf{M}} = \overline{\pi^{\mathsf{T}};\mathsf{M}}$, we get

$$\mathfrak{total} := \overline{\mathsf{L};\overline{\pi^{\mathsf{T}};\mathsf{M}}}^{\mathsf{T}} : [X \leftrightarrow Y] \leftrightarrow \mathbf{1}$$

as model of the set of all total relations of type $X \leftrightarrow Y$. Altogether, $\mathfrak{univ} \cap \mathfrak{total}$ models the set of all mappings of type $X \leftrightarrow Y$. If we interchange π and ρ in the specification of $\mathfrak{univ}$, then the result

$$\mathfrak{inj} := \overline{\mathsf{L};((\rho;\rho^{\mathsf{T}} \cap \overline{\pi;\pi^{\mathsf{T}}});\mathsf{M} \cap \overline{\mathsf{M}})}^{\mathsf{T}} : [X \leftrightarrow Y] \leftrightarrow \mathbf{1},$$

models the set of all *injective relations* of type $X \leftrightarrow Y$, and if we replace π by ρ in the specification of $\mathfrak{total}$, then the result

$$\mathfrak{surj} := \overline{\mathsf{L};\overline{\rho^{\mathsf{T}};\mathsf{M}}}^{\mathsf{T}} : [X \leftrightarrow Y] \leftrightarrow \mathbf{1}$$

models the set of all *surjective relations* of type $X \leftrightarrow Y$.

Also the computation of vertex colourings of undirected graphs, which in [17] is treated as an application of vector predicates, easily can be solved with our technique. Suppose $R : X \leftrightarrow X$ to be the symmetric and irreflexive relation of an undirected graph G and let Y be a set of colours. Then the mapping $c : X \to Y$ is a *vertex colouring* if from $R_{x,y}$ it follows $c(x) \neq c(y)$, for all $x, y \in X$. Hence, we have for all $s : X \times Y \leftrightarrow \mathbf{1}$, provided $rel(s)$ is a mapping, that

$$rel(s)\text{vertexcolouring} \iff \forall u, v : rel(s)_{u_1,u_2} \wedge rel(s)_{v_1,v_2} \wedge R_{u_1,v_1} \to u_2 \neq v_2,$$

where the variables u, v of the universal quantifications range over the set $X \times Y$. Similar to the case of univalence, the right-hand-side of this logical description can be shown as equivalent to $(\pi; R; \pi^{\mathsf{T}} \cap \rho; \rho^{\mathsf{T}}); s \subseteq \overline{s}$ and the vector

$$\mathfrak{colour} := \mathfrak{univ} \cap \mathfrak{total} \cap \overline{\mathsf{L}; ((\pi; R; \pi^{\mathsf{T}} \cap \rho; \rho^{\mathsf{T}}); \mathsf{M} \cap \mathsf{M})}^{\mathsf{T}} : [X \leftrightarrow Y] \leftrightarrow \mathbf{1},$$

hence, models the set of all vertex colourings of G with colours from Y (or $|Y|$-colourings) because of Theorem 1.

Now, suppose $X = Y$. Then the following vector models the set of all *permutations* on the set X:

$$\mathfrak{perm} := \mathfrak{univ} \cap \mathfrak{total} \cap \mathfrak{inj} \cap \mathfrak{surj} : [X \leftrightarrow X] \leftrightarrow \mathbf{1},$$

If we want to model the set of all *involutions* (*self-inverse permutations*) on X, we only need a further vector $\mathfrak{sym} : [X \leftrightarrow X] \leftrightarrow \mathbf{1}$ that models the set of all symmetric relations of type $X \leftrightarrow X$. Then $\mathfrak{perm} \cap \mathfrak{sym}$ solves the task. To get a specification of $\mathfrak{sym}$, let $s : X \times X \leftrightarrow \mathbf{1}$ be given. Then we can calculate as follows, where the variables u and v of all quantifications range over the set $X \times X$:

$$\begin{array}{ll}
& rel(s)\text{symmetric} \\
\Longleftrightarrow & \forall u : rel(s)_{u_1,u_2} \to rel(s)_{u_2.u_1} \qquad \text{logical description} \\
\Longleftrightarrow & \forall u : rel(s)_{u_1,u_2} \to \exists v : u_1 = v_2 \wedge u_2 = v_1 \wedge rel(s)_{v_1.v_2} \\
\Longleftrightarrow & \forall u : s_u \to \exists v : u_1 = v_2 \wedge u_2 = v_1 \wedge s_v \qquad \text{def.} rel(s) \\
\Longleftrightarrow & \forall u : s_u \to \exists v : (\pi; \rho^{\mathsf{T}})_{u,v} \wedge (\rho; \pi^{\mathsf{T}})_{u,v} \wedge s_v \qquad \text{def. } \pi \text{ and } \rho \\
\Longleftrightarrow & \forall u : s_u \to \exists v : (\pi; \rho^{\mathsf{T}} \cap \rho; \pi^{\mathsf{T}})_{u,v} \wedge s_v \qquad \text{def. intersection} \\
\Longleftrightarrow & \forall u : s_u \to ((\pi; \rho^{\mathsf{T}} \cap \rho; \pi^{\mathsf{T}}); s)_u \qquad \text{def. composition} \\
\Longleftrightarrow & s \subseteq (\pi; \rho^{\mathsf{T}} \cap \rho; \pi^{\mathsf{T}}); s \qquad \text{def. inclusion}
\end{array}$$

Both sides of the inclusion are from $\mathsf{VE}(s)$. Hence, Theorem 1 is applicable and leads to the following specification:

$$\mathfrak{sym} := \overline{\mathsf{L}; (\mathsf{M} \cap \overline{(\pi; \rho^{\mathsf{T}} \cap \rho; \pi^{\mathsf{T}}); \mathsf{M}})}^{\mathsf{T}} : [X \leftrightarrow X] \leftrightarrow \mathbf{1}$$

In a similar way a vector $\mathfrak{irrefl} : [X \leftrightarrow X] \leftrightarrow \mathbf{1}$ can be developed, which models the set of all irreflexive relations of type $X \leftrightarrow X$. Combining it with $\mathfrak{perm}$ allows to model the set of all *proper (i.e., fixed-point-free) permutations* by $\mathfrak{perm} \cap \mathfrak{irrefl}$. And $\mathfrak{perm} \cap \mathfrak{sym} \cap \mathfrak{irrefl}$ models the set of all *proper involutions.*

7 Implementation and Results of Practical Experiments

As already mentioned, RELVIEW is a ROBDD-based specific purpose computer algebra system for the manipulation and visualization of relations and relational prototyping and programming. It is written in the C programming language and makes full use of the X-windows GUI. The main purpose of the tool is the evaluation of relation-algebraic expressions. These are constructed from the relations of its workspace using pre-defined operations and tests, user-defined relational functions, and user-defined relational programs.

In RelView a relational function is defined as is customary in mathematics, i.e., as $f(X_1, \dots, X_n) = \mathfrak{R}$, where the X_i are variables for relations and the right-hand side $\mathfrak{R}$ is a relation-algebraic expression over the variables and the relations of the workspace. A relational program in RelView is much like a function procedure in Pascal-like programming languages, except that it is not able to modify the tool's workspace (that is, its applications are free of side effects) and it only uses relations as data type. A program starts with a head line containing the program name and the list of formal parameters, which stand for relations. Then the declaration of the local relational domains, functions, and variables follows. Domain declarations can be used to introduce projection relations $\pi : X \times Y \leftrightarrow X$ and $\rho : X \times Y \leftrightarrow Y$ in the case of direct products $X \times Y$ and injection relations $\iota : X \leftrightarrow X + Y$ and $\kappa : Y \leftrightarrow X + Y$ in the case of disjoint unions $X + Y$. The third part of a program is the body, a while-program over relations. As each RelView-program computes a value, finally, its last part is the return-clause, which consists of the keyword `RETURN` and a relation-algebraic expression, whose value after the execution of the body is the result.

Each of the vector expressions we have presented in the Sects. 4 and 6 immediately can be translated into RelView-code and, hence, evaluated by the tool. To give an impression, in the following we show a RelView-implementation of $\mathfrak{stab}$, where we use a relational program instead of a relational function to avoid the two-fold computation of a membership relation:

```
stab(R)
  DECL M
  BEG  M = epsi(R)
       RETURN -(L1n(R)*(R*M & M))^
  END.
```

In this RelView-program '^' denotes transposition, '-' complement, '&' intersection and '*' composition. By means of the pre-defined operations `L1n` and `epsi` transposed universal vectors and membership relations, respectively, are computed, where the types of the results are determined by those of the arguments (see [24] for details).

In Sect. 1 we have mentioned that the use of ROBDDs in RelView often leads to an amazing computational power. To prove this fact, in the following we present some results of practical experiments. They have been carried out with the newest version of RelView (Version 8.1, released September 2012 and available via [24]) on an ordinary desktop computer, with an AMD Phenom II X4 955 processor, 8GB 1600Mhz DDR3 RAM and running Linux.

The table of Fig. 1 shows the average running times (in seconds) of our first series of experiments. They concern the computation of all cliques of randomly generated symmetric and irreflexive relations $R : X \leftrightarrow X$, where $|X|$ ranges from 50 to 500 and the 'density' of the graph from 10 % to 50 % of all possible

$\|X\|$	50	100	150	200	250	300	350	400	450	500
density 10%	0.0	0.0	0.1	0.2	0.8	1.7	3.1	6.5	11.1	17.4
density 20%	0.0	0.0	0.4	1.9	5.0	13.0	18.3	96.1	187.2	421.1
density 30%	0.0	0.0	1.2	11.3	52.5	100.4	346.2	–	–	–
density 40%	0.0	0.7	6.5	112.0	693.7	1 502.2	–	–	–	–
density 50%	0.0	3.7	40.9	–	–	–	–	–	–	–

Fig. 1. Running times for computing cliques

edges. All times result from 5 experiments per size. In each case the times for the random generation of the relations are included, but they are negligible. To obtain from the vector $\mathfrak{clique}$ the vector model $min_{\mathsf{ST}}(\mathfrak{clique})$ of the set of all maximum cliques required at most some seconds. For instance, in one experiment with $|X| = 250$ and 40 % as density the running time was 13.21 s, where $\mathfrak{clique}$ had $1.91 \cdot 10^6$ 1-entries, $min_{\mathsf{ST}}(\mathfrak{clique})$ had 29 1-entries and 9 was the size of each maximum clique.

Because of lack of memory, in the case of 30 % density we have not been able to deal with $|X| > 350$ vertices. If, however, we restricted us to more sparse graphs, larger numbers of vertices could be treated successfully. For example, $|X| = 700$ and a density of 10 % led to approximately 70 s running time and, for the same density, $|X| = 1000$ led to roughly 12 min.

In case of absorbant sets, stable sets and vertex covers RelView was able to cope with the same number of vertices. However, in contrast with the computation of cliques, increasing densities now decreased the running times. For example, if $|X| = 250$, then a density of 90 % led to running times about 1 s, whereas a density of 60 % led to approximately 700 s.

Also when computing specific relations the RelView tool showed an amazing potential. For all sets X and Y such that $|X| \leq 50$ and $|Y| \leq 50$, the evaluation of $\mathfrak{univ}$, $\mathfrak{total}$ and $\mathfrak{irrefl}$ required only a fraction of a second. In case $X = Y$ an evaluation of $\mathfrak{inj}$ was able up to $|X| = 11$ (502.3 s running time) and that of $\mathfrak{surj}$ up to $|X| = 15$ (2 567.8 s running time), such that we have been able to determine for the latter size that 481 066 515 734 of the 15! ($\approx 1.30 \cdot 10^{12}$) permutations are proper.

When experimenting with the computation of vertex colourings, besides randomly generated symmetric and irreflexive relations we applied the RelView-version of $\mathfrak{colour}$ also to the relations of certain well-known specific graphs. An example is the *5-regular Clebsch graph*, which may be constructed from the 4-dimensional hypercube graph by adding edges between all pairs of opposite vertices. Hence, if $H : X \leftrightarrow X$ is the (symmetric and irreflexive) relation of the 4-dimensional hypercube graph, then

$$C_5 := H \cup (H^4 \cap \overline{\bigcup_{i=0}^{3} H^i}) : X \leftrightarrow X$$

is the relation of the 5-regular Clebsch graph. The RelView-pictures of Fig. 2 show C_5 and a 4-colouring $F : X \leftrightarrow \{f_1, f_2, f_3, f_4\}$ of the 5-regular Clebsch

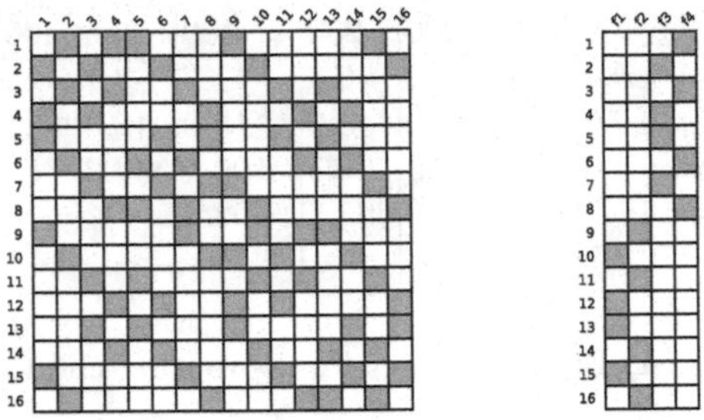

Fig. 2. Relation and a 4-colouring of the 5-regular Clebsch graph

graph as labeled Boolean matrices, where the labels indicate the elements of the source and the target, a black square means a 1-entry, and a white square means a 0-entry. The RelView-picture of Fig. 3 depicts C_5 as a graph. In the latter the 32 edges of the 4-dimensional hypercube graph as well as the 4 colours assigned by F to the vertices are emphasized. The chromatic number of the 5-regular Clebsch graph is 4. RelView required 1.49 s to compute its 28 560 4-coulourings and to determine that there exists no vertex colouring with less than 4 colours.

Originally, in [20] J.J. Seidel introduced the notation 'Clebsch graph' for the graph with relation $\overline{C_5} \cap \overline{\mathsf{I}} : X \leftrightarrow X$, that nowadays often is called the *10-regular Clebsch graph*. The chromatic number of the 10-regular Clebsch graph is 8. Because of the larger number of colours, when compared with the 5-regular Clebsch graph, by means of the RelView-version of colour we only were able to compute that there is no k-colouring of the 10-regular Clebsch graph such that $k \leq 7$. Here the case $k = 7$ required 2 438.97 s running time. To compute the chromatic number of the 10-regular Clebsch graph and a 8-colouring we implemented the recursive contraction-algorithm of Zykov (cf. [23]) and also the well-known greedy colouring-algorithm in RelView. Both programs delivered the desired result in 25.25 and 0 s, respectively. But till now we do not know the number of 8-coulourings of the 10-regular Clebsch graph.

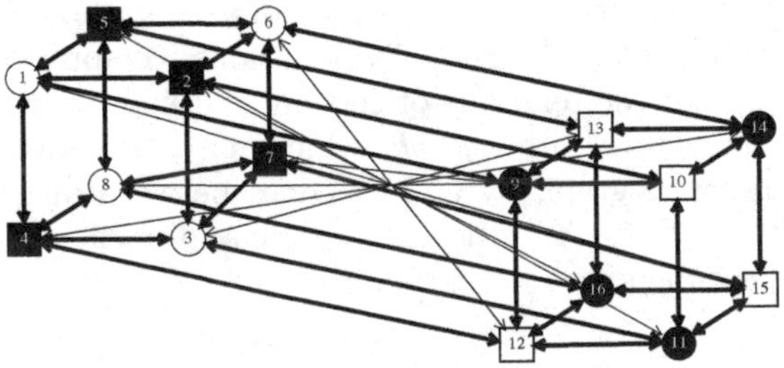

Fig. 3. Construction of the 5-regular Clebsch graph

8 Concluding Remarks

On closer examination, each of our relational algorithms for a given problem does not apply specific knowledge about it. We have only modeled the problem in the language of relation algebra, then translated the resulting formula (inclusions or fixed points equations) into a single relation-algebraic expression, and finally left the execution of the latter to a clever implementation of relation algebra. Under this point of view we have applied a *declarative problem solving* approach. And, in fact, the RelView-program `stab` of Sect. 7 satisfies many of the characteristics of declarative programs. The experiments of Sect. 7 show again that in specific situations declarative solutions can compete with specifically tailored imperative ones.

Concerning efficiency, of course, in general our approach can not compete with specific algorithms implemented in a conventional imperative programming language. But we believe that it has two decisive pros. First, we have obtained all relational problem solutions by developing them from the original logical descriptions. We regard this formal and goal-oriented development of algorithms from formal descriptions, which are correct by construction, as the first main advantage. It is also beneficial that the very formal calculations allow the use of theorem-proving systems. Presently we work with Prover9 and Coq.

As the second main advantage we regard the support by RelView. All solutions of the Sects. 4 and 6 lead to very short and concise RelView-programs. These are nothing else than formulations in a specific syntax. They are easy to alter in case of slightly changed specifications. Combining this with RelView's possibilities for visualization, animation and the random generation of relations (also with specific properties) allows prototyping of specifications, testing of hypotheses, experimentation with new concepts, exploration of new ideas, generation of examples and counterexamples etc., while avoiding unnecessary overhead. This makes RelView very useful for scientific research. In this regard the ROBDD-implementation of relations proved to be of immense help since it allows to treat also non-trivial examples. To mention two applications, in [11] RelView-experiments helped in the solution of chessboard problems, especially concerning bishops, and in [14] with the help of RelView all up-closed multirelations are computed and this led to an appropriate definition of an approximation order for modeling computations which also may be infinite.

Acknowledgement. I want to thank the reviewers for carefully reading the paper and for their very detailed and valuable comments. They helped to improve the paper.

References

1. Banks, J.: Sophisticated voting outcomes and agenda control. Soc. Choice Welfare **2**, 295–306 (1985)
2. Berghammer, R., Gritzner, T., Schmidt, G.: Prototyping relational specifications using higher-order objects. In: Heering, J., Meinke, K., Möller, B., Nipkow, T. (eds.) Higher Order Algebra, Logic and Term Rewriting. LNCS, pp. 56–75. Springer, Heidelberg (1994)

3. Berghammer, R., von Ulke, K.C.: Relation-algebraic analysis of Petri nets with RelView. In: Margaria, T., Steffen, B. (eds.) Second International Workshop. LNCS, pp. 49–69. Springer, Heidelberg (1996)
4. Berghammer, R., Neumann, F.: RelView – An OBDD-based computer algebra system for relations. In: Ganzha, V.G., Mayr, E.W., Vorozhtsov, E.V. (eds.) CASC 2005. LNCS, vol. 3718, pp. 40–51. Springer, Heidelberg (2005)
5. Berghammer, R.: Relation-algebraic computation of fixed points with applications. J. Logic Algebraic Program. **66**, 112–126 (2006)
6. Berghammer, R., Fronk, A.: Exact computation of minimum feedback vertex sets with relational algebra. Fundamenta Informaticae **70**, 301–316 (2006)
7. Berghammer, R., Milanese, U.: Relational approach to boolean logic problems. In: MacCaull, W., Winter, M., Düntsch, I. (eds.) RelMiCS 2005. LNCS, vol. 3929, pp. 48–59. Springer, Heidelberg (2006)
8. Berghammer, R.: Applying relation algebra and RelView to solve problems on orders and lattices. Acta Informatica **45**, 211–236 (2008)
9. Berghammer, R., Braßel, B.: Computing and visualizing closure objects using relation algebra and RelView. In: Gerdt, V.P., Mayr, E.W., Vorozhtsov, E.V. (eds.) CASC 2009. LNCS, vol. 5743, pp. 29–44. Springer, Heidelberg (2009)
10. Berghammer, R., Bolus, S., Rusinowska, A., de Swart, H.: A relation-algebraic approach to simple games. Eur. J. Oper. Res. **210**, 68–80 (2011)
11. Berghammer, R.: Relation-algebraic modeling and solution of chessboard independence and domination problems. J. Logic Algebraic Program. **81**, 625–642 (2012)
12. Berghammer, R.: Computing and visualizing Banks sets of dominance relations using relation algebra and RelView. J. Logic Algebraic Program. **82**, 223–236 (2013)
13. Ganter, B., Wille, R.: Formal Concept Analysis: Mathematical Foundations. Springer, Berlin (1999)
14. Guttmann, W.: Multirelations with infinite computations. J. Logic Algebraic Program. **83**, 194–211 (2014)
15. Jerrum, M.R., Sinclair, J.A.: Approximating the permanent. SIAM J. Comput. **18**, 1149–1178 (1989)
16. Kehden, B., Neumann, F.: A relation-algebraic view on evolutionary algorithms for some graph problems. In: Gottlieb, J., Raidl, G.R. (eds.) EvoCOP 2006. LNCS, vol. 3906, pp. 147–158. Springer, Heidelberg (2006)
17. Kehden, B.: Evaluating sets of search points using relational algebra. In: Schmidt, R.A. (ed.) RelMiCS/AKA 2006. LNCS, vol. 4136, pp. 266–280. Springer, Heidelberg (2006)
18. Schmidt, G., Ströhlein, T.: Discrete Mathematics for Computer Scientists. EATCS Monographs on Theoretical Computer Science. Springer, Heidelberg (1993)
19. Schmidt, G.: Relational Mathematics. Encyclopedia of Mathematics and its Applications. Cambridge University Press, Cambridge (2010)
20. Seidel, J.J.: Strongly regular graphs with $(-1, 1, 0)$ adjacency matrix having eigenvalue 3. Linear Algebra Appl. **1**, 281–298 (1968)
21. Tarski, A.: On the calculus of relations. J. Symbolic Logic **6**, 73–89 (1941)
22. Tarski, A., Givant, S.: A Formalization of Set Theory Without Variables. AMS Colloquium Publications, Rhode Island (1987)
23. Zykov, A.A.: On some properties of linear complexes (in Russian). Math. Sbornik **24**, 163–188 (1949)
24. RelView homepage: http://www.informatik.uni-kiel.de/~progsys/relview/

Turing-Completeness Totally Free

Conor McBride(✉)

University of Strathclyde, Glasgow, Scotland, UK
conor@strictlypositive.org

Abstract. In this paper, I show that general recursive definitions can be represented in the free monad which supports the 'effect' of making a recursive call, without saying how these calls should be executed. Diverse semantics can be given within a total framework by suitable monad morphisms. The Bove-Capretta construction of the domain of a general recursive function can be presented datatype-generically as an instance of this technique. The paper is literate Agda, but its key ideas are more broadly transferable.

1 Introduction

Advocates of Total Functional Programming [21], such as myself, can prove prone to a false confession, namely that the price of functions which provably function is the loss of Turing-completeness. In a total language, to construct $f : S \to T$ is to promise a canonical T eventually, given a canonical S. The alleged benefit of general recursion is just to inhibit such strong promises. To make a weaker promise, simply construct a total function of type $S \to GT$ where G is a suitable monad.

The literature and lore of our discipline are littered with candidates for G [10,19,22], and this article will contribute another—the *free* monad with one operation $f : S \to T$. To work in such a monad is to *write* a general recursive function without prejudice as to how it might be *executed*. We are then free, in the technical sense, to choose any semantics for general recursion we like by giving a suitable *monad morphism* to another notion of partial computation. For example, Venanzio Capretta's partiality monad [10], also known as the *completely iterative* monad on the operation $yield : 1 \to 1$, which might never deliver a value, but periodically offers its environment the choice of whether to interrupt computation or to continue.

Meanwhile, Ana Bove gave, with Capretta, a method for defining the *domain predicate* of a general recursive function simultaneously with the delivery of a value for every input in the domain [8]. When recursive calls are nested, their technique gives a paradigmatic example of defining a datatype and its interpretation by *induction-recursion* in the sense of Peter Dybjer and Anton Setzer [11,12]. Dybjer and Setzer further gave a coding scheme which renders first class the characterising data for inductive-recursive definitions. In this article, I show how to compute from the free monadic presentation of a general recursive function the code for its domain predicate. By doing so, I implement the

R. Hinze and J. Voigtländer (Eds.): MPC 2015, LNCS 9129, pp. 257–275, 2015.
DOI:10.1007/978-3-319-19797-5_13

Bove-Capretta method once for all, systematically delivering (but not, of course, discharging) the proof obligation required to strengthen the promise from partial $f : S \to G\,T$ to the total $f : S \to T$.

Total functional languages remain *logically* incomplete in the sense of Gödel. There are termination proof obligations which we can formulate but not discharge within any given total language, even though the relevant programs—notably the language's own evaluator—are total. Translated across the Curry-Howard correspondence, the argument for general recursion asserts that logical inconsistency is a price worth paying for logical completeness, notwithstanding the loss of the language's value as a carrier of checkable *evidence*. Programmers are free to maintain that such dishonesty is essential to their capacity to earn a living, but a new generation of programming technology enables some of us to offer and deliver a higher standard of guarantee. *Faites vos jeux!*

2 The General Free Monad

Working[1], in Agda, we may define a free monad which is both general purpose, in the sense of being generated by the strictly positive functor of your choice, but also suited to the purpose of supporting general recursion.

```
data General (S : Set) (T : S → Set) (X : Set) : Set where
  !!    : X → General S T X
  _??_ : (s : S) → (T s → General S T X) → General S T X
infixr 5 _??_
```

At each step, we either output an X, or we make the request s ?? k, for some $s : S$, where k explains how to continue once a response in $T\ s$ has been received. That is, values in General $S\ T\ X$ are request-response trees with X-values at the leaves; each internal node is labelled by a request and branches over the possible meaningful responses. The key idea in this paper is to represent recursive calls as just such request-response interactions, and recursive definitions by just such trees. General is a standard inductive definition, available in many dialects of type theory. Indeed, it closely resembles the archetypal such definition: Martin-Löf's 'W-type' [17], W $S\ T$, of well-ordered trees is the leafless special case, General $S\ T$ 0; the reverse construction, General $S\ T\ X$ = W $(X + S)$ [const 0, T][2], just attaches value leaves as an extra sort of childless node.

General datatypes come with a catamorphism, or 'fold' operator[3].

```
fold : ∀ {l S T X} {Y : Set l} →
         (X → Y) → ((s : S) → (T s → Y) → Y) →
         General S T X → Y
```

[1] http://github.com/pigworker/Totality.
[2] I write brackets for case analysis over a sum.
[3] Whenever I intend a monoidal accumulation, I say 'crush', not 'fold'.

```
fold r c (!! x)    = r x
fold r c (s ?? k)  = c s λ t → fold r c (k t)
```

Agda uses curly braces in types to declare arguments which are suppressed by default; corresponding curly braces are used in abstractions and applications to override that default. Note that unlike in Haskell, Agda's λ-abstractions scope as far to the right as possible and may thus stand without parentheses as the last argument to a function such as c. It is idiomatic Agda to drop those parentheses, especially when we think of the function as a defined form of quantification.

The 'bind' operation for the monad General S T substitutes computations for values to build larger computations. It is, of course, a fold.

```
_>>=G_ : ∀ {S T X Y} →
          General S T X → (X → General S T Y) → General S T Y
g >>=G k  = fold k _??_ g
infixl 4 _>>=G_
```

We then acquire what Gordon Plotkin and John Power refer to as a *generic effect* [20]—the presentation of an individual request-response interaction:

```
call : ∀ {S T} (s : S) → General S T (T s)
call s = s ?? !!
```

Now we may say how to give a recursive definition for a function. For each argument $s : S$, we must build a request-response tree from individual calls, ultimately delivering a value in $T\ s$. We may thus define the 'general recursive Π-type',

```
PiG : (S : Set) (T : S → Set) → Set
PiG S T = (s : S) → General S T (T s)
```

to be a type of functions delivering the recursive *strategy* for computing a $T\ s$ from some $s : S$.

For example, given the natural numbers,

```
data Nat : Set where
  zero : Nat
  suc  : Nat → Nat
```

the following obfuscated identity function will not pass Agda's syntactic check for guardedness of recursion.

```
fusc : Nat → Nat
fusc zero     = zero
fusc (suc n)  = suc ( fusc ( fusc n))
```

However, we can represent its definition without such controversy.

```
fusc : PiG Nat λ _ → Nat
fusc zero     = !! zero
fusc (suc n)  = call n >>=G λ fn → call fn >>=G λ ffn → !! (suc ffn)
```

Firstly, note that PiG S λ s $\rightarrow$ T is a defined form of quantification, needing no parentheses. Secondly, observe that each call is but a *placeholder* for a recursive call to fusc. Our code tells us just how to expand the recursion *once*. Note that fusc's *nested* recursive calls make use of the way $\gg\!=_{\mathsf{G}}$ allows values from earlier effects to influence the choice of later effects: nested recursion demands the full monadic power of General.

Even so, it is fair to object that the 'monadified' definition is ugly compared to its direct but not obviously terminating counterpart, with more intermediate naming. Monadic programming is ugly in general, not just in General! Haskell may reserve direct style for the particular monad of nontermination and finite failure, necessitating a more explicit notation for other monads, but languages like Bauer and Pretnar's *Eff* [6] give us grounds to hope for better. They let us write in direct style for whatever effectful interface is locally available, then obtain the computation delivered by the appropriate Moggi-style translation into an explicitly monadic kernel [18]. Just such a translation would give us our monadic fusc from its direct presentation.

By choosing the General monad, we have not committed to any notion of 'infinite computation'. Rather, we are free to work with a variety of monads M which might represent the execution of a general recursive function, by giving a *monad morphism* from General S T to M, mapping each request to something which tries to deliver its response. Correspondingly, we shall need to define these concepts more formally.

3 Monads and Monad Morphisms, More or Less

This section is a formalisation of material which is largely standard. The reader familiar with monad morphisms should feel free to skim for notation without fear of missing significant developments. To save space and concentrate on essentials, I have suppressed some proofs. The online sources for this paper omit nothing.

Let us introduce the notion of a Kleisli structure on sets, as Altenkirch and Reus called it, known to Altenkirch, Chapman and Uustalu as a 'relative' monad [4,5]. Where Haskell programmers might expect a type class and ML programmers a module signature, Agda requires me to be more specific and give the type of records which pack up the necessary equipment. Although the 'notion of computation' is given by a functor taking sets of values to sets of computations, that functor need not preserve the level in the predicative set-theoretic hierarchy[4] at which we work, and we shall need this flexibility for the Bove-Capretta

[4] Agda, Coq, etc., stratify types by 'size' into sets-of-values, sets-of-sets-of-values, and so on, ensuring consistency by forbidding the quantification over 'large' types by 'small'.

construction which represents computations as descriptions of datatypes. The upshot is that we are obliged to work polymorphically in *levels*, i for values and j for computations.

```
record Kleisli {i j} (M : Set i → Set j) : Set (lsuc i ⊔ j) where
  field
    return : ∀ {X} →     X → M X
    _>>=_ : ∀ {A B} → M A → (A → M B) → M B
  _◇_ : ∀ {A B C : Set i} →
        (B → M C) → (A → M B) → (A → M C)
  (f ◇ g) a = g a >>= f
  infixl 4 _>>=_ _◇_
```

Given the fields return and >>=, we may equip ourselves with Kleisli composition in the usual way, replacing each value emerging from g with the computation indicated by f. The Kleisli record type thus packs up operations which quantify over all level i value sets and target specific level j computation sets, so the lowest level we can assign is the maximum of lsuc i and j. Of course, we have

```
GeneralK : ∀ {S T} → Kleisli (General S T)
GeneralK = record {return = !!; _>>=_ = _>>=G_}
```

The 'Monad laws' amount to requiring that return and ◇ give us a category.

```
record KleisliLaws {i j} {M : Set i → Set j} (KM : Kleisli M)
  : Set (lsuc i ⊔ j) where
  open Kleisli KM
  field
    .idLeft  : ∀ {A B} (g : A → M B) → return ◇ g ≡ g
    .idRight : ∀ {A B} (f : A → M B) → f ◇ return ≡ f
    .assoc   : ∀ {A B C D}
                 (f : C → M D) (g : B → M C) (h : A → M B) →
                                        (f ◇ g) ◇ h ≡ f ◇ (g ◇ h)
```

The dots before the field names make those fields unavailable for computational purposes. Indeed, any declaration marked with a dot introduces an entity which exists for reasoning only. Correspondingly, it will not interfere with the effectiveness of computation if we postulate an extensional equality and calculate with functions.

```
postulate
  .ext : ∀ {i j} {A : Set i} {B : A → Set j} {f g : (a : A) → B a} →
           ((a : A) → f a ≡ g a) → f ≡ g
```

Indeed, it is clear that the above laws, expressing equations between functions rather than between the concrete results of applying them, can hold only in an extensional setting.

Extensionality gives us, for example, that anything satisfying the defining equations of a fold is a fold.

$$\begin{aligned}&.\mathsf{foldUnique} : \forall \{l\ S\ T\ X\}\ \{Y : \mathsf{Set}\ l\}\ (f : \mathsf{General}\ S\ T\ X \to Y)\ r\ c \to\\ &\quad (\forall\ x \to f\ (!!\ x) \equiv r\ x) \to (\forall\ s\ k \to f\ (s\ ??\ k) \equiv c\ s\ (f \cdot k)) \to\\ &\quad f \equiv \mathsf{fold}\ r\ c\end{aligned}$$

Specifically, the identity function is a fold, giving idLeft for General $S\ T$.

Meanwhile, computations are often constructed using $\gg\!=_{\mathsf{G}}$ (defined by specialising fold to the constructor ??) and interpreted recursively by some other fold. Correspondingly, a crucial lemma in the development is the following *fusion* law which eliminates an intermediate $\gg\!=_{\mathsf{G}}$ in favour of its direct interpretation.

$$\begin{aligned}&.\mathsf{foldFusion} : \forall \{l\ S\ T\ X\ Y\}\ \{Z : \mathsf{Set}\ l\}\\ &\quad (r : Y \to Z)\ (c : (s : S) \to (T\ s \to Z) \to Z)\\ &\quad (f : X \to \mathsf{General}\ S\ T\ Y) \to\\ &\quad (\mathsf{fold}\ r\ c \cdot \mathsf{fold}\ f\ \_??\_) \equiv \mathsf{fold}\ (\mathsf{fold}\ r\ c \cdot f)\ c\end{aligned}$$

Indeed, the associativity of Kleisli composition amounts to the fusion of two layers of $\gg\!=_{\mathsf{G}}$, and thus comes out as a special case. The above two results readily give us the KleisliLaws for the GeneralK operations on General $S\ T$.

$$.\mathsf{GeneralKLaws} : \forall \{S\ T\} \to \mathsf{KleisliLaws}\ (\mathsf{GeneralK}\ \{S\}\ \{T\})$$

Now, let us consider when a polymorphic function $m : \forall \{X\} \to M\ X \to N\ X$ is a *monad morphism* in this setting. Given Kleisli M and Kleisli N, $m \cdot -$ should map return and $\diamond$ from M to N.

$$\begin{aligned}&\textbf{record}\ \mathsf{Morphism}\ \{i\ j\ k\}\ \{M : \mathsf{Set}\ i \to \mathsf{Set}\ j\}\ \{N : \mathsf{Set}\ i \to \mathsf{Set}\ k\}\\ &\qquad\qquad\qquad (KM : \mathsf{Kleisli}\ M)\ (KN : \mathsf{Kleisli}\ N)\\ &\quad (m : \forall \{X\} \to M\ X \to N\ X) : \mathsf{Set}\ (\mathsf{lsuc}\ i \sqcup j \sqcup k)\ \textbf{where}\\ &\quad \textbf{module}\ -_M = \mathsf{Kleisli}\ KM;\ \textbf{module}\ -_N = \mathsf{Kleisli}\ KN\\ &\quad \textbf{field}\\ &\qquad .\mathsf{respectI} : \{X : \mathsf{Set}\ i\} \to\\ &\qquad\qquad m \cdot \mathsf{return}_M\ \{X\} \equiv \mathsf{return}_N\ \{X\}\\ &\qquad .\mathsf{respectC} : \{A\ B\ C : \mathsf{Set}\ i\}\ (f : B \to M\ C)\ (g : A \to M\ B) \to\\ &\qquad\qquad m \cdot (f \diamond_M g) \equiv (m \cdot f) \diamond_N (m \cdot g)\end{aligned}$$

The proofs, idMorph and compMorph, that monad morphisms are closed under identity and composition, are straightforward and omitted.

Now, General $S\ T$ is the free monad on the functor $\Sigma\ S\ \lambda\ s \to T\ s \to -$ which captures a single request-response interaction. The fact that it is the free construction, turning functors into monads, tells us something valuable about the ways in which its monad morphisms arise. Categorically, what earns the construction the designation 'free' is that it is left adjoint to the 'forgetful' map which turns monads back into functors by ignoring the additional structure given by return and $\gg=$. We seek monad morphisms from General S T to some other M, i.e., some

$$m : \forall \{X\} \rightarrow \mathsf{General}\ S\ T\ X \rightarrow M\ X$$

satisfying the above demands for respect. The free-forgetful adjunction amounts to the fact that such monad morphisms from General S T are in one-to-one correspondence with the functor morphisms (i.e., polymorphic functions) from the underlying request-response functor to M considered only as a functor. That is, every such m is given by and gives a function of type

$$\forall \{X\} \rightarrow (\Sigma\ S\ \lambda\ s \rightarrow T\ s \rightarrow X) \rightarrow M\ X$$

with no conditions to obey. Uncurrying the dependent pair and reordering gives us the equivalent type

$$(s : S) \rightarrow \forall \{X\} \rightarrow (T\ s \rightarrow X) \rightarrow M\ X$$

which the Yoneda lemma tells us amounts to

$$(s : S) \rightarrow M\ (T\ s)$$

That is, the monad morphisms from General S T to M are exactly given by the 'M-acting versions' of our function. Every such morphism is given by instantiating the parameters of the following definition.

```
morph : ∀ {l S T} {M : Set → Set l} (KM : Kleisli M)
          (h : (s : S) → M (T s))
          {X} → General S T X → M X
morph KM h = fold return (_>>=_ · h) where open Kleisli KM
```

We may show that morph makes Morphisms.

```
.morphMorphism : ∀ {l S T} {M : Set → Set l}
  (KM : Kleisli M) (KLM : KleisliLaws KM) →
  (h : (s : S) → M (T s)) →
  Morphism (GeneralK {S} {T}) KM (morph KM h)
```

Moreover, just as the categorical presentation would have us expect, morph give us the *only* monad morphisms from General S T, by the uniqueness of fold.

```
.morphOnly : ∀ {l S T}
  {M : Set → Set l} (KM : Kleisli M) (KLM : KleisliLaws KM) →
  (m : {X : Set} → General S T X → M X) →
  Morphism GeneralK KM m →
  {X : Set} → m {X} ≡ morph KM (m · call) {X}
```

I do not think like, or of myself as, a category theorist. I do, by instinct and training, look to inductive definitions as the basis of a general treatment: syntax as the precursor to any semantics, in this case any monad semantics for programs which can make recursive calls. The categorical question 'What is the *free* monad for call?' poses that problem and tells us to look no further for a solution than General S T and morph.

4 General Recursion with the General Monad

General strategies are finite: they tell us how to expand one request in terms of a bounded number recursive calls. The operation which expands each such request is a monad endomorphism—exactly the one generated by our f : PiG S T itself, replacing each call s node in the tree by the whole tree given by f s.

$$\begin{array}{l}\mathsf{expand} : \forall \{S\ T\ X\} \to \mathsf{PiG}\ S\ T \to \mathsf{General}\ S\ T\ X \to \mathsf{General}\ S\ T\ X \\ \mathsf{expand}\ f\ =\ \mathsf{morph}\ \mathsf{GeneralK}\ f\end{array}$$

You will have noticed that call : PiG S T, and that expand call just replaces one request with another, acting as the identity. As a recursive strategy, taking $f = \lambda\ s \to \mathsf{call}\ s$ amounts to the often valid but seldom helpful 'definition':

$$\mathsf{f}\ s = \mathsf{f}\ s$$

By way of example, let us consider the evolution of state machines. We shall need Boolean values, equipped with conditional expression:

$$\begin{array}{l}\textbf{data}\ \mathsf{Bool} : \mathsf{Set}\ \textbf{where}\ \mathsf{tt}\ \mathsf{ff} : \mathsf{Bool} \\ \mathsf{if\_then\_else\_} : \{X : \mathsf{Set}\} \to \mathsf{Bool} \to X \to X \to X \\ \mathsf{if}\ \mathsf{tt}\ \mathsf{then}\ t\ \mathsf{else}\ f\ =\ t \\ \mathsf{if}\ \mathsf{ff}\ \mathsf{then}\ t\ \mathsf{else}\ f\ =\ f\end{array}$$

Now let us construct the method for computing the halting state of a machine, given its initial state and its one-step transition function.

$$\begin{array}{l}\mathsf{halting} : \forall \{S\} \to (S \to \mathsf{Bool}) \to (S \to S) \to \mathsf{PiG}\ S\ \lambda\ \_ \to S \\ \mathsf{halting}\ \mathit{stop}\ \mathit{step}\ \mathit{start}\ \textbf{with}\ \mathit{stop}\ \mathit{start} \\ \ldots \qquad\qquad \mid\ \mathsf{tt}\ =\ !!\ \mathit{start} \\ \ldots \qquad\qquad \mid\ \mathsf{ff}\ =\ \mathsf{call}\ (\mathit{step}\ \mathit{start})\end{array}$$

For Turing machines, S should pair a machine state with a tape, *stop* should check if the machine state is halting, and *step* should look up the current state and tape-symbol in the machine description then return the next state and tape. We can clearly explain how any old Turing machine computes without stepping beyond the confines of total programming, and without making any rash promises about what values such a computation might deliver, or when.

5 The Petrol-Driven Semantics

It is one thing to describe a general-recursive computation but quite another to perform it. A simple way to give an arbitrary total approximation to partial computation is to provide an engine which consumes one unit of petrol for each recursive call it performs, then specify the initial fuel supply. The resulting program is primitive recursive, but makes no promise to deliver a value. Let us construct it as a monad morphism. We shall need the usual model of *finite* failure, allowing us to give up when we are out of fuel.

```
data Maybe (X : Set) : Set where
  yes : X → Maybe X
  no  : Maybe X
```

Maybe is monadic in the usual failure-propagating way.

```
MaybeK : Kleisli Maybe
MaybeK = record {return = yes
                ; _>>=_ = λ {(yes a) k → k a; no k → no}}
```

The proof MaybeKL : KleisliLaws MaybeK is a matter of elementary case analysis.

We may directly construct the monad morphism which executes a general recursion impatiently.

```
already : ∀ {S T X} → General S T X → Maybe X
already = morph MaybeK λ s → no
```

That is, !! becomes yes and ?? becomes no, so the recursion delivers a value only if it has terminated already. Now, if we have some petrol, we can run an engine which expands the recursion for a while, beforehand.

```
engine : ∀ {S T} (f : PiG S T) (n : Nat)
         {X} → General S T X → General S T X
engine f zero    = id
engine f (suc n) = engine f n · expand f
```

We gain the petrol-driven (or step-indexed, if you prefer) semantics by composition.

```
petrol : ∀ {S T} → PiG S T → Nat → (s : S) → Maybe (T s)
petrol f n = already · engine f n · f
```

If we consider Nat with the usual order and Maybe X ordered by no $<$ yes x, we can readily check that petrol $f\ n\ s$ is monotone in n: supplying more fuel can only (but sadly not strictly) increase the risk of successfully delivering output.

An amusing possibility in a system such as Agda, supporting the partial evaluation of incomplete expressions, is to invoke petrol with ? as the quantity of fuel. We are free to refine the ? with suc ? and resume evaluation repeatedly for as long as we are willing to wait in expectation of a yes. Whilst this may be a clunky way to signal continuing consent for execution, compared to the simple maintenance of the electricity supply, it certainly simulates the conventional experience of executing a general recursive program.

What, then, is the substance of the often repeated claim that a total language capable of this construction is not Turing-complete? Barely this: there is more to delivering the run time execution semantics of programs than the pure evaluation of expressions. A pedant might quibble that the *language* is Turing-incomplete because it takes the *system* in which you use it to execute arbitrary recursive computations for as long as you are willing to tolerate. Such an objection has merit only in that it speaks against casually classifying a *language* as Turing-complete or otherwise, without clarifying the variety of its semanticses and the relationships between them.

Whilst we are discussing the semantics of total languages, emphatically in the plural, it is worth remembering that we expect dependently typed languages to come with at least *two*: a run time execution semantics which computes only with closed terms, and an evaluation semantics which the typechecker applies to open terms. It is quite normal for general recursive languages to have a total type checking algorithm and indeed to make use of restricted evaluation in the course of code generation.

6 Capretta's Coinductive Semantics, via Abel and Chapman

Coinduction in dependent type theory remains a vexed issue: we are gradually making progress towards a presentation of productive programming for infinite data structures, but we can certainly not claim that we have a presentation which combines honesty, convenience and compositional. The state of the art is the current Agda account due to Andrea Abel and colleagues, based on the notion of *copatterns* [3] which allow us to define lazy data by specifying observations of them, and on *sized types* [1] which give a more flexible semantic account of productivity at the cost of additional indexing.

Abel and Chapman [2] give a development of normalization for simply typed λ-calculus, using Capretta's Delay monad [10] as a showcase for copatterns and sized types. I will follow their setup, then construct a monad morphism from General. The essence of their method is to define Delay as the data type of *observations* of lazy computations, mutually with the record type, Delay^∞, of those lazy computations themselves. We gain a useful basis for reasoning about infinite behaviour.

```
mutual
  data Delay (i : Size) (X : Set) : Set where
    now   : X            → Delay i X
    later : Delay∞ i X   → Delay i X
  record Delay∞ (i : Size) (X : Set) : Set where
    coinductive; constructor ⟨_⟩
    field force : {j : Size < i} → Delay j X
```

Abel explains that Size, here, is a special type which characterizes the *observation depth* to which one may iteratively force the lazy computation. Values of type Size cannot be inspected by programs, but corecursive calls must reduce size, so cannot be used to satisfy the topmost observation. That is, we must deliver the outermost now or later without self-invocation. Pleasingly, corecursive need not be *syntactically* guarded by constructors, because their sized types document their legitimate use. For example, we may define the *anamorphism*, or unfold, constructing a Delay X from a coalgebra for the underlying functor $X + -$.

```
data _+_ (S T : Set) : Set where
  inl : S → S + T
  inr : T → S + T
[_,_] : {S T X : Set} → (S → X) → (T → X) → S + T → X
[f, g] (inl s) = f s
[f, g] (inr t) = g t
mutual
  unfold   : ∀ {i X Y} → (Y → X + Y) → Y → Delay i X
  unfold f y = [now, later · unfold∞ f] (f y)
  unfold∞ : ∀ {i X Y} → (Y → X + Y) → Y → Delay∞ i X
  force (unfold∞ f y) = unfold f y
```

Syntactically, the corecursive call unfold$^\infty$ f is inside a composition inside a defined case analysis operator, but the type of later ensures that the recursive unfold$^\infty$ has a smaller size than that being forced.

Based on projection, copatterns favour products over sum, which is why most of the motivating examples are based on streams. As soon as we have a choice, mutual recursion becomes inevitable as we alternate between the projection and case analysis. However, thus equipped, we can build a Delay X value by stepping a computation which can choose either to deliver an X or to continue.

Capretta explored the use of Delay as a monad to model general recursion, with the >>= operator concatenating sequences of laters. By way of example, he gives an interpretation of the classic language with an operator seeking the minimum number satisfying a test. Let us therefore equip Delay with a >>= operator. It can be given as an unfold, but the direct definition with sized types is more straightforward. Abel and Chapman give us the following definition.

```
mutual
  _>>=D_ : ∀ {i A B} →
              Delay i A → (A → Delay i B) → Delay i B
  now a >>=D f = f a
  later a' >>=D f = later (a' >>=∞D f)
  _>>=∞D_ : ∀ {i A B} →
              Delay∞ i A → (A → Delay i B) → Delay∞ i B
  force (a' >>=∞D f) = force a' >>=D f
```

and hence our purpose will be served by taking

```
DelayK : {i : Size} → Kleisli (Delay i)
DelayK = record {return = now; _>>=_ = _>>=D_}
```

Abel and Chapman go further and demonstrate that these definitions satisfy the monad laws up to strong bisimilarity, which is the appropriate notion of equality for coinductive data but sadly not the propositional equality which Agda makes available. I shall not recapitulate their proof.

It is worth noting that the Delay monad is an example of a *completely iterative* monad, a final coalgebra $\nu\ Y.\ X + F\ Y$, where the free monad, General, is an initial algebra [14]. For Delay, take $F\ Y\ =\ Y$, or isomorphically, $F\ Y\ =\ 1 \times 1\ \rightarrow\ Y$, representing a trivial request-response interaction. That is Delay represents processes which must always eventually *yield*, allowing their environment the choice of whether or not to resume them. We have at least promised to obey control-C! Of course, naïve $\gg\!=_{\mathsf{D}}$ is expensive, especially when left-nested, but the usual efficient semantics, based on interruption rather than yielding, can be used at run time.

By way of connecting the Capretta semantics with the petrol-driven variety, we may equip every Delay process with a monotonic engine.

$$
\begin{array}{llcl}
\mathsf{engine} : \mathsf{Nat} \rightarrow \forall \{X\} \rightarrow \mathsf{Delay}\ \_\ X \rightarrow \mathsf{Maybe}\ X \\
\mathsf{engine}\ \_ & (\mathsf{now}\ x) & = & \mathsf{yes}\ x \\
\mathsf{engine}\ \mathsf{zero} & (\mathsf{later}\ \_) & = & \mathsf{no} \\
\mathsf{engine}\ (\mathsf{suc}\ n) & (\mathsf{later}\ d) & = & \mathsf{engine}\ n\ (\mathsf{force}\ d)
\end{array}
$$

Note that engine n is not a monad morphism unless n is zero.

$$
\begin{array}{lcl}
\mathsf{engine}\ \mathit{1}\ (\mathsf{later}\ \langle\, \mathsf{now}\ \mathsf{tt}\,\rangle \gg\!= \lambda\ v \rightarrow \mathsf{later}\ \langle\, \mathsf{now}\ v\,\rangle) & = & \mathsf{no} \\
\mathsf{engine}\ \mathit{1}\ (\mathsf{later}\ \langle\, \mathsf{now}\ \mathsf{tt}\,\rangle) \gg\!= \lambda\ v \rightarrow \mathsf{engine}\ \mathit{1}\ (\mathsf{later}\ \langle\, \mathsf{now}\ v\,\rangle) & = & \mathsf{yes}\ \mathsf{tt}
\end{array}
$$

Meanwhile, given a petrol-driven process, we can just keep trying more and more fuel. This is one easy way to write the minimization operator.

$$
\begin{array}{l}
\mathsf{tryMorePetrol} : \forall \{i\ X\} \rightarrow (\mathsf{Nat} \rightarrow \mathsf{Maybe}\ X) \rightarrow \mathsf{Delay}\ i\ X \\
\mathsf{tryMorePetrol}\ \{\_\}\ \{X\}\ f\ =\ \mathsf{unfold\ try\ zero}\ \mathbf{where} \\
\quad \mathsf{try} : \mathsf{Nat} \rightarrow X + \mathsf{Nat} \\
\quad \mathsf{try}\ n\ \mathbf{with}\ f\ n \\
\quad \ldots \quad \mid \quad \mathsf{yes}\ x\ =\ \mathsf{inl}\ x \\
\quad \ldots \quad \mid \quad \mathsf{no}\ \ \ \ =\ \mathsf{inr}\ (\mathsf{suc}\ n) \\
\mathsf{minimize} : (\mathsf{Nat} \rightarrow \mathsf{Bool}) \rightarrow \mathsf{Delay}\ \_\ \mathsf{Nat} \\
\mathsf{minimize}\ \mathit{test}\ =\ \mathsf{tryMorePetrol}\ \lambda\ n \rightarrow \mathsf{if}\ \mathit{test}\ n\ \mathsf{then\ yes}\ n\ \mathsf{else\ no}
\end{array}
$$

Our request-response characterization of general recursion is readily mapped onto Delay. Sized types allow us to give the monad morphism directly, corecursively interpreting each recursive call.

$$
\begin{array}{l}
\mathbf{mutual} \\
\quad \mathsf{delay} \ : \forall \{i\ S\ T\}\ (f : \mathsf{PiG}\ S\ T)\ \{X\} \rightarrow \mathsf{General}\ S\ T\ X \rightarrow \mathsf{Delay} i X \\
\quad \mathsf{delay}\ f\ =\ \mathsf{morph\ DelayK}\ \lambda\ s \rightarrow \mathsf{later}\ (\mathsf{delay}^{\infty}\ f\ (f\ s)) \\
\quad \mathsf{delay}^{\infty} : \forall \{i\ S\ T\}\ (f : \mathsf{PiG}\ S\ T)\ \{X\} \rightarrow \mathsf{General}\ S\ T\ X \rightarrow \mathsf{Delay}^{\infty}\ i\ X \\
\quad \mathsf{force}\ (\mathsf{delay}^{\infty}\ f\ g)\ =\ \mathsf{delay}\ f\ g
\end{array}
$$

We can now transform our General functions into their coinductive counterparts.

```
lazy : ∀ {S T} → PiG S T → (s : S) → Delay _ (T s)
lazy f  =  delay f · f
```

Although my definition of delay is a monad morphism by construction, it is quite possible to give extensionally the same operation as an unfold, thus removing the reliance on sized types but incurring the obligation to show that delay respects return and $\gg=_G$ upto strong bisimulation. Some such manoeuvre would be necessary to port this section's work to Coq's treatment of coinduction [15]. There is certainly no deep obstacle to the treatment of general recursion via coinduction in Coq.

7 A Little λ-Calculus

By way of a worked example, let us implement the untyped λ-calculus. We can equip ourselves with de Bruijn-indexed terms, using finite sets to police scope, as did Altenkirch and Reus [5]. It is sensible to think of Fin n as the type of natural numbers strictly less than n.

```
data Fin : Nat → Set where
  zero : {n : Nat}          → Fin (suc n)
  suc  : {n : Nat} → Fin n → Fin (suc n)
```

I have taken the liberty of parametrizing terms by a type X of inert constants.

```
data Λ (X : Set) (n : Nat) : Set where
  κ    : X                  → Λ X n
  #    : Fin n              → Λ X n
  λ    : Λ X (suc n)        → Λ X n
  _$_  : Λ X n → Λ X n      → Λ X n
infixl 5 _$_
```

In order to evaluate terms, we shall need a suitable notion of environment. Let us make sure they have the correct size to enable projection.

```
data Vec (X : Set) : Nat → Set where
  ⟨⟩  : Vec X zero
  _‚_ : {n : Nat} → Vec X n → X → Vec X (suc n)
proj : ∀ {X n} → Vec X n → Fin n → X
proj (_‚ x) zero      = x
proj (γ‚ _) (suc n)   = proj γ n
```

Correspondingly, a *value* is either a constant applied to other values, or a function which has got stuck for want of its argument.

```
data Val (X : Set) : Set where
  κ : X → {n : Nat} → Vec (Val X) n                → Val X
  λ : {n : Nat} → Vec (Val X) n → Λ X (suc n) → Val X
```

Now, in general, we will need to evaluate *closures*—open terms in environments.

```
data Closure (X : Set) : Set where
  _⊢_ : {n : Nat} → Vec (Val X) n → Λ X n → Closure X
infixr 4 _⊢_
```

We can now give the evaluator, ⟦_⟧ as a General recursive strategy to compute a value from a closure. Application is the fun case: a carefully placed **let** shadows the evaluator with an appeal to call, so the rest of the code looks familiar, yet it is no problem to invoke the non-structural recursion demanded by a β-redex.

```
⟦_⟧ : {X : Set} → PiG (Closure X) λ _ → Val X
⟦ γ ⊢ κ x   ⟧      = !! (κ x ⟨⟩)     -- Constants are inert.
⟦ γ ⊢ # i   ⟧      = !! (proj γ i)  -- Variables index the environment.
⟦ γ ⊢ λ b   ⟧      = !! (λ γ b)     -- Unapplied functions get stuck.
⟦ γ ⊢ f $ s ⟧      =
  let ⟦_⟧ : PiG (Closure _) λ _ → Val _; ⟦_⟧ = call in   -- shadow ⟦_⟧
  ⟦ γ ⊢ s ⟧  ≫=G λ v →     -- evaluate the argument, then
  ⟦ γ ⊢ f ⟧  ≫=G λ {        -- evaluate the function, then, inspecting its value,
    (κ x vs) → !! (κ x (vs ‚ v));   -- either grow an inert constant application,
    (λ δ b)  → ⟦ δ ‚ v ⊢ b ⟧     }  -- or grow a closure from a stuck function.
```

Thus equipped, lazy ⟦_⟧ is the Delayed version. Abel and Chapman give a Delayed interpreter (for typed terms) directly, exercising some craft in negotiating size and mutual recursion [2]. The General method makes that craft rather more systematic.

There is still a little room for programmer choice, however. Where a recursive call happens to be structurally decreasing, e.g. when evaluating the function and its argument, we are not *forced* to appeal to the call oracle, but could instead compute by structural recursion the expansion of an evaluation to those of its redexes. Indeed, that is the choice which Abel and Chapman make. It is not yet clear which course is preferable in practice.

8 An Introduction or Reimmersion in Induction-Recursion

I have one more semantics for general recursion to show you, constructing for any given$f$: PiG S T its *domain*. The domain is an inductively defined predicate, classifying the arguments which give rise to call trees whose paths are finite. As Ana Bove observed, the fact that a function is defined on its domain is a structural recursion—the tricky part is to show that the domain predicate holds [7]. However, to support nested recursion, we need to define the domain predicate and the resulting output *mutually*. Bove and Capretta realised that such mutual definitions are just what we get from Dybjer and Setzer's notion of *induction-recursion* [8,12], giving rise to the 'Bove-Capretta method' of modelling general recursion and generating termination proof obligations.

We can make the Bove-Capretta method generic, via the universe encoding for (indexed) inductive-recursive sets shown consistent by Dybjer and Setzer. The idea is that each node of data is a record with some ordinary fields coded by σ, and some places for recursive substructures coded by δ, with ι coding the end.

```
data IR {l} {S : Set} (I : S → Set l) (O : Set l) : Set (l ⊔ lsuc lzero) where
  ι : (o : O)                                  → IR I O
  σ : (A : Set) (K : A → IR I O)               → IR I O
  δ : (B : Set) (s : B → S)
      (K : (i : (b : B) → I (s b)) → IR I O) → IR I O
```

Now, in the indexed setting, we have S sorts of recursive substructure, and for each $s : S$, we know that an 'input' substructure can be interpreted as a value of type $I\ s$. Meanwhile, O is the 'output' type in which we must interpret the whole node. I separate inputs and outputs when specifying individual nodes, but the connection between them will appear when we tie the recursive knot.

When we ask for substructures with δ branching over B, we must say which sort each must take via $s : B \to S$, and then K learns the interpretations of those substructures *before* we continue. It is that early availability of the interpretation which allows Bove and Capretta to define the domain predicate for nested recursions: the interpretation is exactly the value of the recursive call that is needed to determine the argument of the enclosing recursive call.

Eventually, we must signal 'end of node' with ι and specify the output. As you can see, σ and δ pack up Sets, so IR codes are certainly large: the interpretation types I and O can be still larger. Induction-recursive definitions add convenience rather than expressive strength when I and O happen to be small and can be translated to ordinary datatype families indexed by the interpretation under just those circumstances [16]. For our application, they need only be as large as the return type of the function whose domain we define, so the translation applies, yielding not the *domain* of the function but the relational presentation of its *graph*. The latter is less convenient than the domain for termination proofs, but is at least expressible in systems without induction-recursion, such as Coq.

Now, to interpret these codes as record types, we need the usual notion of *small* dependent pair types, for IR gives small types with possibly large interpretations.

```
record Σ (S : Set) (T : S → Set) : Set where
  constructor _,_
  field fst : S; snd : T fst
```

By way of abbreviation, let me also introduce the notion of a sort-indexed family of maps, between sort-indexed families of sets.

$$
\begin{aligned}
&\_\dot{\rightarrow}\_ : \forall \{l\} \{S : \mathsf{Set}\} (X : S \rightarrow \mathsf{Set}) (I : S \rightarrow \mathsf{Set}\ l) \rightarrow \mathsf{Set}\ l \\
&X \dot{\rightarrow} I = \forall \{s\} \rightarrow X\ s \rightarrow I\ s
\end{aligned}
$$

If we know what the recursive substructures are and how to interpret them, we can say what nodes consist of, namely tuples made with Σ and 1.

$$
\begin{aligned}
&[\![\_]\!]_{\mathsf{Set}} : \forall \{l\ S\ I\ O\} (C : \mathsf{IR}\ \{l\}\ I\ O) (X : S \rightarrow \mathsf{Set}) (i : X \dot{\rightarrow} I) \\
&\qquad\qquad \rightarrow \mathsf{Set} \\
&[\![\ \iota\ o\ ]\!]_{\mathsf{Set}}\ X\ i = 1 \\
&[\![\ \sigma\ A\ K\ ]\!]_{\mathsf{Set}}\ X\ i = \Sigma\ A\ \lambda\ a \rightarrow [\![\ K\ a\ ]\!]_{\mathsf{Set}}\ X\ i \\
&[\![\ \delta\ B\ s\ K\ ]\!]_{\mathsf{Set}}\ X\ i = \Sigma\ ((b : B) \rightarrow X\ (s\ b))\ \lambda\ r \rightarrow [\![\ K\ (i \cdot r)\ ]\!]_{\mathsf{Set}}\ X\ i
\end{aligned}
$$

Moreover, we can read off their output by applying i at each δ until we reach $\iota\ o$.

$$
\begin{aligned}
&[\![\_]\!]_{\mathsf{out}} : \forall \{l\ S\ I\ O\} (C : \mathsf{IR}\ \{l\}\ I\ O) (X : S \rightarrow \mathsf{Set}) (i : X \dot{\rightarrow} I) \\
&\qquad\qquad \rightarrow [\![\ C\ ]\!]_{\mathsf{Set}}\ X\ i \rightarrow O \\
&[\![\ \iota\ o\ ]\!]_{\mathsf{out}}\ X\ i\ \langle\rangle = o \\
&[\![\ \sigma\ A\ K\ ]\!]_{\mathsf{out}}\ X\ i\ (a, t) = [\![\ K\ a\ ]\!]_{\mathsf{out}}\ X\ i\ t \\
&[\![\ \delta\ B\ s\ K\ ]\!]_{\mathsf{out}}\ X\ i\ (r, t) = [\![\ K\ (i \cdot r)\ ]\!]_{\mathsf{out}}\ X\ i\ t
\end{aligned}
$$

Now we can tie the recursive knot, following Dybjer and Setzer's recipe. Again, I make use of Abel's sized types to convince Agda that decode terminates.

```
mutual
  data μ {l} {S} {I} (F : (s : S) → IR {l} I (I s)) (j : Size) (s : S) : Set
    where ⟨_⟩ : {k : Size < j} → ⟦ F s ⟧Set (μ F k) decode → μ F j s
  decode : ∀ {l} {S} {I} {F} {j} → μ {l} {S} {I} F j ∸→ I
  decode {F = F} {s = s} ⟨ n ⟩ = ⟦ F s ⟧out (μ F _) decode n
```

Of course, you and I can see from the definition of $[\![\_]\!]_{\mathsf{out}}$ that the recursive uses of decode will occur only at substructures, but without sized types, we should need to inline $[\![\_]\!]_{\mathsf{out}}$ to expose that guardedness to Agda.

Now, as Ghani and Hancock observe, IR I is a (relative) monad [13].[5] Indeed, it is the free monad generated by σ and δ. Its $\gg\!=$ operator is perfectly standard, concatenating dependent record types. I omit the unremarkable proofs of the laws.

```
IRK : ∀ {l} {S} {I : S → Set l} → Kleisli (IR I)
IRK {l} {S} {I} = record {return = ι; _>>=_ = _>>=I_} where
  _>>=I_ : ∀ {X Y} → IR I X → (X → IR I Y) → IR I Y
  ι x       >>=I K' = K' x
  σ A K     >>=I K' = σ A λ a → K a >>=I K'
  δ B s K   >>=I K' = δ B s λ f → K f >>=I K'
```

Now, the Bove-Capretta method amounts to a monad morphism from General S T to IR T. That is, the domain predicate is indexed over S, with

[5] They observe also that $[\![\_]\!]_{\mathsf{Set}}$ and $[\![\_]\!]_{\mathsf{out}}$ form a monad morphism.

domain evidence for a given s decoded in $T\ s$. We may generate the morphism as usual from the treatment of a typical call s, demanding the single piece of evidence that s is also in the domain, then returning at once its decoding.

$$\begin{aligned}&\mathsf{callInDom} : \forall \{l\ S\ T\} \to (s : S) \to \mathsf{IR}\ \{l\}\ T\ (T\ s)\\&\mathsf{callInDom}\ s\ =\ \delta\ 1\ (\mathsf{const}\ s)\ \lambda\ t \to \iota\ (t\ \langle\rangle)\\&\mathsf{DOM} : \forall \{S\ T\} \to \mathsf{PiG}\ S\ T \to (s : S) \to \mathsf{IR}\ T\ (T\ s)\\&\mathsf{DOM}\ f\ s\ =\ \mathsf{morph\ IRK\ callInDom}\ (f\ s)\end{aligned}$$

Now, to make a given f : $\mathsf{PiG}\ S\ T$ total, it is sufficient to show that its domain predicate holds for all $s : S$.

$$\begin{aligned}&\mathsf{total} : \forall \{S\ T\}\ (f : \mathsf{PiG}\ S\ T)\ (\mathit{allInDom} : (s : S) \to \mu\ (\mathsf{DOM}\ f)\ \_\ s) \to\\&\qquad (s : S) \to T\ s\\&\mathsf{total}\ f\ \mathit{allInDom}\ =\ \mathsf{decode} \cdot \mathit{allInDom}\end{aligned}$$

The absence of σ from callInDom tells us that domain evidence contains at most zero bits of data and is thus 'collapsible' in Edwin Brady's sense [9], thus enabling total f to be compiled for run time execution exactly as the naïve recursive definition of f.

9 Discussion

We have seen how to separate the business of saying what it is to *be* a recursive definition from the details of what it means to *run* one. The former requires only that we work in the appropriate free monad to give us an interface permitting the recursive calls we need to make. Here, I have considered only recursion at a fixed arity, but the method should readily extend to partially applied recursive calls, given that we need only account for their *syntax* in the first instance. It does not seem like a big stretch to expect that the familiar equational style of recursive definition could be translated monadically, much as we see in the work on algebraic effects.

The question, then, is not what is *the* semantics for general recursion, but rather how to make use of recursive definitions in diverse ways by giving appropriate monad morphisms—that is, by explaining how each individual call is to be handled. We have seen a number of useful possibilities, not least the Bove-Capretta domain construction, by which we can seek to establish the totality of our function and rescue it from its monadic status.

However, the key message of this paper is that the status of general recursive definitions is readily negotiable within a total framework. There is no need to give up on the ability either to execute potentially nonterminating computations or to be trustably total. There is no difference between what you can *do* with a partial language and what you can *do* with a total language: the difference is in what you can *promise*, and it is the partial languages which fall short.

References

1. Abel, A.: Type-based termination: a polymorphic lambda-calculus with sized higher-order types. Ph.D. thesis, Ludwig Maximilians University Munich (2007)
2. Abel, A., Chapman, J.: Normalization by evaluation in the delay monad: a case study for coinduction via copatterns and sized types. In: Levy, P., Krishnaswami, N. (eds.) Workshop on Mathematically Structured Functional Programming 2014, vol. 153 of EPTCS, pp. 51–67 (2014)
3. Abel, A., Pientka, B., Thibodeau, D., Setzer, A.: Copatterns: programming infinite structures by observations. In: Giacobazzi, R., Cousot, R. (eds.) ACM Symposium on Principles of Programming Languages, POPL 2013, ACM, pp. 27–38 (2013)
4. Altenkirch, T., Chapman, J., Uustalu, T.: Monads need not be endofunctors. In: Ong, L. (ed.) FOSSACS 2010. LNCS, vol. 6014, pp. 297–311. Springer, Heidelberg (2010)
5. Altenkirch, T., Reus, B.: Monadic presentations of lambda terms using generalized inductive types. In: Flum, J., Rodríguez-Artalejo, M. (eds.) CSL 1999. LNCS, vol. 1683, pp. 453–468. Springer, Heidelberg (1999)
6. Bauer, A., Pretnar, M.: Programming with algebraic effects and handlers. J. Log. Algebr. Meth. Program. **84**(1), 108–123 (2015)
7. Bove, A.: Simple general recursion in type theory. Nord. J. Comput. **8**(1), 22–42 (2001)
8. Bove, A., Capretta, V.: Nested general recursion and partiality in type theory. In: Boulton, R.J., Jackson, P.B. (eds.) TPHOLs 2001. LNCS, vol. 2152, pp. 121–135. Springer, Heidelberg (2001)
9. Brady, E., McBride, C., McKinna, J.: Inductive families need not store their indices. In: Berardi, S., Coppo, M., Damiani, F. (eds.) TYPES 2003. LNCS, vol. 3085, pp. 115–129. Springer, Heidelberg (2004)
10. Capretta, V.: General recursion via coinductive types. Log. Meth. Comput. Sci. **1**(2), 1–28 (2005)
11. Dybjer, P., Setzer, A.: A finite axiomatization of inductive-recursive definitions. In: Girard, J.-Y. (ed.) TLCA 1999. LNCS, vol. 1581, pp. 129–146. Springer, Heidelberg (1999)
12. Dybjer, P., Setzer, A.: Indexed induction-recursion. In: Kahle, R., Schroeder-Heister, P., Stärk, R.F. (eds.) PTCS 2001. LNCS, vol. 2183, pp. 93–113. Springer, Heidelberg (2001)
13. Ghani, N., Hancock, P.: Containers, monads and induction recursion. Math. Struct. Comput. Sci. FirstView: 1–25, 2 (2015)
14. Ghani, N., Lüth, C., De Marchi, F., Power, J.: Algebras, coalgebras, monads and comonads. Electr. Notes Theor. Comput. Sci. **44**(1), 128–145 (2001)
15. Giménez, E.: Codifying guarded definitions with recursive schemes. In: Smith, J., Dybjer, Peter, Nordström, Bengt (eds.) TYPES 1994. LNCS, vol. 996, pp. 39–59. Springer, Heidelberg (1994)
16. Hancock, P., McBride, C., Ghani, N., Malatesta, L., Altenkirch, T.: Small Induction Recursion. In: Hasegawa, M. (ed.) TLCA 2013. LNCS, vol. 7941, pp. 156–172. Springer, Heidelberg (2013)
17. Martin-Löf, P.: Constructive mathematics and computer programming. In: Cohen, L.J., Los, J., Pfeiffer, H., Podewski, K.-P. (eds.) LMPS 6, pp. 153–175. North-Holland (1982)
18. Moggi, E.: Computational lambda-calculus and monads. In: Proceedings of the Fourth Annual Symposium on Logic in Computer Science (LICS 1989), pp. 14–23. IEEE Computer Society, Pacific Grove, California, USA, 5–8 June 1989

19. Moggi, E.: Notions of computation and monads. Inf. Comput. **93**(1), 55–92 (1991)
20. Plotkin, G.D., Power, J.: Algebraic operations and generic effects. Appl. Categorical Struct. **11**(1), 69–94 (2003)
21. Turner, D.A.: Total functional programming. J. Univ. Comput. Sci. **10**(7), 751–768 (2004)
22. Uustalu, T.: Partiality is an effect. Talk Given at Dagstuhl Seminar on Dependently Typed Programming. http://www.ioc.ee/~tarmo/tday-veskisilla/uustalu-slides.pdf (2004)

Auto in Agda

Programming Proof Search Using Reflection

Wen Kokke(✉) and Wouter Swierstra

Universiteit Utrecht, Utrecht, The Netherlands
wen.kokke@gmail.com, w.s.swierstra@uu.nl

Abstract. As proofs in type theory become increasingly complex, there is a growing need to provide better proof automation. This paper shows how to implement a Prolog-style resolution procedure in the dependently typed programming language Agda. Connecting this resolution procedure to Agda's reflection mechanism provides a first-class proof search tactic for first-order Agda terms. As a result, writing proof automation tactics need not be different from writing any other program.

1 Introduction

Writing proof terms in type theory is hard and often tedious. Interactive proof assistants based on type theory, such as Agda (Norell, 2007) or Coq (2004), take very different approaches to facilitating this process.

The Coq proof assistant has two distinct language fragments. Besides the programming language Gallina, there is a separate tactic language for writing and programming proof scripts. Together with several highly customizable tactics, the tactic language Ltac can provide powerful proof automation (Chlipala, 2013). Having to introduce a separate tactic language, however, seems at odds with the spirit of type theory, where a single language is used for both proof and computation. Having a separate language for programming proofs has its drawbacks: programmers need to learn another language to automate proofs, debugging Ltac programs can be difficult, and the resulting proof automation may be inefficient (Braibant, 2012).

Agda does not have Coq's segregation of proof and programming language. Instead, programmers are encouraged to automate proofs by writing their own solvers (Norell, 2009b). In combination with Agda's reflection mechanism (Agda development team, 2013, van der Walt and Swierstra, 2013), developers can write powerful automatic decision procedures (Allais, 2010). Unfortunately, not all proofs are easily automated in this fashion. If this is the case, the user is forced to interact with the integrated development environment and manually construct a proof term step by step.

This paper tries to combine the best of both worlds by implementing a library for proof search *within* Agda itself. In other words, we have defined a *program* for the automatic *construction* of *mathematical* proofs. More specifically, this paper makes several novel contributions:

R. Hinze and J. Voigtländer (Eds.): MPC 2015, LNCS 9129, pp. 276–301, 2015.
DOI:10.1007/978-3-319-19797-5_14

- We show how to implement a Prolog interpreter in the style of (Stutterheim, 2013) in Agda (Sect. 3). Note that, in contrast to Agda, resolving a Prolog query need not terminate. Using coinduction, however, we can write an interpreter for Prolog that is *total.*
- Resolving a Prolog query results in a substitution that, when applied to the goal, produces a solution in the form of a term that can be derived from the given rules. We extend our interpreter to also produce a trace of the applied rules, which enables it to produce a proof term that shows the resulting substitution is valid.
- We integrate this proof search algorithm with Agda's *reflection* mechanism (Sect. 4). This enables us to *quote* the type of a lemma we would like to prove, pass this term as the goal of our proof search algorithm, and finally, *unquote* the resulting proof term, thereby proving the desired lemma.

Although Agda already has built-in proof search functionality (Lindblad and Benke, 2004), our approach has several key advantages over most existing approaches to proof automation:

- Our library is highly customizable. We may parametrize our tactics over the search depth, hint database, or search strategy. Each of these is itself a first-class Agda value, that may be inspected or transformed, depending on the user's needs.
- Although we limit ourself in the paper to a simple depth-first search, different proofs may require a different search strategy. Such changes are easily made in our library. To illustrate this point, we will develop a variation of our tactic which allows the user to limit the number of times certain rules may be applied (Sect. 5).
- Users need not learn a new programming language to modify existing tactics or develop tactics of their own. They can use a full-blown programming language to define their tactics, rather than restrict themselves to a domain-specific tactic language such as Ltac.
- Finally, users can use all the existing Agda technology for testing and debugging *programs* when debugging the generation of *proofs.* Debugging complex tactics in Coq requires a great deal of expertise – we hope that implementing tactics as a library will make this process easier.

We will compare our library with the various alternative forms of proof automation in greater depth in Sect. 6, after we have presented our development.

All the code described in this paper is freely available from GitHub[1]. It is important to emphasize that all our code is written in the safe fragment of Agda: it does not depend on any postulates or foreign functions; all definitions pass Agda's termination checker; and all metavariables are resolved.

2 Motivation

Before describing the *implementation* of our library, we will provide a brief introduction to Agda's reflection mechanism and illustrate how the proof automation described in this paper may be used.

[1] See https://github.com/wenkokke/AutoInAgda.

Reflection in Agda

Agda has a *reflection* mechanism[2] for compile time metaprogramming in the style of Lisp (Pitman, 1980), MetaML (Taha and Sheard, 1997), and Template Haskell (Sheard and Peyton Jones, 2002). This reflection mechanism makes it possible to convert a program fragment into its corresponding abstract syntax tree and vice versa. We will introduce Agda's reflection mechanism here with several short examples, based on the explanation in previous work (van der Walt and Swierstra, 2013). A more complete overview can be found in the Agda release notes (Agda development team, 2013) and Van der Walt's thesis (2012).

The type Term : Set is the central type provided by the reflection mechanism. It defines an abstract syntax tree for Agda terms. There are several language constructs for quoting and unquoting program fragments. The simplest example of the reflection mechanism is the quotation of a single term. In the definition of idTerm below, we quote the identity function on Boolean values.

```
idTerm : Term
idTerm = quoteTerm (λ (x : Bool) → x)
```

When evaluated, the idTerm yields the following value:

```
lam visible (var 0 [])
```

On the outermost level, the lam constructor produces a lambda abstraction. It has a single argument that is passed explicitly (as opposed to Agda's implicit arguments). The body of the lambda consists of the variable identified by the De Bruijn index 0, applied to an empty list of arguments.

The **quote** language construct allows users to access the internal representation of an *identifier*, a value of a built-in type Name. Users can subsequently request the type or definition of such names.

Dual to quotation, the **unquote** mechanism allows users to splice in a Term, replacing it with its concrete syntax. For example, we could give a convoluted definition of the K combinator as follows:

```
const : ∀ {A B} → A → B → A
const = unquote (lam visible (lam visible (var 1 [])))
```

The language construct **unquote** is followed by a value of type Term. In this example, we manually construct a Term representing the K combinator and splice it in the definition of const. The **unquote** construct then type-checks the given term, and turns it into the definition λ x → λ y → x.

The final piece of the reflection mechanism that we will use is the **quoteGoal** construct. The usage of **quoteGoal** is best illustrated with an example:

---

[2] Note that Agda's reflection mechanism should not be confused with 'proof by reflection' – the technique of writing a verified decisionprocedure for some class of problems.

```
goalInHole : ℕ
goalInHole = quoteGoal g in { }0
```

In this example, the construct **quoteGoal** g binds the Term representing the *type* of the current goal, ℕ, to the variable g. When completing this definition by filling in the hole labeled 0, we may now refer to the variable g. This variable is bound to def ℕ [], the Term representing the type ℕ.

Using Proof Automation

To illustrate the usage of our proof automation, we begin by defining a predicate Even on natural numbers as follows:

```
data Even : ℕ → Set where
  isEven0   : Even 0
  isEven+2 : ∀ {n} → Even n → Even (suc (suc n))
```

Next we may want to prove properties of this definition:

```
even+ : Even n → Even m → Even (n + m)
even+ isEven0          e2 = e2
even+ (isEven+2 e1) e2 = isEven+2 (even+ e1 e2)
```

Note that we omit universally quantified implicit arguments from the typeset version of this paper, in accordance with convention used by Haskell (Peyton Jones, 2003) and Idris (Brady, 2013).

As shown by Van der Walt and Swierstra (2013), it is easy to decide the Even property for closed terms using proof by reflection. The interesting terms, however, are seldom closed. For instance, if we would like to use the even+ lemma in the proof below, we need to call it explicitly.

```
trivial : Even n → Even (n + 2)
trivial e = even+ e (isEven+2 isEven0)
```

Manually constructing explicit proof objects in this fashion is not easy. The proof is brittle. We cannot easily reuse it to prove similar statements such as Even (n + 4). If we need to reformulate our statement slightly, proving Even (2 + n) instead, we need to rewrite our proof. Proof automation can make propositions more robust against such changes.

Coq's proof search tactics, such as auto, can be customized with a *hint database*, a collection of related lemmas. In our example, auto would be able to prove the trivial lemma, provided the hint database contains at least the constructors of the Even data type and the even+ lemma. In contrast to the construction of explicit proof terms, changes to the theorem statement need not break the proof. This paper shows how to implement a similar tactic as an ordinary function in Agda.

Before we can use our auto function, we need to construct a hint database:

```
hints : HintDB
hints = ε « quote isEven0 « quote isEven+2 « quote even+
```

To construct such a database, we use **quote** to obtain the names of any terms that we wish to include in it and pass them to the right-hand side of the _«_ function, which will insert them into a hint database to the left. Note that ε represents the empty hint database. We will describe the implementation of _«_ in more detail in Sect. 4. For now it should suffice to say that, in the case of even+, after the **quote** construct obtains an Agda Name, _«_ uses the built-in function type to look up the type associated with even+, and generates a derivation rule which states that given two proofs of Even n and Even m, applying the rule even+ will result in a proof of Even (n + m).

Note, however, that unlike Coq, the hint data base is a *first-class* value that can be manipulated, inspected, or passed as an argument to a function.

We now give an alternative proof of the trivial lemma using the auto tactic and the hint database defined above:

```
trivial : Even n → Even (n + 2)
trivial = quoteGoal g in unquote (auto 5 hints g)
```

Or, using the newly added Agda tactic syntax[3]:

```
trivial : Even n → Even (n + 2)
trivial = tactic (auto 5 hints)
```

The notation **tactic** f is simply syntactic sugar for **quoteGoal** g **in unquote** (f g), for some function f.

The central ingredient is a *function* auto with the following type:

```
auto : (depth : ℕ) → HintDB → Term → Term
```

Given a maximum depth, hint database, and goal, it searches for a proof Term that witnesses our goal. If this term can be found, it is spliced back into our program using the **unquote** statement.

Of course, such invocations of the auto function may fail. What happens if no proof exists? For example, trying to prove Even n → Even (n + 3) in this style gives the following error:

```
Exception searchSpaceExhausted !=<
  Even .n -> Even (.n + 3) of type Set
```

When no proof can be found, the auto function generates a dummy term with a type that explains the reason the search has failed. In this example, the search space has been exhausted. Unquoting this term, then gives the type error message above. It is up to the programmer to fix this, either by providing a manual proof or diagnosing why no proof could be found.

[3] Syntax for Agda tactics was added in Agda 2.4.2.

Overview. The remainder of this paper describes how the auto function is implemented. Before delving into the details of its implementation, however, we will give a high-level overview of the steps involved:

1. The **tactic** keyword converts the goal type to an abstract syntax tree, i.e., a value of type Term. In what follows we will use AgTerm to denote such terms, to avoid confusion with the other term data type that we use.
2. Next, we check the goal term. If it has a functional type, we add the arguments of this function to our hint database, implicitly introducing additional lambdas to the proof term we intend to construct. At this point we check that the remaining type and all its original arguments are are first-order. If this check fails, we produce an error message, not unlike the searchSpaceExhausted term we saw above. We require terms to be first-order to ensure that the unification algorithm, used in later steps for proof search, is decidable. If the goal term is first-order, we convert it to our own term data type for proof search, PsTerm.
3. The key proof search algorithm, presented in the next section, then tries to apply the hints from the hint database to prove the goal. This process coinductively generates a (potentially infinite) search tree. A simple bounded depth-first search through this tree tries to find a series of hints that can be used to prove the goal.
4. If such a proof is found, this is converted back to an AgTerm; otherwise, we produce an erroneous term describing that the search space has been exhausted. Finally, the **unquote** keyword type checks the generated AgTerm and splices it back into our development.

The rest of this paper will explain these steps in greater detail.

3 Proof Search in Agda

The following section describes our implementation of proof search á la Prolog in Agda. This implementation abstracts over two data types for names—one for inference rules and one for term constructors. These data types will be referred to as RuleName and TermName, and will be instantiated with concrete types (with the same names) in Sect. 4.

Terms and Unification

The heart of our proof search implementation is the structurally recursive unification algorithm described by (McBride, 2003). Here the type of terms is indexed by the number of variables a given term may contain. Doing so enables the formulation of the unification algorithm by structural induction on the number of free variables. For this to work, we will use the following definition of terms

```
data PsTerm (n : ℕ) : Set where
  var : Fin n → PsTerm n
  con : TermName → List (PsTerm n) → PsTerm n
```

We will use the name PsTerm to stand for *proof search term* to differentiate them from the terms from Agda's *reflection* mechanism, AgTerm. In addition to variables, represented by the finite type Fin n, we will allow first-order constants encoded as a name with a list of arguments.

For instance, if we choose to instantiate TermName with the following Arith data type, we can encode numbers and simple arithmetic expressions:

```
data Arith : Set where
  Suc  : Arith
  Zero : Arith
  Add  : Arith
```

The closed term corresponding to the number one could be written as follows:

```
One : PsTerm 0
One = con Suc (con Zero [] :: [])
```

Similarly, we can use the var constructor to represent open terms, such as x + 1. We use the prefix operator # to convert from natural numbers to finite types:

```
AddOne : PsTerm 1
AddOne = con Add (var (# 0) :: One :: [])
```

Note that this representation of terms is untyped. There is no check that enforces addition is provided precisely two arguments. Although we could add further type information to this effect, this introduces additional overhead without adding safety to the proof automation presented in this paper. For the sake of simplicity, we have therefore chosen to work with this untyped definition.

We shall refrain from further discussion of the unification algorithm itself. Instead, we restrict ourselves to presenting the interface that we will use:

```
unify : (t₁ t₂ : PsTerm m) → Maybe (∃ [n] Subst m n)
```

The unify function takes two terms t_1 and t_2 and tries to compute a substitution—the most general unifier. Substitutions are indexed by two natural numbers m and n. A substitution of type Subst m n can be applied to a PsTerm m to produce a value of type PsTerm n. As unification may fail, the result is wrapped in the Maybe type. In addition, since the number of variables in the terms resulting from the unifying substitution is not known *a priori*, this number is existentially quantified over. For the remainder of the paper, we will write ∃ [x] B to mean a type B with occurrences of an existentially quantified variable x, or ∃ (λ x → B) in full.

Inference Rules

The hints in the hint database will form *inference rules* that we may use to prove a goal term. We represent such rules as records containing a rule name, a list of terms for its premises, and a term for its conclusion:

```
AddBase : Rule 1
AddBase = record {
  name       = "AddBase"
  conclusion = con Add ( con Zero []
                       :: var (# 0)
                       :: var (# 0)
                       :: [])
  premises   = []
  }
AddStep : Rule 3
AddStep = record {
  name       = "AddStep"
  conclusion = con Add ( con Suc (var (# 0) :: [])
                       :: var (# 1)
                       :: con Suc (var (# 2) :: [])
                       :: [])
  premises   = con Add ( var (# 0)
                       :: var (# 1)
                       :: var (# 2)
                       :: [])
             :: []
  }
```

Fig. 1. Agda representation of example rules

```
record Rule (n : ℕ) : Set where
  field
    name       : RuleName
    premises   : List (PsTerm n)
    conclusion : PsTerm n
  arity : ℕ
  arity = length premises
```

Once again the data-type is quantified over the number of variables used in the rule. Note that the number of variables in the premises and the conclusion is the same.

Using our newly defined Rule type we can give a simple definition of addition. In Prolog, this would be written as follows.

```
add(0, X, X).
add(suc(X), Y, suc(Z)) :- add(X, Y, Z).
```

Unfortunately, the named equivalents in our Agda implementation given in Fig. 1 are a bit more verbose. Note that we have, for the sake of this example, instantiated the RuleName and TermName to String and Arith respectively.

A *hint database* is nothing more than a list of rules. As the individual rules may have different numbers of variables, we existentially quantify these:

```
HintDB : Set
HintDB = List (∃ [n] Rule n)
```

Generalised Injection and Raising

Before we can implement some form of proof search, we need to define a pair of auxiliary functions. During proof resolution, we will work with terms and rules containing a different number of variables. We will use the following pair of functions, inject and raise, to weaken bound variables, that is, map values of type Fin n to some larger finite type.

```
inject : ∀ {m} n → Fin m → Fin (m + n)
inject n zero    = zero
inject n (suc i) = suc (inject n i)
raise : ∀ m {n} → Fin n → Fin (m + n)
raise zero    i = i
raise (suc m) i = suc (raise m i)
```

On the surface, the inject function appears to be the identity. When you make all the implicit arguments explicit, however, you will see that it sends the zero constructor in Fin m to the zero constructor of type Fin (m + n). Hence, the inject function maps Fin m into the *first* m elements of the type Fin (m + n). Dually, the raise function maps Fin n into the *last* n elements of the type Fin (m + n) by repeatedly applying the suc constructor.

We can use inject and raise to define similar functions that work on our Rule and PsTerm data types, by mapping them over all the variables that they contain.

Constructing the Search Tree

Our proof search procedure is consists of two steps. First, we coinductively construct a (potentially infinite) search space; next, we will perform a bounded depth-first traversal of this space to find a proof of our goal.

We will represent the search space as a (potentially) infinitely deep, but finitely branching rose tree.

```
data SearchTree (A : Set) : Set where
  leaf : A → SearchTree A
  node : List (∞ (SearchTree A)) → SearchTree A
```

We will instantiate the type parameter A with a type representing proof terms. These terms consist of applications of rules, with a sub-proof for every premise.

```
data Proof : Set where
  con : (name : RuleName) (args : List Proof) → Proof
```

Unfortunately, during the proof search we will have to work with *partially complete* proof terms.

Such partial completed proofs are represented by the PartialProof type. In contrast to the Proof data type, the PartialProof type may contain variables, hence the type takes an additional number as its argument:

```
PartialProof : ℕ → Set
PartialProof m = ∃ [k] Vec (PsTerm m) k × (Vec Proof k → Proof)
```

A value of type PartialProof m records three separate pieces of information:

- a number k, representing the number of open subgoals;
- a vector of length k, recording the subgoals that are still open;
- a function that, given a vector of k proofs for each of the subgoals, will produce a complete proof of the original goal.

Next, we define the following function to help construct partial proof terms:

```
apply : (r : Rule n) → Vec Proof (arity r + k) → Vec Proof (suc k)
apply r xs = new :: rest
  where
    new = con (name r) (toList (take (arity r) xs))
    rest = drop (arity r) xs
```

Given a Rule and a list of proofs of subgoals, this apply function takes the required sub-proofs from the vector, and creates a new proof by applying the argument rule to these sub-proofs. The result then consists of this new proof, together with any unused sub-proofs. This is essentially the 'unflattening' of a rose tree.

We can now finally return to our proof search algorithm. The solveAcc function forms the heart of the search procedure. Given a hint database and the current partially complete proof, it produces a SearchTree containing completed proofs.

```
solveAcc : HintDB → PartialProof (δ + m) → SearchTree Proof
solveAcc rules (0     ,      [] , p) = leaf (p [])
solveAcc rules (suc k, g :: gs, p) = node (map step rules)
```

If there are no remaining subgoals, i.e., the list in the second component of the PartialProof is empty, the search is finished. We construct a proof p [], and wrap this in the leaf constructor of the SearchTree. If we still have open subgoals, we have more work to do. In that case, we will try to apply every rule in our hint database to resolve this open goal—our rose tree has as many branches as there are hints in the hint database. The real work is done by the step function, locally defined in a where clause, that given the rule to apply, computes the remainder of the SearchTree.

Before giving the definition of the step function, we will try to provide some intuition. Given a rule, the step function will try to unify its conclusion with the current subgoal g. When this succeeds, the premises of the rule are added to the list of open subgoals. When this fails, we return a node with no children, indicating that applying this rule can never prove our current goal.

Carefully dealing with variables, however, introduces some complication, as the code for the step function illustrates:

```
step : ∃ [δ] (Rule δ) → ∞ (SearchTree Proof)
step (δ, r)
  with unify (inject δ g) (raise m (conclusion r))
... | nothing        = ♯ node []
... | just (n, mgu)  = ♯ solveAcc prf
  where
  prf : PartialProof n
  prf = arity r + k, gs′, (p ∘ apply r)
    where
    gs′ : Vec (Goal n) (arity r + k)
    gs′ = map (sub mgu) (raise m (fromList (premises r)) ++ inject δ gs)
```

Note that we use the function sub to apply a substitution to a term. This function is defined by McBride (2003).

The rule given to the step function may have a number of free variables of its own. As a result, all goals have to be injected into a larger domain which includes all current variables *and* the new rule's variables. The rule's premises and conclusion are then also raised into this larger domain, to guarantee freshness of the rule variables.

The definition of the step function attempts to unify the current subgoal g and conclusion of the rule r. If this fails, we can return node [] immediately. If this succeeds, however, we build up a new partial proof, prf. This new partial proof, once again, consists of three parts:

- the number of open subgoals is incremented by arity r, i.e., the number of premises of the rule r.
- the vector of open subgoals gs is extended with the premises of r, after weakening the variables of appropriately.
- the function producing the final Proof object will, given the proofs of the premises of r, call apply r to create the desired con node in the final proof object.

The only remaining step, is to kick-off our proof search algorithm with a partial proof, consisting of a single goal.

```
solve : (goal : PsTerm m) → HintDB → SearchTree Proof
solve g rules = solveAcc (1, g :: [], head)
```

Searching for Proofs

After all this construction, we are left with a simple tree structure, which we can traverse in search of solutions. For instance, we can define a bounded depth-first traversal.

```
dfs : (depth : ℕ) → SearchTree A → List A
dfs zero     _          = []
dfs (suc k) (leaf x)    = return x
dfs (suc k) (node xs)   = concatMap (λ x → dfs k (♭ x)) xs
```

It is fairly straightforward to define other traversal strategies, such as a breadth-first search. Similarly, we could define a function which traverses the search tree aided by some heuristic. We will explore further variations on search strategies in Sect. 5.

4 Adding Reflection

To complete the definition of our auto function, we still need to convert between Agda's built-in AgTerm data type and the data type required by our unification and resolution algorithms, PsTerm. Similarly, we will need to transform the Proof produced by our solve function to an AgTerm that can be unquoted. These are essential pieces of plumbing, necessary to provide the desired proof automation. While not conceptually difficult, this does expose some of the limitations and design choices of the auto function. If you are unfamiliar with the precise workings of the Agda reflection mechanism, you may want to skim this section.

The first thing we will need are concrete definitions for the TermName and RuleName data types, which were parameters to the development presented in the previous section. It would be desirable to identify both types with Agda's Name type, but unfortunately Agda does not assign a name to the function space type operator, _ → _; nor does Agda assign names to locally bound variables. To address this, we define two new data types TermName and RuleName.

First, we define the TermName data type.

```
data TermName : Set where
  name : Name → TermName
  pvar : ℕ → TermName
  impl : TermName
```

The TermName data type has three constructors. The name constructor embeds Agda's built-in Name in the TermName type. The pvar constructor describes locally bound variables, represented by their De Bruijn index. Note that the pvar constructor has nothing to do with PsTerm's var constructor: it is not used to construct a Prolog variable, but rather to be able to refer to a local variable as a Prolog constant. Finally, impl explicitly represents the Agda function space.

We define the RuleName type in a similar fashion.

```
data RuleName : Set where
  name : Name → RuleName
  rvar : ℕ → RuleName
```

The rvar constructor is used to refer to Agda variables as rules. Its argument i corresponds to the variable's De Bruijn index – the value of i can be used directly as an argument to the var constructor of Agda's Term data type.

As we have seen in Sect. 2, the auto function may fail to find the desired proof. Furthermore, the conversion from AgTerm to PsTerm may also fail for various reasons. To handle such errors, we will work in the Error monad defined below:

```
Error : (A : Set a) → Set a
Error A = Message ⊎ A
```

Upon failure, the auto function will produce an error message. The corresponding Message type simply enumerates the possible sources of failure:

```
data Message : Set where
  searchSpaceExhausted : Message
  unsupportedSyntax    : Message
```

The meaning of each of these error messages will be explained as we encounter them in our implementation below.

Finally, we will need one more auxiliary function to manipulate bound variables. The match function takes two bound variables of types Fin m and Fin n and computes the corresponding variables in Fin (m ⊔ n) – where m ⊔ n denotes the maximum of m and n:

```
match : Fin m → Fin n → Fin (m ⊔ n) × Fin (m ⊔ n)
```

The implementation is reasonably straightforward. We compare the numbers n and m, and use the inject function to weaken the appropriate bound variable. It is straightforward to use this match function to define similar operations on two terms or a term and a list of terms.

Constructing Terms

We now turn our attention to the conversion of an AgTerm to a PsTerm. There are two problems that we must address.

First of all, the AgTerm type represents all (possibly higher-order) terms, whereas the PsTerm type is necessarily first-order. We mitigate this problem by allowing the conversion to 'fail', by producing a term of the type Exception, as we saw in the introduction.

Secondly, the AgTerm data type uses natural numbers to represent variables. The PsTerm data type, on the other hand, represents variables using a finite type Fin n, for some n. To convert between these representations, the function keeps track of the current depth, i.e. the number of Π-types it has encountered, and uses this information to ensure a correct conversion. We sketch the definition of the main function below:

```
convert : (binders : ℕ) → AgTerm → Error (∃ PsTerm)
convert b (var i [])     = inj₂ (convertVar b i)
convert b (con n args) = convertName n ∘ convert b ⟨$⟩ args
```

```
convert b (def n args)  =  convertName n ∘ convert b ⟨$⟩ args
convert b (pi (arg (arg-info visible _) (el _ t₁)) (el _ t₂))
  with convert b t₁ | convert (suc b) t₂
... | inj₁ msg      | _           = inj₁ msg
... | _             | inj₁ msg    = inj₁ msg
... | inj₂ (n₁, p₁) | inj₂ (n₂, p₂)
  with match p₁ p₂
... | (p₁′, p₂′) = inj₂′ (n₁ ⊔ n₂, con impl (p₁′ :: p₂′ :: []))
convert b (pi (arg _ _) (el _ t₂))  =  convert (suc b) t₂
convert b _                 = inj₁ unsupportedSyntax
```

We define special functions, convertVar and name2term, to convert variables and constructors or defined terms respectively. The arguments to constructors or defined terms are processed using the convertChildren function defined below. The conversion of a pi node binding an explicit argument proceeds by converting the domain and then codomain. If both conversions succeed, the resulting terms are matched and a PsTerm is constructed using impl. Implicit arguments and instance arguments are ignored by this conversion function. Sorts, levels, or any other Agda feature mapped to the constructor unknown of type Term triggers a failure with the message unsupportedSyntax.

The convertChildren function converts a list of Term arguments to a list of Prolog terms, by stripping the arg constructor and recursively applying the convert function. We only give its type signature here, as the definition is straightforward:

```
convertChildren : ℕ → List (Arg Term) → Error (∃ (List ∘ PsTerm))
```

To convert between an AgTerm and PsTerm we simply call the convert function, initializing the number of binders encountered to 0.

```
agda2term : AgTerm → Error (∃ PsTerm)
agda2term t = convert 0 t
```

Constructing Rules

Our next goal is to construct rules. More specifically, we need to convert a list of quoted Names to a hint database of Prolog rules. To return to our example in Sect. 2, the definition of even+ had the following type:

```
even+ : Even n → Even m → Even (n + m)
```

We would like to construct a value of type Rule that expresses how even+ can be used. In Prolog, we might formulate the lemma above as the rule:

```
even(add(M,N)) :- even(M), even(N).
```

In our Agda implementation, we can define such a rule manually:

```
Even+  : Rule 2
Even+  = record {
  name       = name even+
  conclusion = con (name (quote Even)) (
               con (name (quote _+_)) (var (# 0) :: var (# 1) :: [])
               :: []
               )
  premises   =    con (name (quote Even)) (var (# 0) :: [])
               :: con (name (quote Even)) (var (# 1) :: [])
               :: []
  }
```

In the coming subsection, we will show how to generate the above definition from the Name representing even+.

This generation of rules is done in two steps. First, we will convert a Name to its corresponding PsTerm:

```
name2term : Name → Error (∃ PsTerm)
name2term = agda2term ∘ unel ∘ type
```

The type construct maps a Name to the AgTerm representing its type; the unel function discards any information about sorts; the agda2term was defined previously.

In the next step, we process this PsTerm. The split function, defined below, splits a PsTerm at every top-most occurrence of the function symbol impl. Note that it would be possible to define this function directly on the AgTerm data type, but defining it on the PsTerm data type is much cleaner as we may assume that any unsupported syntax has already been removed.

```
split : PsTerm n → ∃ (λ k → Vec (PsTerm n) (suc k))
split (con impl (t₁ :: t₂ :: [])) = Product.map suc (_::_ t₁) (split t₂)
split t = (0, t :: [])
```

Using all these auxiliary functions, we now define the name2rule function below that constructs a Rule from an Agda Name.

```
name2rule : Name → Error (∃ Rule)
name2rule nm with name2term nm
... | inj₁ msg            = inj₁ msg
... | inj₂ (n, t)         with split t
... | (k, ts)             with initLast ts
... | (prems, concl, _) = inj₂ (n, rule (name nm) concl (toList prems))
```

We convert a name to its corresponding PsTerm, which is converted to a vector of terms using split. The last element of this vector is the conclusion of the rule; the prefix constitutes the premises. We use the initLast function from the Agda standard library, to decompose this vector accordingly.

Constructing Goals

Next, we turn our attention to converting a goal AgTerm to a PsTerm. While we could use the agda2term function to do so, there are good reasons to explore other alternatives.

Consider the example given in Sect. 2. The goal AgTerm we wish to prove is Even n → Even (n + 2). Calling agda2term would convert this to a PsTerm, where the function space has been replaced by the constructor impl. Instead, however, we would like to *introduce* arguments, such as Even n, as assumptions to our hint database.

In addition, we cannot directly reuse the implementation of convert that was used in the construction of terms. The convert function maps every AgTerm variable is mapped to a Prolog variable *that may still be instantiated.* When considering the goal type, however, we want to generate *skolem constants* for our variables. To account for this difference we have two flavours of the convert function: convert and convert4Goal. Both differ only in their implementation of convertVar.

```
agda2goal × premises : AgTerm → Error (∃ PsTerm × HintDB)
agda2goal × premises t with convert4Goal 0 t
... | inj₁ msg            = inj₁ msg
... | inj₂ (n, p)         with split p
... | (k, ts)             with initLast ts
... | (prems, goal, _)    = inj₂ ((n, goal), toPremises k prems)
```

Fortunately, we can reuse many of the other functions we have defined above, and, using the split and initLast functions, we can get our hands on the list of premises prems and the desired return type goal. The only missing piece of the puzzle is a function, toPremises, which converts a list of PsTerms to a hint database containing rules for the arguments of our goal.

```
toPremises : ∀ {k} → ℕ → Vec (PsTerm n) k → HintDB
toPremises i [] = []
toPremises i (t :: ts) = (n, rule (rvar i) t []) :: toPremises (suc i) ts
```

The toPremises converts every PsTerm in its argument list to a rule, using the argument's De Bruijn index as its rule name.

Reification of Proof Terms

Now that we can compute Prolog terms, goals and rules from an Agda Term, we are ready to call the resolution mechanism described in Sect. 3. The only remaining problem is to convert the witness computed by our proof search back to an AgTerm, which can be unquoted to produce the desired proof. This is done by the reify function that traverses its argument Proof; the only interesting question is how it handles the variables and names it encounters.

The Proof may contain two kinds of variables: locally bound variables, rvar i, or variables storing an Agda Name, name n. Each of these variables is treated differently in the reify function.

```
reify : Proof → AgTerm
reify (con (rvar i) ps) = var i []
reify (con (name n) ps) with definition n
... | function x    = def n (toArg ∘ reify ⟨$⟩ ps)
... | constructor'  = con n (toArg ∘ reify ⟨$⟩ ps)
... | _             = unknown
  where
    toArg : AgTerm → Arg AgTerm
    toArg = arg (arg-info visible relevant)
```

Any references to locally bound variables are mapped to the var constructor of the AgTerm data type. These variables correspond to usage of arguments to the function being defined. As we know by construction that these arguments are mapped to rules without premises, the corresponding Agda variables do not need any further arguments.

If, on the other hand, the rule being applied is constructed using a name, we do disambiguate whether the rule name refers to a function or a constructor. The definition function, defined in Agda's reflection library, tells you how a name was defined (i.e. as a function name, constructor, etc). For the sake of brevity, we restrict the definition here to only handle defined functions and data constructors. It is easy enough to extend with further branches for postulates, primitives, and so forth.

We will also need to wrap additional lambdas around the resulting term, due to the premises that were introduced by the agda2goal × premises function. To do so, we define the intros function that repeatedly wraps its argument term in a lambda.

```
intros : AgTerm → AgTerm
intros = introsAcc (length args)
  where
    introsAcc : ℕ → AgTerm → AgTerm
    introsAcc zero t = t
    introsAcc (suc k) t = lam visible (introsAcc k t)
```

Hint Databases

Users to provide hints, i.e., rules that may be used during resolution, in the form of a *hint database*. These hint databases consist of an (existentially quantified) a list of rules. We can add new hints to an existing database using the insertion operator, «, defined as follows:

```
_«_ : HintDB → Name → HintDB
db « n with name2rule n
```

```
db « n | inj₁ msg = db
db « n | inj₂ r   = db ++ [r]
```

If the generation of a rule fails for whatever reason, no error is raised, and the rule is simply ignored. Our actual implementation requires an implicit proof argument that all the names in the argument list can be quoted successfully. If you define such proofs to compute the trivial unit record as evidence, Agda will fill them in automatically in every call to the _«_ function on constant arguments. This simple form of proof automation is pervasive in Agda programs (Oury and Swierstra, 2008, Swierstra, 2010).

This is the simplest possible form of hint database. In principle, there is no reason not to define alternative versions that assign priorities to certain rules or limit the number of times a rule may be applied. We will investigate some possibilities for extensible proof search in Sect. 5.

It is worth repeating that hint databases are first-class objects. We can combine hints databases, filter certain rules from a hint database, or manipulate them in any way we wish.

Error Messages

Lastly, we need to decide how to report error messages. Since we are going to return an AgTerm, we need to transform the Message type we saw previously into an AgTerm. When unquoted, this term will cause a type error, reporting the reason for failure. To accomplish this, we introduce a dependent type, indexed by a Message:

```
data Exception : Message → Set where
  throw : (msg : Message) → Exception msg
```

The message passed as an argument to the throw constructor, will be recorded in the Exception's type, as we intended.

Next, we define a function to produce an AgTerm from a Message. We could construct such terms by hand, but it is easier to just use Agda's **quoteTerm** construct:

```
quoteError : Message → Term
quoteError searchSpaceExhausted =
  quoteTerm (throw searchSpaceExhausted)
quoteError unsupportedSyntax    =
  quoteTerm (throw unsupportedSyntax)
```

Putting it all Together

Finally, we can present the definition of the auto function used in the examples in Sect. 2:

```
auto : ℕ → HintDB → AgTerm → AgTerm
auto depth rules goalType
```

```
  with agda2goal × premises goalType
... | inj₁ msg = quoteError msg
... | inj₂ ((n, g), args)
  with dfs depth (solve g (args ++ rules))
... | [] = quoteError searchSpaceExhausted
... | (p :: _) = intros (reify p)
```

The auto function takes an AgTerm representing the goal type, splits it into PsTerms representing the goal g and a list of arguments, args. These arguments are added to the initial hint database. Calling the solve function with this hint database and the goal g, constructs a proof tree, that we traverse up to the given depth in search of a solution. If this proof search succeeds, the Proof is converted to an AgTerm, a witness that the original goal is inhabited. There are two places where this function may fail: the conversion to a PsTerm may fail because of unsupported syntax; or the proof search may not find a result.

5 Extensible Proof Search

As we promised in the previous section, we will now explore several variations and extensions to the auto tactic described above.

Custom Search Strategies

The simplest change we can make is to abstract over the search strategy used by the auto function. In the interest of readability we will create a simple alias for the types of search strategies. A Strategy represents a function which searches a SearchTree up to depth, and returns a list of the leaves (or Proofs) found in the SearchTree in an order which is dependent on the search strategy.

```
Strategy = (depth : ℕ) → SearchTree A → List A
```

The changed type of the auto function now becomes.

```
auto : Strategy → ℕ → HintDB → AgTerm → AgTerm
```

This will allow us to choose whether to pass in dfs, breadth-first search or even a custom user-provided search strategy.

Custom Hint Databases

In addition, we have developed a variant of the auto tactic described in the paper that allows users to define their own type of hint database, provided they can implement the following interface:

```
HintDB  : Set
Hint    : ℕ → Set
```

```
getHints : HintDB → Hints
getRule  : Hint k → Rule k
getTr    : Hint k → (HintDB → HintDB)
```

Besides the obvious types for hints and rules, we allow hint databases to evolve during the proof search. The user-defined getTr function describes a transformation that may modify the hint database after a certain hint has been applied.

Using this interface, we can implement many variations on proof search. For instance, we could implement a 'linear' proof search function that removes a rule from the hint database after it has been applied. Alternatively, we may want to assign priorities to our hints. To illustrate one possible application of this interface, we will describe a hint database implementation that limits the usage of certain rules. Before we do so, however, we need to introduce a motivating example.

Example: Limited Usage of Hints

We start by defining the following sublist relation, taken from the Agda tutorial (Norell, 2009a):

```
data _⊆_ : List A → List A → Set where
  stop : [] ⊆ []
  drop : xs ⊆ ys →      xs ⊆ y :: ys
  keep : xs ⊆ ys → x :: xs ⊆ x :: ys
```

It is easy to show that the sublist relation is both reflexive and transitive—and using these simple proofs, we can build up a small hint database to search for proofs on the sublist relation.

```
hintdb : HintDB
hintdb = ε « quote drop « quote keep « quote ⊆-refl « quote ⊆-trans
```

Our auto tactic quickly finds a proof for the following lemma:

```
lemma₁ : ws ⊆ 1 :: xs → xs ⊆ ys → ys ⊆ zs → ws ⊆ 1 :: 2 :: zs
lemma₁ = tactic (auto dfs 10 hintdb)
```

The following lemma, however, is false.

```
lemma₂ : ws ⊆ 1 :: xs → xs ⊆ ys → ys ⊆ zs → ws ⊆ 2 :: zs
lemma₂ = tactic (auto dfs 10 hintdb)
```

Indeed, this example does not type check and our tactic reports that the search space is exhausted. As noted by (Chlipala, 2013) when examining tactics in Coq, auto will nonetheless spend a considerable amount of time trying to construct a proof. As the trans rule is always applicable, the proof search will construct a search tree up to the full search depth—resulting in an exponental running time.

We will use a variation of the auto tactic to address this problem. Upon constructing the new hint database, users may assign limits to the number of times certain hints may be used. By limiting the usage of transitivity, our tactic will fail more quickly.

To begin with, we choose the representation of our hints: a pair of a rule and a 'counter' that records how often the rule may still be applied:

```
record Hint (k : ℕ) : Set where
  field
    rule    : Rule k
    counter : Counter
```

These counter values will either be a natural number n, representing that the rule can still be used at most n times; or ⊤, when the usage of the rule is unrestricted.

```
Counter : Set
Counter = ℕ ⊎ ⊤
```

Next, we define a decrementing function, decrCounter, that returns nothing when a rule can no longer be applied:

```
decrCounter : Counter → Maybe Counter
decrCounter (inj₁ 0)  = nothing
decrCounter (inj₁ 1)  = nothing
decrCounter (inj₁ x)  = just (inj₁ (pred x))
decrCounter (inj₂ tt) = just (inj₂ tt)
```

Given a hint h, the transition function will now simply find the position of h in the hint database and decrement the hint's counter, removing it from the database if necessary.

We can redefine the default insertion function (_«_) to allow unrestricted usage of a rule. However, we will define a new insertion function which will allow the user to limit the usage of a rule during proof search:

```
_« [_] _ : HintDB → ℕ → Name → HintDB
db « [0] _ = db
db « [x] n with (name2rule n)
db « [x] n | inj₁ msg   = db
db « [x] n | inj₂ (k, r) = db ++ [k, record {rule = r, counter = inj₁ x}]
```

We now revisit our original hint database and limit the number of times transitivity may be applied:

```
hintdb : HintDB
hintdb = ε «     quote drop
           «     quote keep
           «     quote refl
           « [2] quote trans
```

If we were to search for a proof of lemma_2 now, our proof search fails sooner. *A fortiori*, if we use this restricted database when searching for a proof of lemma_1, the auto function succeeds sooner, as we have greatly reduced the search space. Of course, there are plenty of lemmas that require more than two applications of transitivity. The key insight, however, is that users now have control over these issues – something which is not even possible in current implementations of auto in Coq.

6 Discussion

The auto function presented here is far from perfect. This section not only discusses its limitations, but compares it to existing proof automation techniques in interactive proof assistants.

Restricted Language Fragment. The auto function can only handle first-order terms. Even though higher-order unification is not decidable in general, we believe that it should be possible to adapt our algorithm to work on second-order goals. Furthermore, there are plenty of Agda features that are not supported or ignored by our quotation functions, such as universe polymorphism, instance arguments, and primitive functions.

Even for definitions that seem completely first-order, our auto function can fail unexpectedly. Consider the following definition of the pair type:

```
_×_ : (A B : Set) → Set
A × B = Σ A (λ _ → B)
pair : {A B : Set} → A → B → A × B
pair x y = x, y
```

Here a (non-dependent) pair is defined as a special case of the dependent pair type Σ. Now consider the following trivial lemma:

```
andIntro : (A : Set) → (B : Set) → A × B
```

Somewhat surprisingly, trying to prove this lemma using our auto function, providing the pair function as a hint, fails. The **quoteGoal** construct always returns the goal in normal form, which exposes the higher-order nature of A × B. Converting the goal (A × (λ _ → B)) to a PsTerm will raise the ‘exception’ unsupportedSyntax; the goal type contains a lambda which causes the proof search to fail before it has even started.

Refinement. The auto function returns a complete proof term or fails entirely. This is not always desirable. We may want to return an incomplete proof, that still has open holes that the user must complete. The difficulty lies with the current implementation of Agda's reflection mechanism, as it cannot generate an incomplete Term.

In the future, it may be interesting to explore how to integrate proof automation using the reflection mechanism better with Agda's IDE. For instance, we could create an IDE feature which replaces a call to auto with the proof terms that it generates. As a result, reloading the file would no longer need to recompute the proof terms.

Metatheory. The auto function is necessarily untyped because the interface of Agda's reflection mechanism is untyped. Defining a well-typed representation of dependent types in a dependently typed language remains an open problem, despite various efforts in this direction (Chapman, 2009, Danielsson, 2006, Devriese and Piessens, 2013, McBride, 2010). If we had such a representation, however, we could use the type information to prove that when the auto function succeeds, the resulting term has the correct type. As it stands, a bug in our auto function could potentially produce an ill-typed proof term, that only causes a type error when that term is unquoted.

Variables. The astute reader will have noticed that the tactic we have implemented is closer to Coq's eauto tactic than the auto tactic. The difference between the two tactics lies in the treatment of unification variables: eauto may introduce new variables during unification; auto will never do so. It would be fairly straightforward to restrict our tactic to only apply hints when all variables known. A suitable instantiation algorithm, which we could use instead of the more general unification algorithm in this paper, has already been developed in previous work (Swierstra and van Noort, 2013).

Technical Limitations. The auto tactic relies on the unification algorithm and proof search mechanism we have implemented ourselves. These are all run *at compile time*, using the reflection mechanism to try and find a suitable proof term. It is very difficult to say anything meaningful about the performance of the auto tactic, as Agda currently has no mechanism for debugging or profiling programs run at compile time. We hope that further advancement of the Agda compiler and associated toolchain can help provide meaningful measurements of the performance of auto. Similarly, a better (static) debugger would be invaluable when trying to understand why a call to auto failed to produce the desired proof.

Related Work

There are several other interactive proof assistants, dependently typed programming languages, and alternative forms of proof automation in Agda. In the remainder of this section, we will briefly compare the approach taken in this paper to these existing systems.

Coq. Coq has rich support for proof automation. The Ltac language and the many primitive, customizable tactics are extremely powerful (Chlipala, 2013). Despite Coq's success, it is still worthwhile to explore better methods for proof automation. Recent work on Mtac (Ziliani et al., 2013) shows how to add a typed language for proof automation on top of Ltac. Furthermore, Ltac itself is not designed to be a general purpose programming language. It can be difficult to abstract over certain patterns and debugging proof automation is not easy. The programmable proof automation, written using reflection, presented here may not be as mature as Coq's Ltac language, but addresses these issues.

More recently, Malecha et al. [2014] have designed a higher-order reflective programming language (MirrorCore) and an associated tactic language (Rtac). MirrorCore defines a unification algorithm – similar to the one we have implemented in this paper. Alternative implementations of several familiar Coq tactics, such as eauto and setoid_rewrite, have been developed using Rtac. The authors have identified several similar advantages of 'programming' tactics, rather than using built-in primitives, that we mention in this paper, such as manipulating and assembling first-class hint databases.

Idris. The dependently typed programming language Idris also has a collection of tactics, inspired by some of the more simple Coq tactics, such as rewrite, intros, or exact. Each of these tactics is built-in and implemented as part of the Idris system. There is a small Haskell library for tactic writers to use that exposes common commands, such as unification, evaluation, or type checking. Furthermore, there are library functions to help handle the construction of proof terms, generation of fresh names, and splitting sub-goals. This approach is reminiscent of the HOL family of theorem provers (Gordon and Melham, 1993) or Coq's plug-in mechanism. An important drawback is that tactic writers need to write their tactics in a different language to the rest of their Idris code; furthermore, any changes to tactics requires a recompilation of the entire Idris system.

Agsy. Agda already has a built-in 'auto' tactic that outperforms the auto function we have defined here (Lindblad and Benke, 2004). It is nicely integrated with the IDE and does not require the users to provide an explicit hint database. It is, however, implemented in Haskell and shipped as part of the Agda system. As a result, users have very few opportunities for customization: there is limited control over which hints may (or may not) be used; there is no way to assign priorities to certain hints; and there is a single fixed search strategy. In contrast to the proof search presented here, where we have much more fine grained control over all these issues.

Conclusion

The proof automation presented in this paper is not as mature as some of these alternative systems. Yet we strongly believe that this style of proof automation is worth pursuing further.

The advantages of using reflection to program proof tactics should be clear: we do not need to learn a new programming language to write new tactics; we can use existing language technology to debug and test our tactics; and we can use all of Agda's expressive power in the design and implementation of our tactics. If a particular problem domain requires a different search strategy, this can be implemented by writing a new traversal over a SearchTree. Hint databases are first-class values. There is never any built-in magic; there are no compiler primitives beyond Agda's reflection mechanism.

The central philosophy of Martin-Löf type theory is that the construction of programs and proofs is the same activity. Any external language for proof automation renounces this philosophy. This paper demonstrates that proof automation is not inherently at odds with the philosophy of type theory. Paraphrasing Martin-Löf (1985), it no longer seems possible to distinguish the discipline of *programming* from the *construction* of mathematics.

Acknowledgements. We would like to thank the Software Technology Reading Club at the Universiteit Utrecht, and all our anonymous reviewers for their helpful feedback – we hope we have done their feedback justice.

References

Agda development team. Agda release notes documenting the reflection mechanism. The Agda Wiki (2013). http://wiki.portal.chalmers.se/agda/agda.php?n=Main.Version-2-2-8 and http://wiki.portal.chalmers.se/agda/agda.php?n=Main.Version-2-3-0. Accessed 9 Feb 2013

Allais, G.: Proof automatization using reflection (implementations in Agda). MSc Intern report, University of Nottingham (2010)

Brady, E.: Idris, a general-purpose dependently typed programming language: design and implementation. J. Funct. Program. **9**, 23:552–593 (2013). doi:10.1017/S095679681300018X

Braibant, T.: Emancipate yourself from LTac (2012). http://gallium.inria.fr/blog/your-first-coq-plugin/

Chapman, J.: Type checking and normalisation. Ph.D. thesis, University of Nottingham (2009)

Chlipala, A.: Certified Programming with Dependent Types. MIT Press, Cambridge (2013)

Danielsson, N.A.: A formalisation of a dependently typed language as an inductive-recursive family. In: Altenkirch, T., McBride, C. (eds.) TYPES 2006. LNCS, vol. 4502, pp. 93–109. Springer, Heidelberg (2007). doi:10.1145/2500365.2500575

Coq development team. The Coq proof assistant reference manual. Logical Project (2004)

Devriese, D., Piessens, F.: Typed syntactic meta-programming. In: Proceedings of the 2013 ACM SIGPLAN International Conference on Functional Programming (ICFP 2013). ACM, September 2013. doi:10.1145/2500365.2500575

Gordon, M.J.C., Melham, T.F.: Introduction to HOL: A Theorem Proving Environment for Higher Order Logic. Cambridge University Press, New York (1993)

Lindblad, F., Benke, M.: A tool for automated theorem proving in Agda. In: Filliâtre, J.-C., Paulin-Mohring, C., Werner, B. (eds.) TYPES 2004. LNCS, vol. 3839, pp. 154–169. Springer, Heidelberg (2006)

Malecha, G., Chlipala, A., Braibant, T.: Compositional computational reflection. In: Klein, G., Gamboa, R. (eds.) ITP 2014. LNCS, vol. 8558, pp. 374–389. Springer, Heidelberg (2014)

Martin-Löf, P.: Constructive mathematics and computer programming. In: Proceedings of a Discussion Meeting of the Royal Society of London on Mathematical Logic and Programming Languages, pp. 167–184. Prentice-Hall Inc. (1985)

McBride, C.: First-order unification by structural recursion. J. Funct. Program. **11**, 13:1061–1075 (2003). doi:10.1017/S0956796803004957

McBride, C.: Outrageous but meaningful coincidences: dependent type-safe syntax and evaluation. In: Proceedings of the 6th ACM SIGPLAN Workshop on Generic Programming, WGP 2010, pp. 1–12. ACM, New York (2010). doi:10.1145/1863495.1863497

Norell, U.: Towards a practical programming language based on dependent type theory. Ph.D. thesis, Department of Computer Science and Engineering, Chalmers University of Technology (2007)

Norell, U.: Dependently typed programming in Agda. In: Koopman, P., Plasmeijer, R., Swierstra, D. (eds.) AFP 2008. LNCS, vol. 5832, pp. 230–266. Springer, Heidelberg (2009a)

Norell, U.: Playing with Agda. Invited talk at TPHOLS (2009b)

Oury, N., Swierstra, W.: The power of Pi. In: Proceedings of the 13th ACM SIGPLAN International Conference on Functional Programming, ICFP 2008, pp. 39–50 (2008). doi:10.1145/1411204.1411213

Jones, S.P. (ed.): Haskell 98 Language and Libraries: the Revised report. Cambridge University Press, Cambridge (2003)

Pitman, K.M.: Special forms in LISP. In: Proceedings of the 1980 ACM Conference on LISP and Functional Programming, pp. 179–187. ACM (1980)

Sheard, T., Jones, S.P.: Template meta-programming for Haskell. In: Proceedings of the 2002 ACM SIGPLAN Workshop on Haskell, pp 1–16 (2002). doi:10.1145/581690.581691

Stutterheim, J., Swierstra, W., Swierstra, D.: Forty hours of declarative programming: teaching prolog at the junior college Utrecht. In: Proceedings First International Workshop on Trends in Functional Programming in Education, Electronic Proceedings in Theoretical Computer Science, University of St. Andrews, Scotland, UK, 11 June 2012, vol. 106, pp 50–62 (2013)

Swierstra, W.: More dependent types for distributed arrays. High.-Ord. Symbol. Comput. **23**(4), 489–506 (2010)

Swierstra, W., van Noort, T.: A library for polymorphic dynamic typing. J. Funct. Program. **23**, 229–248 (2013). doi:10.1017/S0956796813000063

Taha, W., Sheard, T.: Multi-stage programming with explicit annotations. In: Proceedings of the 1997 ACM SIGPLAN Symposium on Partial Evaluation and Semantics-Based Program Manipulation, PEPM 1997 (1997). doi:10.1145/258993.259019

van der Walt, P., Swierstra, W.: Engineering proof by reflection in Agda. In: Hinze, R. (ed.) IFL 2012. LNCS, vol. 8241, pp. 157–173. Springer, Heidelberg (2013)

van der Walt, P.: Reflection in Agda. Master's thesis, Department of Computer Science, Utrecht University, Utrecht, The Netherlands (2012). http://igitur-archive.library.uu.nl/student-theses/2012-1030-200720/UUindex.html

Ziliani, B., Dreyer, D., Krishnaswami, N.R., Nanevski, A., Vafeiadis, V.: Mtac: a monad for typed tactic programming in Coq. In: Proceedings of the 18th ACM SIGPLAN International Conference on Functional Programming, ICFP 2013, pp 87–100 (2013). doi:10.1145/2500365.2500579

Fusion for Free

Efficient Algebraic Effect Handlers

Nicolas Wu[1(✉)] and Tom Schrijvers[2(✉)]

[1] Department of Computer Science, University of Bristol, Bristol, UK
[2] Department of Computer Science, KU Leuven, Leuven, Belgium
nicolas.wu@bristol.ac.uk, tom.schrijvers@cs.kuleuven.be

Abstract. Algebraic effect handlers are a recently popular approach for modelling side-effects that separates the syntax and semantics of effectful operations. The shape of syntax is captured by functors, and free monads over these functors denote syntax trees. The semantics is captured by algebras, and effect handlers pass these over the syntax trees to interpret them into a semantic domain.

This approach is inherently modular: different functors can be composed to make trees with richer structure. Such trees are interpreted by applying several handlers in sequence, each removing the syntactic constructs it recognizes. Unfortunately, the construction and traversal of intermediate trees is painfully inefficient and has hindered the adoption of the handler approach.

This paper explains how a sequence of handlers can be fused into one, so that multiple tree traversals can be reduced to a single one and no intermediate trees need to be allocated. At the heart of this optimization is keeping the notion of a free monad abstract, thus enabling a change of representation that opens up the possibility of fusion. We demonstrate how the ensuing code can be inlined at compile time to produce efficient handlers.

1 Introduction

Free monads are currently receiving a lot of attention. They are at the heart of *algebraic effect handlers*, a new purely functional approach for modelling side effects introduced by Plotkin and Power [15]. Much of their appeal stems from the separation of the syntax and semantics of effectful operations. This is both conceptually simple and flexible, as multiple different semantics can be provided for the same syntax.

The syntax of the primitive side-effect operations is captured in a signature functor. The free monad over this functor assembles the syntax for the individual operations into an abstract syntax tree for an effectful program. The semantics of the individual operations is captured in an algebra, and an effect handler folds the algebra over the syntax tree of the program to interpret it into a semantic domain.

A particular strength of the approach is its inherent modularity. Different signature functors can be composed to make trees with richer structure. Such

R. Hinze and J. Voigtländer (Eds.): MPC 2015, LNCS 9129, pp. 302–322, 2015.
DOI:10.1007/978-3-319-19797-5_15

trees are interpreted by applying several handlers in sequence, each removing the syntactic constructs it recognizes.

Unfortunately, the construction and traversal of intermediate trees is rather costly. This inefficiency is perceived as a serious weakness of effect handlers, especially when compared to the traditional approach of composing effects with monad transformers. While several authors address the cost of constructing syntax trees with free monads, efficiently applying multiple handlers in sequence has received very little attention. As far as we know, only Kammar et al. [9] provide an efficient implementation. Unfortunately, this implementation does not come with an explanation. Hence it is underappreciated and ill-understood.

In this paper we close the gap and explain how a sequence of algebraic effect handlers can be effectively fused into a single handler. Central to the paper are the many interpretations of the word *free*. Interpreting *free* monads as the initial objects of the more general term algebras and, in particular, term monads provide an essential change of perspective where *free* theorems enable fusion. The codensity monad facilitates the way, turning any term algebra into a monadic one for *free*, and with an appropriate code setup in Haskell the GHC compiler takes care of fusion at virtually no cost[1] to the programmer. The result is an effective implementation that compares well to monad transformers.

2 Algebraic Effect Handlers

The idea of the algebraic effect handlers approach is to consider the free monad over a particular functor as an abstract syntax tree (AST) for an effectful computation. The functor is used to generate the nodes of a free structure whose leaves correspond to variables. This can be defined as an inductive datatype *Free f* for a given functor *f*.

data *Free f a* **where**
  Var :: *a* → *Free f a*
  Con :: *f* (*Free f a*) → *Free f a*

The nodes are *con*structed by *Con*, and the *var*iables are given by *Var*.

Since a value of type *Free f a* is an inductive structure, we can define a *fold* for it by providing a function *gen* that deals with generation of values from *Var x*, and an algebra *alg* that is used to recursively collapse an operation *Con op*.

fold :: *Functor f* ⇒ (*f b* → *b*) → (*a* → *b*) → (*Free f a* → *b*)
fold alg gen (*Var x*) = *gen x*
fold alg gen (*Con op*) = *alg* (*fmap* (*fold alg gen*) *op*)

Algebraic effect handlers give a semantics to the syntax tree: one way of doing this is by using a *fold*.

[1] Yes, almost for *free*!

The behaviour of folds when composed with other functions is described by fusion laws. The first law describes how certain functions that are precomposed with a fold can be incorporated into a new fold:

$$fold\ alg\ gen \cdot fmap\ h = fold\ alg\ (gen \cdot h) \tag{1}$$

The second law shows how certain functions that are postcomposed with a fold can be incorporated into a new fold:

$$k \cdot fold\ alg\ gen = fold\ alg'\ (k \cdot gen) \tag{2}$$

this is subject to the condition that $k \cdot alg = alg' \cdot fmap\ k$.

The monadic instance of the free monad is witnessed by the following:

```
instance Functor f ⇒ Monad (Free f) where
  return x = Var x
  m >>= f  = fold Con f m
```

Variables are the way of providing a return for the monad, and extending a syntax tree by means of a function f corresponds to applying that function to the variables found at the leaves of the tree.

2.1 Nondeterminism

A functor supplies the abstract syntax for the primitive effectful operations in the free monad. For instance, the *Nondet* functor provides the *Or k k* syntax for a binary nondeterministic choice primitive. The parameter to the constructor of type k marks the recursive site of syntax, which indicates where the continuation is after this syntactic fragment has been evaluated.

```
data Nondet k where
  Or :: k → k → Nondet k
instance Functor Nondet where
  fmap f (Or x y) = Or (f x) (f y)
```

This allows us to express the syntax tree of a computation that nondeterministically returns *True* or *False*.

```
coin :: Free Nondet Bool
coin = Con (Or (Var True) (Var False))
```

The syntax is complemented by semantics in the form of effect handlers—functions that replace the syntax by values from a semantic domain. Using a *fold* for the free monad is a natural way of expressing such functions. For instance, here is a handler that interprets *Nondet* in terms of lists of possible outcomes:

$$handle_{Nondet_{[]}} :: Free\ Nondet\ a \to [a]$$
$$handle_{Nondet_{[]}} = fold\ alg_{Nondet_{[]}}\ gen_{Nondet_{[]}}$$

where $alg_{Nondet_{[]}}$ is the *Nondet*-algebra that interprets terms constructed by *Or* operations and $gen_{Nondet_{[]}}$ interprets variables.

$$
\begin{aligned}
&alg_{Nondet_{[]}} :: Nondet\ [a] \to [a] \\
&alg_{Nondet_{[]}}\ (Or\ l_1\ l_2) = l_1 +\!\!+\ l_2 \\
&gen_{Nondet_{[]}} :: a \to [a] \\
&gen_{Nondet_{[]}}\ x = [x]
\end{aligned}
$$

The variables of the syntax tree are turned into singleton lists by $gen_{Nondet_{[]}}$, and choices between alternatives are put together by $alg_{Nondet_{[]}}$, which appends lists.

As an example, we can interpret the *coin* program:

$$
\begin{aligned}
&> handle_{Nondet_{[]}}\ coin \\
&[True, False]
\end{aligned}
$$

This particular interpretation gives us a list of the possible outcomes.

Generalizing away from the details, handlers are usually presented in the form

$$hdl :: \forall a\ .\ Free\ F\ a \to H\ a$$

where F and H are arbitrary functors determined by the handler.

2.2 Handler Composition

There are many useful scenarios that involve the (function) composition of effect handlers. We now consider two classes of this kind of composition.

Effect Composition. A first important class of scenarios is where multiple effects are combined in the same program. To this end we compose signatures and handlers.

The coproduct functor $f + g$ makes it easy to compose functors:

$$
\begin{aligned}
&\textbf{data}\ (+)\ f\ g\ a\ \textbf{where} \\
&\quad Inl :: f\ a \to (f + g)\ a \\
&\quad Inr :: g\ a \to (f + g)\ a \\
&\textbf{instance}\ (Functor\ f, Functor\ g) \Rightarrow Functor\ (f + g)\ \textbf{where} \\
&\quad fmap\ f\ (Inl\ s) = Inl\ (fmap\ f\ s) \\
&\quad fmap\ f\ (Inr\ s) = Inr\ (fmap\ f\ s)
\end{aligned}
$$

The free monad of a coproduct functor is a tree where each node can be built from syntax from either f or g.

Composing handlers is easy too: if the handlers are written in a compositional style, then function composition does the trick. A compositional handler for the functor F has a signature of the form:

$$hdl :: \forall g\ a\ .\ Free\ (F + g)\ a \to H_1\ (Free\ g\ (G_1\ a))$$

This processes only the F-nodes in the AST and leaves the g-nodes as they are. Hence the result of the compositional handler is a new (typically smaller) AST with only g-nodes. The variables of type $G_1\ a$ in the resulting AST are derived from the variables of type a in original AST as well as from the processed operations. Moreover, the new AST may be embedded in a context H_1.

For instance, the compositional nondeterminism handler is defined as follows, where $F = Nondet$, $G_1 = [\,]$ and implicitly $H_1 = \mathsf{Id}$.

$$\begin{array}{l} handle_{Nondet} :: Functor\ g \Rightarrow Free\ (Nondet + g)\ a \to Free\ g\ [a] \\ handle_{Nondet} = fold\ (alg_{Nondet} \triangledown Con)\ gen_{Nondet} \end{array}$$

Here the variables are handled with the monadified version of $gen_{Nondet_{[]}}$, given by gen_{Nondet}.

$$\begin{array}{l} gen_{Nondet} :: Functor\ g \Rightarrow a \to Free\ g\ [a] \\ gen_{Nondet}\ x = Var\ [x] \end{array}$$

The g nodes are handled by a Con algebra, which essentially leaves them untouched. The $Nondet$ nodes are handled by the alg_{Nondet} algebra, which is a monadified version of $alg_{Nondet_{[]}}$.

$$\begin{array}{l} alg_{Nondet} :: Functor\ g \Rightarrow Nondet\ (Free\ g\ [a]) \to Free\ g\ [a] \\ alg_{Nondet}\ (Or\ ml_1\ ml_2) = \\ \quad \mathbf{do}\ \{l_1 \leftarrow ml_1;\ l_2 \leftarrow ml_2;\ Var\ (l_1 +\!\!+ l_2)\} \end{array}$$

The junction combinator $(\triangledown)$ composes the algebras for the two kinds of nodes.

$$\begin{array}{l} (\triangledown) :: (f\ b \to b) \to (g\ b \to b) \to ((f + g)\ b \to b) \\ (\triangledown)\ alg_f\ alg_g\ (Inl\ s) = alg_f\ s \\ (\triangledown)\ alg_f\ alg_g\ (Inr\ s) = alg_g\ s \end{array}$$

In the definition of $handlerNondet$, we use $alg_{Nondet} \triangledown Con$. Since the functor in question is $Nondet + g$, the values constructed by $Nondet$ are handled by alg_{Nondet}, and values constructed by g are left untouched: the fold unwraps one level of Con, but then replaces it with a Con again.

A second example of an effect signature is that of state, whose primitive operations Get and Put respectively query and modify the implicit state.

$$\begin{array}{l} \mathbf{data}\ State\ s\ k\ \mathbf{where} \\ \quad Put :: s \to k \to State\ s\ k \\ \quad Get :: (s \to k) \to State\ s\ k \\ \mathbf{instance}\ Functor\ (State\ s)\ \mathbf{where} \\ \quad fmap\ f\ (Put\ s\ k) = Put\ s\ (f\ k) \\ \quad fmap\ f\ (Get\ k) \quad = Get\ (f \cdot k) \end{array}$$

The compositional handler for state is as follows, where $F = State\ s$, $H_1 = s \rightarrow -$ and implicitly $G_1 = \mathsf{Id}$.

$$handle_{State} :: Functor\ g \Rightarrow Free\ (State\ s + g)\ a \rightarrow (s \rightarrow Free\ g\ a)$$
$$handle_{State} = fold\ (alg_{State} \triangledown con_{State})\ gen_{State}$$

This time the variable and constructor cases are defined as:

$$gen_{State} :: Functor\ g \Rightarrow a \rightarrow (s \rightarrow Free\ g\ a)$$
$$gen_{State}\ x\ s = Var\ x$$

$$alg_{State} :: Functor\ g \Rightarrow State\ s\ (s \rightarrow Free\ g\ a) \rightarrow (s \rightarrow Free\ g\ a)$$
$$alg_{State}\ (Put\ s'\ k)\ s = k\ s'$$
$$alg_{State}\ (Get\ k)\quad s = k\ s\ s$$

Using gen_{State}, a variable x is replaced by a version that ignores any incoming state parameter s. Any stateful constructs are handled by alg_{State}, where a continuation k proceeds by using the appropriate state: if the syntax is a $Put\ s'\ k$, then the new state is s', otherwise the syntax is $Get\ k$, in which case the state is left unchanged and passed as s.

Finally, any syntax provided by g is adapted by con_{State} to take the extra state parameter s into account:

$$con_{State} :: Functor\ g \Rightarrow g\ (s \rightarrow Free\ g\ a) \rightarrow (s \rightarrow Free\ g\ a)$$
$$con_{State}\ op\ s = Con\ (fmap\ (\lambda m \rightarrow m\ s)\ op)$$

This feeds the state s to the continuations of the operation.

To demonstrate effect composition we can put *Nondet* and *State* together and handle them both. Before we do so, we also need a base case for the composition, which is the empty signature *Void*.

data *Void k*

instance *Functor Void*

The *Void* handler only provides a variable case since the signature has no constructors. In fact, a *Free Void* term can only be a $Var\ x$, so x is immediately output using the identity function.

$$handle_{Void} :: Free\ Void\ a \rightarrow a$$
$$handle_{Void} = fold\ \bot\ id$$

Finally, we can put together a composite handler for programs that feature both nondeterminism and state. The signature of such programs is the composition of the three basic signatures:

type $\Sigma = Nondet + (State\ Int + Void)$

The handler is the composition of the three handlers, working from the left-most functor in the signature:

$$handle_{\Sigma} :: Free\ \Sigma\ a \rightarrow Int \rightarrow [a]$$
$$handle_{\Sigma}\ prog = handle_{Void} \cdot (handle_{State} \cdot handle_{Nondet})\ prog$$

Effect Delegation. Another important class of applications are those where a handler expresses the complex semantics of particular operations in terms of more primitive effects.

For instance, the following logging handler for state records every update of the state by means of the *Writer* effect.

$$\begin{array}{l}
\mathit{handle}_{\mathit{LogState}} :: \mathit{Free}\ (\mathit{State}\ s)\ a \to s \to \mathit{Free}\ (\mathit{Writer}\ \mathit{String} + \mathit{Void})\ a \\
\mathit{handle}_{\mathit{LogState}} = \mathit{fold}\ \mathit{alg}_{\mathit{LogState}}\ \mathit{gen}_{\mathit{State}} \\
\mathit{alg}_{\mathit{LogState}} :: \mathit{State}\ s\ (s \to \mathit{Free}\ (\mathit{Writer}\ \mathit{String} + \mathit{Void})\ a) \\
\qquad \to s \to \mathit{Free}\ (\mathit{Writer}\ \mathit{String} + \mathit{Void})\ a \\
\mathit{alg}_{\mathit{LogState}}\ (\mathit{Put}\ s'\ k)\ s = \mathit{Con}\ (\mathit{Inl}\ (\mathit{Tell}\ \texttt{"put"}\ (k\ s'))) \\
\mathit{alg}_{\mathit{LogState}}\ (\mathit{Get}\ k)\quad s = k\ s\ s
\end{array}$$

The syntax of the *Writer* effect is captured by the following functor, where w is a parameter that represents the type of values that are written to the log:

$$\begin{array}{l}
\mathbf{data}\ \mathit{Writer}\ w\ k\ \mathbf{where} \\
\quad \mathit{Tell} :: w \to k \to \mathit{Writer}\ w\ k \\
\mathbf{instance}\ \mathit{Functor}\ (\mathit{Writer}\ w)\ \mathbf{where} \\
\quad \mathit{fmap}\ f\ (\mathit{Tell}\ w\ k) = \mathit{Tell}\ w\ (f\ k)
\end{array}$$

A semantics can be given by the following handler, where w is constrained to be a member of the *Monoid* typeclass.

$$\begin{array}{l}
\mathit{handle}_{\mathit{Writer}} :: (\mathit{Functor}\ g, \mathit{Monoid}\ w) \Rightarrow \mathit{Free}\ (\mathit{Writer}\ w + g)\ a \to \mathit{Free}\ g\ (w, a) \\
\mathit{handle}_{\mathit{Writer}} = \mathit{fold}\ (\mathit{alg}_{\mathit{Writer}} \nabla \mathit{Con})\ \mathit{gen}_{\mathit{Writer}}
\end{array}$$

The variables are evaluated by pairing with the unit of the monoid given by *mempty* before being embedded into the monad m_2.

$$\begin{array}{l}
\mathit{gen}_{\mathit{Writer}} :: (\mathit{Monad}\ m_2, \mathit{Monoid}\ w) \Rightarrow a \to m_2\ (w, a) \\
\mathit{gen}_{\mathit{Writer}}\ x = \mathit{return}\ (\mathit{mempty}, x)
\end{array}$$

When a $\mathit{Tell}\ w_1\ k$ operation is encountered, the continuation k is followed by a state where w_1 is appended using *mappend* to any generated logs.

$$\begin{array}{l}
\mathit{alg}_{\mathit{Writer}} :: (\mathit{Monad}\ m_2, \mathit{Monoid}\ w) \Rightarrow \mathit{Writer}\ w\ (m_2\ (w, a)) \to m_2\ (w, a) \\
\mathit{alg}_{\mathit{Writer}}\ (\mathit{Tell}\ w_1\ k) = k \gg\!= \lambda(w_2, x) \to \mathit{return}\ (w_1\ `\mathit{mappend}`\ w_2, x)
\end{array}$$

To see this machinery in action, consider the following program that makes use of state:

$$\begin{array}{l}
\mathit{program} :: \mathit{Int} \to \mathit{Free}\ (\mathit{State}\ \mathit{Int})\ \mathit{Int} \\
\mathit{program}\ n \\
\quad \mid n \leq 0 \qquad\ \ = \mathit{Con}\ (\mathit{Get}\ \mathit{var}) \\
\quad \mid \mathit{otherwise} = \mathit{Con}\ (\mathit{Get}\ (\lambda s \to \mathit{Con}\ (\mathit{Put}\ (s + n)\ (\mathit{program}\ (n - 1)))))
\end{array}$$

This is then simply evaluated by running handlers in sequence.

$$\begin{array}{l} example :: Int \rightarrow (String, Int) \\ example\ n = (handle_{Void} \cdot handle_{Writer} \cdot handle_{LogState}\ (program\ n))\ 0 \end{array}$$

To fully interpret a stateful program, we must first run $handle_{LogState}$, which interprets the *Tell* operations by generating a tree with *Writer String* syntax. This generated syntax is then handled with the $handle_{Writer}$ handler.

3 Fusion

The previous composition examples lead us to the main challenge of this paper: The composition of two handlers produces an intermediate abstract syntax tree. How can we fuse the two handlers into a single one that does not involve an intermediate tree?

More concretely, given two handlers of the form:

$$\begin{array}{l} handler_1 :: Free\ F_1\ a \rightarrow H_1\ (Free\ F_2\ (G_1\ a)) \\ handler_1 = fold\ alg_1\ gen_1 \\ handler_2 :: Free\ F_2\ a \rightarrow H_2\ a \\ handler_2 = fold\ alg_2\ gen_2 \end{array}$$

where F_1 and F_2 are signature functors and H_1, G_1 and H_2 are arbitrary functors, our goal is to obtain a combined handler

$$\begin{array}{l} pipeline_{12} :: Free\ F_1\ a \rightarrow H_1\ (H_2\ (G_1\ a)) \\ pipeline_{12} = fold\ alg_{12}\ gen_{12} \end{array}$$

such that

$$fmap\ handler_2 \cdot handler_1 = pipeline_{12} \tag{3}$$

3.1 Towards Proper Builders

The fact that $handler_1$ *builds* an AST over functor F_2 and that $handler_2$ *folds* over this AST suggests a particular kind of fusion known as shortcut fusion or fold/build fusion [4].

One of the two key ingredients for this kind of fusion is already manifestly present: the *fold* in $handler_2$. Yet, the structure and type of $handler_1$ do not necessarily force it to be a proper builder: fold/build fusion requires that the builder creates the F_2-AST from scratch by generating all the *Var* and *Con* constructors itself. Indeed, in theory, $handler_1$ could produce the F_2-AST out of ready-made components supplied by (fields of) a colluding F_1 functor.

In order to force $handler_1$ to be a proper builder, we require it to be implemented against a builder interface rather than a concrete representation. This builder interface is captured in the typeclass *TermMonad* (explained below), and then, with the following constraint polymorphic signature, $handler_1$ is guaranteed to build properly:

$$handler_1 :: (TermMonad\ m_2\ F_2) \Rightarrow Free\ F_1\ a \rightarrow H_1\ (m_2\ (G_1\ a))$$

Term Algebras. The concept of a *term algebra* provides an abstract interface for the primitive ways to build an AST: the two constructors *Con* and *Var* of the free monad. We borrow the nomenclature from the literature on universal algebras [2].

A *term algebra* is an f-algebra $con :: f\ (h\ a) \rightarrow h\ a$ with a carrier $h\ a$. The values in $h\ a$ are those generated by the set of variables a with a valuation function $var :: a \rightarrow h\ a$, as well as those that arise out of repeated applications of the algebra. This is modelled by the typeclass *TermAlgebra h f* as follows:

```
class Functor f ⇒ TermAlgebra h f | h → f where
  var :: ∀a . a → h a
  con :: ∀a . f (h a) → h a
```

The function *var* is used to embed a *var*iable into the term, and the function *con* is used to *con*struct a term from existing ones. This typeclass is well-defined only when $h\ a$ is indeed generated by *var* and *con*.

The most trivial instance of this typeclass is of course that of the free monad.

```
instance Functor f ⇒ TermAlgebra (Free f) f where
  var = Var
  con = Con
```

Term Monads. There are two additional convenient ways to build an AST: the monadic primitives *return* and ($\gg\!=$).

A monad m is a *term monad* for a functor f, if it there is a term algebra for f whose carrier is $m\ a$. We can model this relationship as a typeclass with no members.

```
class (Monad m, TermAlgebra m f) ⇒ TermMonad m f | m → f
instance (Monad m, TermAlgebra m f) ⇒ TermMonad m f
```

Again, the free monad is the obvious instance of this typeclass.

```
instance Functor f ⇒ TermMonad (Free f) f
```

Its monadic primitives are implemented in terms of *fold*, *con* and *var*. In the abstract builder interface *TermMonad* we only partially expose this fact, by means of the following two laws. Firstly, the *var* operation should coincide with the monad's *return*.

$$var = return \tag{4}$$

Secondly, the monad's bind ($\gg\!=$) should distribute through *con*.

$$con\ op \gg\!= k = con\ (fmap\ (\gg\!= k)\ op) \tag{5}$$

This law states that a term constructed by an operation, where the term is followed by a continuation k, is equivalent to constructing a term from an operation where the operation is followed by k. In other words, the arguments of an operation correspond to its continuations.

Examples. All the compositional handlers of Sect. 2.2 can be easily expressed in terms of the more abstract *TermMonad* interface. For example, the revised nondeterminism handler looks as follows.

$$\begin{array}{l}
\mathit{handle}'_{\mathit{Nondet}} :: \mathit{TermMonad}\ m\ g \Rightarrow \mathit{Free}\ (\mathit{Nondet} + g)\ a \rightarrow m\ [a] \\
\mathit{handle}'_{\mathit{Nondet}} = \mathit{fold}\ (\mathit{alg}'_{\mathit{Nondet}} \nabla \mathit{con})\ \mathit{gen}'_{\mathit{Nondet}} \\
\mathit{gen}'_{\mathit{Nondet}} :: \mathit{TermMonad}\ m\ g \Rightarrow a \rightarrow m\ [a] \\
\mathit{gen}'_{\mathit{Nondet}}\ x = \mathit{var}\ [x] \\
\mathit{alg}'_{\mathit{Nondet}} :: \mathit{TermMonad}\ m\ g \Rightarrow \mathit{Nondet}\ (m\ [a]) \rightarrow m\ [a] \\
\mathit{alg}'_{\mathit{Nondet}}\ (\mathit{Or}\ \mathit{ml}_1\ \mathit{ml}_2) = \\
\quad \mathbf{do}\ \{l_1 \leftarrow \mathit{ml}_1\,;\ l_2 \leftarrow \mathit{ml}_2\,;\ \mathit{var}\ (l_1 +\!\!+\ l_2)\}
\end{array}$$

Notice that not much change has been necessary. We have generalized away from *Free g* into a type *m* that is constrained by *TermMonad m g*.

3.2 Parametricity: Fold/Build Fusion for Free

Hinze et al. [6] state that we get fold/build fusion for free from the *free theorem* [21,22] of the builder's polymorphic type. Hence, let us consider what the new type of $\mathit{handler}_1$ buys us.

Theorem 1. *Assume that* F_1, F_2, H_1, G_1 *and* A *are closed types, with* F_1, F_2 *and* H_1 *functors. Given a function* h *of type* $\forall m\ .\ (\mathit{TermMonad}\ m\ F_2) \Rightarrow \mathit{Free}\ F_1\ A \rightarrow H_1\ (m\ (G_1\ A))$, *two term monads* M_1 *and* M_2 *and a term monad morphism* $\alpha :: \forall a\ .\ M_1\ a \rightarrow M_2\ a$, *then:*

$$\mathit{fmap}\ \alpha\ \cdot\ h_{M_1} = h_{M_2} \tag{6}$$

where the subscripts of h *denote the instantiations of the polymorphic type variable* m.

If $\mathit{handler}_2$ is a term monad morphism, then we can use the free theorem to determine $\mathit{pipeline}_{12}$ in one step, starting from Eq. (3).

$$\begin{array}{l}
\quad \mathit{fmap}\ \mathit{handler}_2\ \cdot\ \mathit{handler}_1 = \mathit{pipeline}_{12} \\
\equiv\ \{\text{Parametricity (6), assuming } \mathit{handler}_2 \text{ is a term monad morphism}\} \\
\quad \mathit{handler}_1 = \mathit{pipeline}_{12}
\end{array}$$

Unfortunately, $\mathit{handler}_2$ is not a term monad morphism for the simple reason that H_2 is just an arbitrary functor that does not necessarily have a monad structure. Hence, in general H_2 is only term algebra.

$$\begin{array}{l}
\mathbf{instance}\ \mathit{TermAlgebra}\ H_2\ F_2\ \mathbf{where} \\
\quad \mathit{var} = \mathit{gen}_2 \\
\quad \mathit{con} = \mathit{alg}_2
\end{array}$$

Fortunately, we can turn any term algebra into a term monad, thanks to the codensity monad, which is what we explore in the next section.

3.3 Codensity: *TermMonads* from *TermAlgebras*

The Codensity Monad. It is well-known that the codensity monad *Cod* turns any (endo)functor h into a monad (in fact, h need not even be a functor at all). It simply instantiates the generalised monoid of endomorphisms (e.g., see [16]) in the category of endofunctors.

$$
\begin{aligned}
&\textbf{newtype}\ \mathit{Cod}\ h\ a = \mathit{Cod}\ \{\mathit{unCod} :: \forall x\ .\ (a \to h\ x) \to h\ x\} \\
&\textbf{instance}\ \mathit{Monad}\ (\mathit{Cod}\ h)\ \textbf{where} \\
&\quad \mathit{return}\ x = \mathit{Cod}\ (\lambda k \to k\ x) \\
&\quad \mathit{Cod}\ m \gg\!= f = \mathit{Cod}\ (\lambda k \to m\ (\lambda a \to \mathit{unCod}\ (f\ a)\ k))
\end{aligned}
$$

TermMonad Construction. Given any term algebra h for functor f, we have that *Cod h* is also a term algebra for f.

$$
\begin{aligned}
&\textbf{instance}\ \mathit{TermAlgebra}\ h\ f \Rightarrow \mathit{TermAlgebra}\ (\mathit{Cod}\ h)\ f\ \textbf{where} \\
&\quad \mathit{var} = \mathit{return} \\
&\quad \mathit{con} = \mathit{alg}_{\mathit{Cod}}\ \mathit{con} \\
&\mathit{alg}_{\mathit{Cod}} :: \mathit{Functor}\ f \Rightarrow (\forall x\ .\ f\ (h\ x) \to h\ x) \to (f\ (\mathit{Cod}\ h\ a) \to \mathit{Cod}\ h\ a) \\
&\mathit{alg}_{\mathit{Cod}}\ \mathit{alg}\ \mathit{op} = \mathit{Cod}\ (\lambda k \to \mathit{alg}\ (\mathit{fmap}\ (\lambda m \to \mathit{unCod}\ m\ k)\ \mathit{op}))
\end{aligned}
$$

Moreover, *Cod h* is also a term monad for f, even if h is not.

$$\textbf{instance}\ \mathit{TermAlgebra}\ h\ f \Rightarrow \mathit{TermMonad}\ (\mathit{Cod}\ h)\ f$$

The definition of *var* makes it easy to see that it satisfies the first term monad law in Eq. (4). The proof for the second law, Eq. (5) is less obvious:

$$
\begin{aligned}
&\quad \mathit{con}\ \mathit{op} \gg\!= f \\
&\equiv \{\text{unfold } (\gg\!=)\} \\
&\quad \mathit{Cod}\ (\lambda k \to \mathit{unCod}\ (\mathit{con}\ \mathit{op})\ (\lambda a \to \mathit{unCod}\ (f\ a)\ k)) \\
&\equiv \{\text{unfold } \mathit{con}\} \\
&\quad \mathit{Cod}\ (\lambda k \to \mathit{unCod}\ (\mathit{alg}_{\mathit{Cod}}\ \mathit{con}\ \mathit{op})\ (\lambda a \to \mathit{unCod}\ (f\ a)\ k)) \\
&\equiv \{\text{unfold } \mathit{alg}_{\mathit{Cod}}\} \\
&\quad \mathit{Cod}\ (\lambda k \to \mathit{unCod}\ (\mathit{Cod}\ (\lambda k \to \mathit{con}\ (\mathit{fmap}\ (\lambda m \to \mathit{unCod}\ m\ k)\ \mathit{op})))\ (\lambda a \to \mathit{unCod}\ (f\ a)\ k)) \\
&\equiv \{\text{apply } \mathit{unCod} \cdot \mathit{Cod} = \mathit{id}\} \\
&\quad \mathit{Cod}\ (\lambda k \to (\lambda k \to \mathit{con}\ (\mathit{fmap}\ (\lambda m \to \mathit{unCod}\ m\ k)\ \mathit{op}))\ (\lambda a \to \mathit{unCod}\ (f\ a)\ k)) \\
&\equiv \{\text{apply } \beta\text{-reduction}\} \\
&\quad \mathit{Cod}\ (\lambda k \to \mathit{con}\ (\mathit{fmap}\ (\lambda m \to \mathit{unCod}\ m\ (\lambda a \to \mathit{unCod}\ (f\ a)\ k))\ \mathit{op})) \\
&\equiv \{\text{apply } \beta\text{-expansion}\} \\
&\quad \mathit{Cod}\ (\lambda k \to \mathit{con}\ (\mathit{fmap}\ (\lambda m \to (\lambda k \to \mathit{unCod}\ m\ (\lambda a \to \mathit{unCod}\ (f\ a)\ k))\ k)\ \mathit{op})) \\
&\equiv \{\text{apply } \mathit{unCod} \cdot \mathit{Cod} = \mathit{id}\} \\
&\quad \mathit{Cod}\ (\lambda k \to \mathit{con}\ (\mathit{fmap}\ (\lambda m \to \mathit{unCod}\ (\mathit{Cod}\ (\lambda k \to \mathit{unCod}\ m\ (\lambda a \to \mathit{unCod}\ (f\ a)\ k)))\ k)\ \mathit{op})) \\
&\equiv \{\text{fold } (\gg\!=)\} \\
&\quad \mathit{Cod}\ (\lambda k \to \mathit{con}\ (\mathit{fmap}\ (\lambda m \to \mathit{unCod}\ (m \gg\!= f)\ k)\ \mathit{op})) \\
&\equiv \{\text{apply } \beta\text{-expansion}\} \\
&\quad \mathit{Cod}\ (\lambda k \to \mathit{con}\ (\mathit{fmap}\ (\lambda m \to (\lambda m \to \mathit{unCod}\ m\ k)\ (m \gg\!= k))\ \mathit{op})) \\
&\equiv \{\text{apply } \beta\text{-expansion}\} \\
&\quad \mathit{Cod}\ (\lambda k \to \mathit{con}\ (\mathit{fmap}\ (\lambda m \to (\lambda m \to \mathit{unCod}\ m\ k)\ ((\lambda m \to m \gg\!= k)\ m))\ \mathit{op})) \\
&\equiv \{\text{fold } (\cdot)\} \\
&\quad \mathit{Cod}\ (\lambda k \to \mathit{con}\ (\mathit{fmap}\ (\lambda m \to ((\lambda m \to \mathit{unCod}\ m\ k) \cdot (\lambda m \to m \gg\!= k))\ m)\ \mathit{op})) \\
&\equiv \{\text{apply } \eta\text{-reduction}\} \\
&\quad \mathit{Cod}\ (\lambda k \to \mathit{con}\ (\mathit{fmap}\ ((\lambda m \to \mathit{unCod}\ m\ k) \cdot (\lambda m \to m \gg\!= k))\ \mathit{op})) \\
&\equiv \{\text{apply } \mathit{fmap}\text{-fission and unfold } (\cdot)\} \\
&\quad \mathit{Cod}\ (\lambda k \to \mathit{con}\ (\mathit{fmap}\ (\lambda m \to \mathit{unCod}\ m\ k)\ (\mathit{fmap}\ (\lambda m \to m \gg\!= k)\ \mathit{op}))) \\
&\equiv \{\text{fold } \mathit{alg}_{\mathit{Cod}}\} \\
&\quad \mathit{alg}_{\mathit{Cod}}\ \mathit{con}\ (\mathit{fmap}\ (\lambda m \to m \gg\!= k)\ \mathit{op}) \\
&\equiv \{\text{fold } \mathit{con} \text{ and } \eta\text{-reduce}\} \\
&\quad \mathit{con}\ (\mathit{fmap}\ (\gg\!= f)\ \mathit{op})
\end{aligned}
$$

3.4 Shifting to Codensity

Now we can write $handler_2$ as the composition of a term monad morphism $handler'_2$ with a post-processing function $runCod\ gen_2$:

$$
\begin{array}{l}
handler_2 :: Free\ F_2\ a \to H_2\ a \\
handler_2 = runCod\ gen_2 \cdot handler'_2 \\
handler'_2 :: Free\ F_2\ a \to Cod\ H_2\ a \\
handler'_2 = fold\ (alg_{Cod}\ alg_2)\ var \\
runCod :: (a \to f\ x) \to Cod\ f\ a \to f\ x \\
runCod\ g\ m = unCod\ m\ g
\end{array}
$$

This decomposition of $handler_2$ hinges on the following property:

$$fold\ alg_2\ gen_2 = runCod\ gen_2 \cdot fold\ (alg_{Cod}\ alg_2)\ var \tag{7}$$

This equation follows from the second fusion law for folds (2), provided that:

$$
\begin{aligned}
gen_2 &= runCod\ gen_2 \cdot var \\
runCod\ gen_2 \cdot alg_{Cod}\ alg_2 &= alg_2 \cdot fmap\ (runCod\ gen_2)
\end{aligned}
$$

The former holds as follows:

$$
\begin{array}{l}
\quad runCod\ gen_2 \cdot var \\
\equiv \{\text{unfold} \cdot \} \\
\quad \lambda x \to runCod\ gen_2\ (var\ x) \\
\equiv \{\text{unfold } runCod \text{ and } var\ \} \\
\quad \lambda x \to unCod\ (Cod\ (\lambda k \to k\ x))\ gen_2 \\
\equiv \{\text{apply } runCod \cdot Cod = id\ \} \\
\quad \lambda x \to (\lambda k \to k\ x)\ gen_2 \\
\equiv \{\beta\text{-reduction}\ \} \\
\quad \lambda x \to gen_2\ x \\
\equiv \{\eta\text{-reduction}\ \} \\
\quad gen_2
\end{array}
$$

and the latter:

$$
\begin{array}{l}
\quad runCod\ gen_2 \cdot alg_{Cod}\ alg_2 \\
\equiv \{\text{unfold} \cdot \} \\
\quad \lambda op \to runCod\ gen_2\ (alg_{Cod}\ alg_2\ op) \\
\equiv \{\text{unfold } runCod \text{ and } alg_{Cod}\ \} \\
\quad \lambda op \to unCod\ (Cod\ (\lambda k \to alg_2\ (fmap\ (\lambda m \to unCod\ m\ k)\ op)))\ gen_2 \\
\equiv \{\text{apply } runCod \cdot Cod = id\ \} \\
\quad \lambda op \to (\lambda k \to alg_2\ (fmap\ (\lambda m \to unCod\ m\ k)\ op))\ gen_2 \\
\equiv \{\beta\text{-reduction}\ \} \\
\quad \lambda op \to alg_2\ (fmap\ (\lambda m \to unCod\ m\ gen_2)\ op) \\
\equiv \{\text{fold } runCod\ \}
\end{array}
$$

$$\begin{aligned}
& \lambda op \rightarrow alg_2\ (fmap\ (\lambda m \rightarrow runCod\ gen_2\ m)\ op) \\
\equiv\ & \{\eta\text{-reduction}\ \} \\
& \lambda op \rightarrow alg_2\ (fmap\ (runCod\ gen_2)\ op) \\
\equiv\ & \{\text{fold}\ \cdot\ \} \\
& alg_2 \cdot fmap\ (runCod\ gen_2)
\end{aligned}$$

3.5 Fusion at Last

Finally, instead of fusing $fmap\ handler_2 \cdot handler_1$ we can fuse $fmap\ handler'_2 \cdot handler_1$ using the free theorem. This yields:

$$pipeline'_{12} = fold\ alg_1\ gen_1$$

Now we can calculate the original fusion:

$$\begin{aligned}
& fmap\ handler_2 \cdot handler_1 \\
\equiv\ & \{\text{decomposition of } handler_2\ \} \\
& fmap\ (runCod\ gen_2 \cdot handler'_2) \cdot handler_1 \\
\equiv\ & \{fmap \text{ fission}\ \} \\
& fmap\ (runCod\ gen_2) \cdot fmap\ handler'_2 \cdot handler_1 \\
\equiv\ & \{\text{free theorem}\ \} \\
& fmap\ (runCod\ gen_2) \cdot handler_1
\end{aligned}$$

In other words, the fused version can be defined as:

$$pipeline_{12} = fmap\ (runCod\ gen_2) \cdot fold\ alg_1\ gen_1$$

Observe that this version only performs a single *fold* and does not allocate any intermediate tree.

3.6 Repeated Fusion

Often a sequence of handlers is not restricted to two. Fortunately, we can easily generalize the above to a pipeline of n handlers

$$fmap^n\ handler_n \cdot \ldots \cdot fmap\ handler_1 \cdot handler_0$$

where $handler_i = fold\ alg_i\ gen_i\ (i \in 1..n)$. This pipeline fusion can start by arbitrarily fusing two consecutive handlers $handler_j$ and $handler_{j+1}$ using the above approach, and then incrementally extending the fused kernel on the left and the right with additional handlers. These two kinds of extensions are explained below.

Fusion on the Right. Suppose that $fmap\ handler_2 \cdot handler_1$ is composed with another handler on the right:

$$\begin{aligned}
& handler_0 :: (TermMonad\ m_1\ F_1) \Rightarrow Free\ F_0\ a \rightarrow H_0\ (m_1\ (G_0\ a)) \\
& handler_0 = fold\ alg_0\ gen_0
\end{aligned}$$

to form the pipeline:

$$\mathit{pipeline}_{012} = \mathit{fmap}\ (\mathit{fmap}\ \mathit{handler}_2 \cdot \mathit{handler}_1) \cdot \mathit{handler}_0$$

Can we perform the fusion twice to obtain a single *fold* and eliminate both intermediate trees? Yes! The first fusion, as before yields:

$$\mathit{pipeline}_{012} = \mathit{fmap}\ (\mathit{fmap}\ (\mathit{runCod}\ \mathit{gen}_2) \cdot \mathit{handler}_1) \cdot \mathit{handler}_0$$

Applying *fmap* fission and regrouping, we obtain:

$$\mathit{pipeline}_{012} = \mathit{fmap}\ (\mathit{fmap}\ (\mathit{runCod}\ \mathit{gen}_2)) \cdot (\mathit{fmap}\ \mathit{handler}_1 \cdot \mathit{handler}_0)$$

Now the right component is another instance of the binary fusion problem, which yields:

$$\mathit{pipeline}_{012} = \mathit{fmap}\ (\mathit{fmap}\ (\mathit{runCod}\ \mathit{gen}_2)) \cdot \mathit{fmap}\ (\mathit{runCod}\ \mathit{gen}_1) \cdot \mathit{handler}_0$$

Fusion on the Left. Suppose that $\mathit{handler}_2$ has the more specialised type:

$$\mathit{handler}_2 :: (\mathit{TermMonad}\ m_3\ F_3) \Rightarrow \mathit{Free}\ F_2\ a \rightarrow H_2\ (m_2\ (G_2\ a))$$

then we can compose $\mathit{fmap}\ \mathit{handler}_2 \cdot \mathit{handler}_1$ on the left with another handler:

$$\begin{aligned}&\mathit{handler}_3 :: \mathit{Free}\ F_3\ a \rightarrow H_3\ a\\ &\mathit{handler}_3 = \mathit{fold}\ \mathit{alg}_3\ \mathit{gen}_3\end{aligned}$$

This yields a slightly more complicated fusion scenario:

$$\mathit{pipeline}_{123} = \mathit{fmap}\ (\mathit{fmap}\ \mathit{handler}_3 \cdot \mathit{handler}_2) \cdot \mathit{handler}_1$$

Of course, we can first fuse $\mathit{handler}_2$ and $\mathit{handler}_3$. That would yield an instance of fusion on the right. However, suppose we first fuse $\mathit{handler}_1$ and $\mathit{handler}_2$, after applying *fmap* fission.

$$\mathit{pipeline}_{123} = \mathit{fmap}\ (\mathit{fmap}\ \mathit{handler}_3) \cdot \mathit{fmap}\ (\mathit{runCod}\ \mathit{gen}_2) \cdot \mathit{handler}_1$$

Now we can shift the carrier of $\mathit{handler}_3$ to codensity and invoke the free theorem on $\mathit{fmap}\ (\mathit{runCod}\ \mathit{gen}_2) \cdot \mathit{handler}_1$. This accomplishes the second fusion.

$$\mathit{pipeline}_{123} = \mathit{fmap}\ (\mathit{fmap}\ (\mathit{runCod}\ \mathit{gen}_3)) \cdot \mathit{fmap}\ (\mathit{runCod}\ \mathit{gen}_2) \cdot \mathit{handler}_1$$

Summary. An arbitrary pipeline of the form:

$$\mathit{fmap}^n\ \mathit{handler}_n \cdot \ldots \cdot \mathit{fmap}\ \mathit{handler}_1 \cdot \mathit{handler}_0$$

where $\mathit{handler}_i = \mathit{fold}\ \mathit{alg}_i\ \mathit{gen}_i\ (i \in 1..n)$ fuses into

$$\mathit{fmap}^n\ (\mathit{runCod}\ g_n) \cdot \ldots \cdot \mathit{fmap}\ (\mathit{runCod}\ \mathit{gen}_1) \cdot \mathit{handler}_0$$

3.7 Fusion All the Way

We are not restricted to fusing handlers, but can fuse all the way, up to and including the expression that builds the initial AST and to which the handlers are applied. Consider for example the *coin* example of Sect. 2.1. The free theorem of *coin*'s type is a variant of Theorem 1:

$$\alpha\ coin = coin$$

where $\alpha :: \forall a\ .\ M_1\ a \rightarrow M_2\ a$ is a term monad morphism between any two term monads M_1 and M_2. We can use this to fuse $handle_{Nondet}\ coin$ into $runCod\ gen_{Nondet}\ coin$. Of course this fusion interacts nicely with the fusion of a pipeline of handlers.

4 Pragmatic Implementation and Evaluation

This section turns the fusion approach into a practical Haskell implementation and evaluates the performance improvement.

4.1 Pragmatic Implementation

Before we can put the fusion approach into practice, we need to consider a few pragmatic implementation aspects.

Inlining with Typeclasses. In a lazy language like Haskell, fusion only leads to a significant performance gain if it is performed statically by the compiler and combined with inlining. In the context of the GHC compiler, the inlining requirement leaves little implementation freedom: GHC is rather reluctant to inline in general recursive code. There is only one exception: GHC is keen to create type-specialised copies of (constraint) polymorphic recursive definitions and to inline the definitions of typeclass methods in the process.

In short, if we wish to get good speed-ups from effect handler fusion, we need to make sure that the effectful programs are polymorphic in the term monad and that all the algebras are held in typeclass instances. For this reason, all handlers should be made instances of *TermAlgebra*.

Explicit Carrier Functors. The carrier functor of the compositional state handler is $s \rightarrow m_2 -$. From the category theory point of view, this is clearly a functor. However, it is neither an instance of the Haskell *Functor* typeclass nor can it be made one in this syntactically invalid form. A new type needs to be created that instantiates the functor typeclass:

```
newtype StateCarrier s m a = SC {unSC :: s → m a}
instance Functor m ⇒ Functor (StateCarrier s m) where
  fmap f x = SC (fmap (fmap f) (unSC x))
```

$$
\begin{array}{l}
\textbf{instance } \mathit{TermMonad}\ m_2\ f \\
\quad \Rightarrow \quad \mathit{TermAlgebra}\ (\mathit{StateCarrier}\ s\ m_2)\ (\mathit{State}\ s + f)\ \textbf{where} \\
\quad \mathit{con} = \mathit{SC} \cdot (\mathit{alg'}_{\mathit{State}} \nabla \mathit{con}_{\mathit{State}}) \cdot \mathit{fmap}\ \mathit{unSC} \\
\quad \mathit{var} = \mathit{SC} \cdot \mathit{gen'}_{\mathit{State}} \\
\mathit{gen'}_{\mathit{State}} :: \mathit{TermMonad}\ m\ f \Rightarrow a \to (s \to m\ a) \\
\mathit{gen'}_{\mathit{State}}\ x = \mathit{const}\ (\mathit{var}\ x) \\
\mathit{alg'}_{\mathit{State}} :: \mathit{TermMonad}\ m\ f \Rightarrow \mathit{State}\ s\ (s \to m\ a) \to (s \to m\ a) \\
\mathit{alg'}_{\mathit{State}}\ (\mathit{Put}\ s'\ k)\ s = \quad k\ s' \\
\mathit{alg'}_{\mathit{State}}\ (\mathit{Get}\ k)\quad s = \quad k\ s\ s
\end{array}
$$

Now the following function is convenient to run a fused state handler.

$$
\begin{array}{l}
\mathit{runStateC} :: \mathit{TermMonad}\ m_2\ f \Rightarrow \mathit{Cod}\ (\mathit{StateCarrier}\ s\ m_2)\ a \to (s \to m_2\ a) \\
\mathit{runStateC} = \mathit{unSC} \cdot \mathit{runCod}\ \mathit{var}
\end{array}
$$

Unique Carrier Functors. Even though the logging state handler has ostensibly[2] the same carrier $s \to m_2 -$ as the regular compositional state handler, we cannot reuse the same functor. The reason is that in the typeclass-based approach the carrier functor must uniquely determine the algebra; two different typeclass instances for the same type are forbidden. Hence, we need to write a new set of definitions as follows:

$$
\begin{array}{l}
\textbf{newtype } \mathit{LogStateCarrier}\ s\ m\ a = \mathit{LSC}\ \{\mathit{unLSC} :: s \to m\ a\} \\
\textbf{instance } \mathit{Functor}\ m \Rightarrow \mathit{Functor}\ (\mathit{LogStateCarrier}\ s\ m)\ \textbf{where} \\
\quad \mathit{fmap}\ f\ x = \mathit{LSC}\ (\mathit{fmap}\ (\mathit{fmap}\ f)\ (\mathit{unLSC}\ x)) \\
\textbf{instance } \mathit{TermMonad}\ m_2\ (\mathit{Writer}\ \mathit{String} + \mathit{Void}) \\
\quad \Rightarrow \mathit{TermAlgebra}\ (\mathit{LogStateCarrier}\ s\ m_2)\ (\mathit{State}\ s)\ \textbf{where} \\
\quad\quad \mathit{var} = \mathit{LSC} \cdot \mathit{gen'}_{\mathit{State}} \\
\quad\quad \mathit{con} = \mathit{LSC} \cdot \mathit{alg}_{\mathit{LogState}} \cdot \mathit{fmap}\ \mathit{unLSC} \\
\mathit{runLogStateC} :: \mathit{TermMonad}\ m_2\ (\mathit{Writer}\ \mathit{String} + \mathit{Void}) \\
\qquad\qquad\qquad \Rightarrow \mathit{Cod}\ (\mathit{LogStateCarrier}\ s\ m_2)\ a \to (s \to m_2\ a) \\
\mathit{runLogStateC} = \mathit{unLSC} \cdot \mathit{runCod}\ \mathit{var}
\end{array}
$$

4.2 Evaluation

To evaluate the impact of fusion we consider several benchmarks implemented in different ways: running handlers over the inductive definition of the free monad (FREE) and to its Church encoding (CHURCH), the fully fused definitions (FUSED), and the conventional definitions of the state monad from MTL (MTL).

The benchmarks are run in the Criterion benchmarking harness using GHC 7.10.1 on a MacBook Pro with a 3 GHz Intel Core i7 processor, 16 GB RAM and Mac OS 10.10.3. All values are in milliseconds, and show the ordinary least-squares regression of the recorded samples; the R^2 goodness of fit is above 0.99 in all instances.

[2] The typeclass constraints on m_2 are different.

Benchmark	FREE	CHURCH	FUSED	MTL
$count_1$				
10^7	1,017	1,311	3	3
10^8	10,250	13,220	29	29
10^9	103,000	129,700	291	295
$count_2$				
10^6	684	746	167	213
10^7	6,937	7,344	1,740	2,157
10^8	102,700	98,010	17,220	20,300
$count_3$				
10^6	559	555	166	205
10^7	5,759	5,561	1,618	2,132
10^8	110,900	94,760	16,300	20,120
grammar	794	763	6	77
pipes	1,325	1,351	43	N/A

The $count_1$ benchmark consists of the simple count-down loop used by Kammar et al. [9].

$$
\begin{array}{l}
count_1 = \\
\quad \mathbf{do}\ i \leftarrow get \\
\quad\quad \mathbf{if}\ i \equiv 0\ \mathbf{then}\ return\ i \\
\quad\quad\quad \mathbf{else}\ put\ (i - 1)\ ;\ count_1
\end{array}
$$

We have evaluated this program with three different initial states of 10^7, 10^8 and 10^9. The results show that all representations scale linearly with the size of the benchmark. However, the fused representation is about 300 times faster than the free monad representations and matches the performance of the traditional monad transformers.

The $count_2$ benchmarks extends the $count_1$ benchmark with a *tell* operation from the *Writer* effect in every iteration of the loop. It is run by sequencing the state and writer handler. The improvement due to fusion is now much less extreme, but still quite significant.

The $count_3$ benchmark is the $count_1$ program, but run with the logging state handler that delegates to the writer handler. The runtimes are slightly better than those of $count_2$.

The *grammar* benchmark implements a simpler parser by layering the state and non-determinism effects. Again fusion has a tremendous impact, even considerably outperforming the MTL implementation.

The *pipes* benchmark consists of the simple producer-consumer pipe used by Kammar et al. [9]. We can see that fusion provides a significant improvement over either free monad representation. There is no sensible MTL implementation to compare with for this benchmark.

The results (in ms) show that the naive approaches based on intermediate trees, either defined inductively or by Church encoding, incur a considerable overhead compared to traditional monads and monad transformers. Yet, thanks to fusion they can easily compete or even slightly outperform the latter.

5 Related Work

5.1 Fusion

Fusion has received much attention, and was first considered as the elimination of trees by Wadler [23] with his so-called deforestation algorithm, which was then later generalized by Chin [3].

From the implementation perspective, Gill et al. first introduced the notion of shortcut fusion to Haskell [4], thus allowing programs written as folds over lists to be fused. Takano and Meijer showed how fusion could be generalized to arbitrary datastructures [19]. The technique of using free theorems to explain certain kinds of fusion was applied by Voigtländer [20].

Work by Hinze et al. [6] builds on recursive coalgebras to show the theory and practice of fusion, limited to the case of folds and unfolds. Later work by Harper [5] provides a pragmatic discussion, that bridges the gap between theory and practice further, by discussing the implementation of fusion in GHC with inline rules. Harper also considers the fusion of Church encodings.

More recently, recursive coalgebras appear in work by Hinze et al. [7] where conjugate hylomorphisms are introduced as a means of unifying all presently known structured recursion schemes. There, the theory behind fusion for general datatypes across all known schemes is described as arising from naturality laws of adjunctions. Special attention is drawn to fusion of the cofree comonad, which is the dual case of free monads we consider here.

5.2 Effect Handlers

Plotkin and Power were the first to explore effect operations [14], and gave an algebraic account of effects [15] and their combination [8]. Subsequently, Plotkin and Pretnar [13] have added the concept of handlers to deal with exceptions. This has led to many implementation approaches.

Lazy Languages. Many implementations of effect handlers focus on the lazy language Haskell.

For the sake of simplicity and without regard for efficiency, Wu et al. [24] use the inductive datatype representation of the free monad for their exposition. They use the *Data Types à la Carte* approach [18] to conveniently inject functors into a co-product; that approach is entirely compatible with this paper. Wu et al. also generalize the free monad to *higher-order functors*; we expect that our fusion approach generalizes to that setting.

Kiselyov et al. [10] provide a Haskell implementation in terms of the free monad too. However, they combine this representation with two optimizations: (1) the codensity monad improves the performance of ($\gg=$), and (2) their *Dynamic*-based open unions have a better time complexity than nested co-products. Due to the use of the codensity monad, this paper also benefits from the former improvement. Moreover, we believe that the latter improvement is unnecessary due to the specialisation and inlining opportunities that are exposed by fusion.

Van der Ploeg and Kiselyov [12] present an implementation of the free monad with good asymptotic complexity for both pattern matching and binding; unfortunately, the constant factors involved are rather high. This representation is mainly useful for effect handlers that cannot be easily expressed as folds, and thus fall outside of the scope of the current paper.

Behind a Template Haskell frontend Kammar et al. [9] consider a range of different Haskell implementations and perform a performance comparison. Their basic representation is the inductive datatype definition, with a minor twist: the functor is split into syntax for the operation itself and a separate continuation. This representation is improved with the codensity monad. Finally, they provide—without explanation—a representation that is very close to the one presented here; their use of the continuation monad instead of the codensity monad is a minor difference.

Both Atkey et al. [1] and Schrijvers et al. [17] study the interleaving of a free monad with an arbitrary monad, i.e., the combination of algebraic effect handlers and conventional monadic effects. We believe that our fusion technique can be adapted for optimizing the free monad aspect of their settings.

Strict Languages. In the absence of lazy evaluation, the inductive datatype definition of the free monad is not practical.

Kammar et al. [9] briefly sketch an implementation based on delimited continuations. Schrijvers et al. [17] show the equivalence between a delimited continuations approach and the inductive datatype; hence the fusion technique presented in this paper is in principle possible. However, in practice, the codensity monad used for fusion is likely not efficient in strict languages. Hence, effective fusion for strict languages remains to be investigated.

5.3 Monad Transformers

Monad transformers, as first introduced by Liang et al. [11], pre-date algebraic effect handlers as a means for modelling compositional effects. Yet, there exists a close connection between both approaches: monad transformers are fused forms of effect handlers. What is particular about their underlying effect handlers is that their carriers are term algebras with a monadic structure, i.e., term monads. This means that the *Cod* construction is not necessary for fusion.

6 Conclusion

We have explained how to fuse algebraic effect handlers by shifting perspective from free monads to term monads. Our benchmarks show that, with a careful code setup in Haskell, this leads to good speed-ups compared to the free monad, and allows algebraic effect handlers to compete with the traditional monad transformers.

Acknowledgments. The authors would like to thank Ralf Hinze for suggesting that they consider the specification that free monads satisfy in terms of adjunctions, and to James McKinna who helped baptize *con* for constructing terms. We are also grateful for the feedback from the members of the IFIP Working Group 2.1 on an early iteration of this work. This work has been partially funded by the Flemish Fund for Scientific Research (FWO).

References

1. Atkey, R., Johann, P., Ghani, N., Jacobs, B.: Interleaving data and effects (2012) (Submitted for publication)
2. Burris, S., Sankappanavar, H.P.: A course in universal algebra, graduate texts in mathematics, vol. 78. Springer, New York (1981)
3. Chin, W.N.: Safe fusion of functional expressions ii: Further improvements. J. Funct. Program. **4**, 515–555 (1994)
4. Gill, A., Launchbury, J., Peyton Jones, S.L.: A short cut to deforestation. In: Proceedings of the Conference on Functional Programming Languages and Computer Architecture, FPCA 1993, pp. 223–232. ACM, New York, NY, USA (1993)
5. Harper, T.: A library writer's guide to shortcut fusion. In: Proceedings of the 4th ACM Symposium on Haskell, Haskell 2011, pp. 47–58. ACM, New York, NY, USA (2011)
6. Hinze, R., Harper, T., James, D.W.H.: Theory and practice of fusion. In: Hage, J., Morazán, M.T. (eds.) IFL. LNCS, vol. 6647, pp. 19–37. Springer, Heidelberg (2011)
7. Hinze, R., Wu, N., Gibbons, J.: Conjugate hylomorphisms - the mother of all structured recursion schemes. In: Proceedings of the 42nd Annual ACM SIGPLAN-SIGACT Symposium on Principles of Programming Languages, POPL 2015, pp. 527–538. ACM, New York, NY, USA (2015)
8. Hyland, M., Plotkin, G.D., Power, J.: Combining effects: sum and tensor. Theor. Comput. Sci. **357**(1–3), 70–99 (2006)
9. Kammar, O., Lindley, S., Oury, N.: Handlers in action. In: Proceedings of the 18th ACM SIGPLAN International Conference on Functional programming, ICFP 2014, pp. 145–158. ACM (2013)
10. Kiselyov, O., Sabry, A., Swords, C.: Extensible effects: an alternative to monad transformers. In: Proceedings of the 2013 ACM SIGPLAN Symposium on Haskell, Haskell 2013, pp. 59–70. ACM (2013)
11. Liang, S., Hudak, P., Jones, M.: Monad transformers and modular interpreters. In: Proceedings of the 22nd ACM SIGPLAN-SIGACT Symposium on Principles of Programming Languages, POPL 1995, pp. 333–343. ACM, New York, NY, USA (1995)
12. Ploeg, A.V.D., Kiselyov, O.: Reflection without remorse: revealing a hidden sequence to speed up monadic reflection. In: Proceedings of the 2014 ACM SIGPLAN Symposium on Haskell, Haskell 2014, pp. 133–144. ACM, New York, NY, USA (2014)

13. Plotkin, G.D., Matija, P.: Handling algebraic effects. Log. Meth. Comput. Sci. **9**(4), 1–36 (2013)
14. Plotkin, G., Power, J.: Notions of computation determine monads. In: Nielsen, M., Engberg, U. (eds.) FOSSACS 2002. LNCS, vol. 2303, pp. 342–356. Springer, Heidelberg (2002)
15. Plotkin, G.D., Power, J.: Algebraic operations and generic effects. Appl. Categorical Struct. **11**(1), 69–94 (2003)
16. Rivas, E., Jaskelioff, M.: Notions of computation as monoids. CoRR abs/1406.4823. (2014). http://arxiv.org/abs/1406.4823
17. Schrijvers, T., Wu, N., Desouter, B., Demoen, B.: Heuristics entwined with handlers combined. In: Proceedings of the 16th International Symposium on Principles and Practice of Declarative Programming, PPDP 2014, pp. 259–270. ACM (2014)
18. Swierstra, W.: Data types à la carte. J. Funct. Program. **18**(4), 423–436 (2008)
19. Takano, A., Meijer, E.: Shortcut deforestation in calculational form. In: Proceedings of the Seventh International Conference on Functional Programming Languages and Computer Architecture, FPCA 1995, pp. 306–313. ACM, New York, NY, USA (1995)
20. Voigtländer, J.: Proving correctness via free theorems: the case of the destroy/build-rule. In: Proceedings of the 2008 ACM SIGPLAN Symposium on Partial Evaluation and Semantics-based Program Manipulation, PEPM 2008, pp. 13–20. ACM, New York, NY, USA (2008)
21. Voigtländer, J.: Free theorems involving type constructor classes: functional pearl. In: Proceedings of the 14th ACM SIGPLAN International Conference on Functional Programming, ICFP 2009, pp. 173–184. ACM, New York, NY, USA (2009)
22. Wadler, P.: Theorems for free!. In: Proceedings of the Fourth International Conference on Functional Programming Languages and Computer Architecture, FPCA 1989, pp. 347–359. ACM, New York, NY, USA (1989)
23. Wadler, P.: Deforestation: transforming programs to eliminate trees. Theor. Comput. Sci. **73**(2), 231–248 (1990)
24. Wu, N., Schrijvers, T., Hinze, R.: Effect handlers in scope. In: Proceedings of the 2014 ACM SIGPLAN Symposium on Haskell, Haskell 2014, pp. 1–12. ACM, New York, NY, USA (2014)

Author Index

Printed in the United States
by Baker & Taylor Publisher Services